# Research Based Undergraduate Science Teaching

*A volume in*
Research in Science Education (RISE)

*Series Editors*

Dennis W. Sunal
*University of Alabama*
Emmett L. Wright
*Kansas State University*

## Research in Science Education

Dennis W. Sunal and Emmett L. Wright, Series Editors

*Reform in Undergraduate Science Teaching for the 21st Century* (2003)
edited by Dennis W. Sunal, Emmett L. Wright, and Jeanelle Bland Day

*The Impact of State and National Standards on K-12 Science Teaching* (2006)
edited by Dennis W. Sunal and Emmett L. Wright

*The Impact of the Laboratory and Technology on Learning and Teaching Science K-16* (2008)
edited by Dennis W. Sunal, Emmett L. Wright, and Cheryl Sundberg

*Teaching Science with Hispanic ELLs in K-16 Classrooms* (2010)
edited by Dennis W. Sunal, Cynthia, S. Sunal, and Emmett L. Wright

*The Role of Public Policy in K-12 Science Education* (2011)
edited by George E. DeBoer

*Research Based Undergraduate Science Teaching* (2014)
edited by Dennis W. Sunal Cynthia, S. Sunal,
Emmett L. Wright, Cheryl Mason, and Dean Zollman

# Research Based Undergraduate Science Teaching

edited by

**Dennis W. Sunal**
**and**
**Cynthia Szymanski Sunal**
*University of Alabama*

**Emmett L. Wright**
*Kansas State University*

**Cheryl L. Mason**
*San Diego State University*

**Dean Zollman**
*Kansas State University*

IAP
INFORMATION AGE PUBLISHING, INC.
Charlotte, NC • www.infoagepub.com

**Library of Congress Cataloging-in-Publication Data**

A CIP record for this book is available from the Library of Congress
http://www.loc.gov

ISBN: 978-1-62396-750-5 (Paperback)
978-1-62396-751-2 (Hardcover)
978-1-62396-752-9 (ebook)



Printed and bound by CPI Group (UK) Ltd, Croydon, CR0 4YY

# CONTENTS

# PREFACE TO THE SERIES

Science education as a professional field has been changing rapidly over the past two decades. Scholars, administrators, practitioners, and students preparing to become teachers of science find it difficult to keep abreast of relevant and applicable knowledge concerning research, leadership, policy, curricula, teaching, and learning that improves science instruction and student science learning. The literature available reports a broad spectrum of diverse science education research, making the search for valid materials on a specific area that is time-consuming and tedious.

Science education professionals at all levels need to be able to access a comprehensive, timely, and valid source of knowledge about the emerging body of research, theory, policy, and practice in their fields. This body of knowledge would inform researchers about emerging trends in research, research procedures, and technological assistance in key areas of science education. It would inform policymakers in need of information about specific areas in which they make key decisions. It would also help practitioners and students become aware of current research knowledge, policy, and best practice in their fields.

For these reasons, the goal of the book series, Research in Science Education, is to provide a comprehensive view of current and emerging knowledge, research strategies, and policy in specific professional fields of science education. This series presents currently unavailable, or difficult to gather, materials from a variety of viewpoints and sources in a usable and organized format.

*Research Based Undergraduate Science Teaching*, pp. ix–x

Each volume in the series presents a juried, scholarly, and accessible review of research, theory, and/or policy in a specific field of science education, K–16. Topics covered in each volume are determined by current issues and trends, as well as generative themes related to up-to-date research findings and accepted theory. Published volumes will include empirical studies, policy analysis, literature reviews, and positing of theoretical and conceptual bases.

# PREFACE

*Research-Based Undergraduate Science Teaching* is Volume 6 of the series, Research in Science Education (RISE). Here, we examine research, theory, and practice concerning issues in teaching science with undergraduates. This RISE volume addresses higher education faculty and all who teach entry-level science with the aim of developing a basic science literacy leading to scientific expertise in our undergraduates. The focus throughout the chapters is on research-based reforms leading to best practices in teaching undergraduate science and engineering.

The goal of this sixth volume of RISE is to provide a research foundation that demonstrates an agenda for strengthening the preparation and enhancement of instructors of undergraduate science leading to short- and long-term impact on student outcomes. This foundation is important in providing faculty across the nation with research-based practices for reforming courses. Our goal was carried out through a series of events over several years that led to the planning and subsequent dissemination of research being conducted by various stakeholders throughout the United States. Researchers were first invited to national conferences supported with funding from the National Science Foundation, from regions of the country that have had a long history of classroom research with undergraduates. Two national conferences with the theme of research-based undergraduate science teaching were held, involving over 150 researchers from 85 higher education institutions. Our website at http://nseus.org provides resource materials from the conference presentations, as well as National Study

*Research Based Undergraduate Science Teaching*, pp. xi–xiii

of Education in Undergraduate Science (NSEUS) background research papers, presentations, and research briefs. These conferences were used as one of the dissemination methods to establish and secure commitments from researchers to a conduct and report research to strengthen our understanding of education in undergraduate science classrooms. The national call for manuscripts requested the inclusion of major priorities and critical research areas, methodological concerns, and concerns and results of implementation of faculty professional development programs and instructional reform in undergraduate science classrooms.

When developing research manuscripts to be reviewed for RISE, Volume 6, researchers were asked to consider the status and effectiveness of current and experimental practices for reforming undergraduate science courses involving all undergraduates, including groups of students who are not always well represented in science, technology, engineering, and mathematics (STEM) education. Manuscripts also were solicited and selected to promote dialogue in the research community for

1. Sharing research on teaching conducted in different disciplinary fields in undergraduate science contexts
2. Addressing the complexities and issues associated with conducting research in undergraduate science teaching
3. Identifying research methodologies for investigating undergraduate science teaching and learning
4. Exploring interventions for reducing course attrition and sustaining science and engineering majors in their disciplinary fields

It has been suggested that the science education research community is isolated from the teaching practices in undergraduate science classrooms, pays little attention to practice in these classrooms, and has little influence on it. But to influence practice, it is important to understand how research-based practice develops and how it is implemented. This volume sheds light on the intersection between research and practice through discussions of research and of practical examples.

This RISE volume provides an overview of research-based reforms leading to best practices in teaching undergraduate science and engineering. It also provides an overview of the preparation and in-service development of science faculty in colleges and universities. Needs and guidelines that should be considered and incorporated in planning professional preparation programs for teaching science in undergraduate classrooms are discussed. The chapters outline and describe in detail research-based best practices for a variety of settings.

In the chapters that follow, a number of researchers, reform developers, and reform implementers discuss a wide range of issues. The volume should be considered as a first step in thinking through what reform in undergraduate science education might look like. To facilitate our thinking about reform, the volume is organized in four parts. Part I focuses on a national study of the impact of reforms in undergraduate science classrooms. Following this introductory section, Part II reports on the development and implementation of five research-based specific reforms in classroom practice. Part III then focuses on the impact of three research-based reforms: student projects, problem-based learning, and research-based instructional strategies. Part IV focuses on the impact of three reforms on professional development in higher education: student learning assistants, graduate teaching assistants, and faculty instructors.

# ACKNOWLEDGMENTS

This sixth volume in the series Research in Science Education (RISE), *Research-Based Undergraduate Science Teaching,* was an outcome made possible by many authors who gave generously of their time and prioritized conducting investigations, writing manuscripts, submitting their work to the scrutiny of others, and persisting through many revisions. Their professional experiences and expertise make this volume possible. Another group of individuals to whom a great deal of credit must go for helping make this volume possible are the reviewers who offered critical feedback on the chapters. They provided the additional insights needed to help craft the contributions that make up this volume.

Collaboration between the authors developed through discussions and communications that originated at the 2011 and 2012 national conferences held at the University of Alabama. *Research-Based Undergraduate Science Teaching I and II* were funded through the National Science Foundation. Therefore, we wish to acknowledge the National Science Foundation and individuals attending the conferences for providing us with a forum to meet, interact, disseminate, and form professional collaboration communities involving individuals with an interest in improving teaching and learning in undergraduate science courses. Special recognition also is given to the students, teachers, and administrators who were concerned enough to take part in the investigations and contribute their thoughts to our discussions.

Dennis W. Sunal, Cynthia S. Sunal,
Emmett L. Wright, Cheryl Mason, and Dean Zollman

December, 2013

*Research Based Undergraduate Science Teaching,* pp. xv–xv

# PART I

## NATIONAL STUDY OF EDUCATION IN UNDERGRADUATE SCIENCE

CHAPTER 1

# NATIONAL STUDY OF EDUCATION IN UNDERGRADUATE SCIENCE

## Purpose and Baseline Data

**Dennis W. Sunal, Cynthia Szymanski Sunal, Erika M. Steele, and Glenda Ogletree**

### ABSTRACT

Creating change in entry-level undergraduate science, technology, engineering, and mathematics (STEM) courses has the potential to impact all students in higher education. Course reform has been seen as a way to facilitate the STEM pipeline by sustaining students in their course programs through graduation. Professional development for higher education faculty has received increasing attention over several decades as a means to create change in higher education. The focus of the national study, the National Study of Education in Undergraduate Science (NSEUS), was on the impact of faculty professional development on changing entry-level undergraduate science courses to more fully engage students in STEM learning. In this chapter we address the problem of the extent to which undergraduate entry-level science course reform affects the science learning outcomes of the students who participate in them. We describe the theoretical basis for the

*Research Based Undergraduate Science Teaching*, pp. 3–34

study, related research, and a baseline national survey of reformed entry-level undergraduate science courses to build the context for the implementation of a national study of education in undergraduate science. Other aspects of the complete study are presented in Chapters 2 and 3.

Professional development for higher education faculty has received increasing attention over several decades as a means to facilitate the science, technology, engineering, and mathematics (STEM) pipeline by sustaining students in their programs through graduation. Reform in entry-level undergraduate STEM courses has the potential to impact all students in higher education. Among students in these courses, elementary preservice teachers play an important part in the STEM pipeline cycle as they will, one day, teach science to their elementary school students, greatly affecting the foundation for science learning laid in kindergarten–6th grade. Two or three entry-level STEM courses often are the only science many students will have in their non-STEM undergraduate major or teacher certification program.

In this chapter we address the problem of the extent to which undergraduate entry-level science courses affect the science learning outcomes of the students who participate in them. Since significant examples of national reports, resources, and professional development efforts have been undertaken over the past 30 years to enable higher education STEM faculty to reform undergraduate courses, it is important that this problem be investigated (National Academy of Sciences, 2007; National Commission on Excellence in Education, 1983; National Research Council [NRC], 1996; Siebert & McIntosh, 2001). The National Study of Education in Undergraduate Science (NSEUS) posed a central research question focused on this problem: "How do undergraduate entry-level science courses, differing in level of reform, affect student learning outcomes?" With this problem and central question in mind we

1. describe a research-based faculty professional development framework designed and implemented nationally to facilitate change in undergraduate courses to enhance science, mathematics, and technology literacy
2. examine previous research on faculty professional development and reform in undergraduate science courses and resulting short- and long-term impacts, including successful reform perspectives
3. summarize findings of a national survey of a population of institutions where STEM reform courses were offered, focusing on successful reform perspectives for developing research supported best practice in undergraduate science

4. describe a national research model and methods for conducting a study designed to determine the short- and long-term impacts of level of reform in undergraduate science courses on students, with special emphasis on preservice teacher education candidates.

Chapters 2 and 3 build on the work presented here, describing methodology and results from the NSEUS study that investigated the relationship between undergraduate science course reform and learning outcomes and the extent of reform needed to produce significantly different learning outcomes from traditional courses.

## BACKGROUND

Over the past several decades, national reports documented the goals of, and concerns about, the teaching of STEM courses in higher education beginning with *A Nation at Risk* (National Commission on Excellence in Education, 1983). A continuing stream of other documents followed, including the *American Association for the Advancement of Science's* (AAAS) *Project 2061: Benchmarks for Science Literacy* (1993), the National Research Council's (NRC) *National Science Education Standards* (1996), the National Science Foundation's (NSF) *Shaping the Future* (1996), the NRC's *Science Teaching Reconsidered: A Handbook* (1997) and *Educating Teachers of Science, Mathematics, and Technology: New Practices for the New Millennium* (2001), Siebert and McIntosh's *College Pathways to the Science Education Standards* (2001), the U.S. Department of Education's *No Child Left Behind* (2001), the NRC's *Evaluating and Improving Undergraduate Teaching in Science, Engineering, and Technology* (2003), the National Academy of Science's *Rising Above the Gathering Storm: Energizing and Employing America for a Brighter Economic Future* (2007), and the NRC's *A Framework for K–12 Science Education: Practices, Crosscutting Concepts, and Core Ideas* (2012). These documents set guidelines providing criteria that can be used to judge whether particular reform actions serve the vision of a scientifically literate society.

Traditional non-reform approaches to teaching undergraduate entry-level science courses do not work effectively with many students, particularly non-science majors. Several past efforts have suggested that teaching and students' learning can be impacted positively by reforms (Sundberg & Monaca, 1994). Recognizing that reform is needed in undergraduate entry-level science courses if all students are to benefit, the National Science Foundation identified the goal for higher education reforms as creating an environment where "all students have access to supportive, excellent undergraduate education in STEM and all students learn these subjects by direct experience with the methods and processes of inquiry" (NSF, 1996,

p. 1). The NRC later called for reforms addressing the same goal, stating, "The ultimate goal of undergraduate education should be for individual faculty and departments to improve the academic growth of students" (2003, p. 14).

Over the past 30 years, science literacy for all was stressed by national governmental agencies including NSF and NRC. K–12 science standards were developed, for example, in *Project 2061: Benchmarks for Science Literacy* (AAAS, 1993), the *National Science Education Standards* (NRC, 1996) and *A Framework for K-12 Science Education: Practices, Crosscutting Concepts, and Core Ideas* (NRC, 2012). Effective K–12 science education reform could not be separated from standards-based reform in undergraduate science (Lederman & Gess-Newsome, 1999). Inservice teachers are heavily influenced by how the undergraduate science courses they take are taught and by the science content included in these courses. A renewed emphasis is needed on the learning approaches, organization, breadth, and depth of content found in undergraduate science. Traditional course formats and contents do not challenge but instead reinforce the fragmented nature of students' knowledge and their relative inability to apply that knowledge within the context of teaching (Lederman & Gess-Newsome, 1999).

Studies of pre- and inservice teacher education programs often note that the development of the desired integration of subject matter and pedagogy is not accomplished (Smith, 1999). Lederman and Gess-Newsome stated,

> Subject matter structures are largely determined in adults in college science courses and it appears that these courses do not yield coherent and integrated subject matter structures ... [and] were not integrated and consistent with the vision of national reforms in science education. [The] investigation made us aware ... that the nature and validity of instruction in college science courses was absolutely critical. (1999, p. 209)

Teacher education courses and programs can accomplish only so much without the support of a reform climate in the undergraduate science courses taken as part of their program by preservice teachers. University faculty beliefs and actions are responsible for *quality of the learning environment* and the *content structure and organization* of undergraduate science courses. These undergraduate science courses, in turn, are critical factors in the development of teachers' pedagogical content knowledge (PCK) and the development of meaningful science learning outcomes among their K–12 students. If students come to the university with poor science backgrounds, the undergraduate science courses provided to preservice teachers are a major factor in this cyclic problematic process.

## THE NATIONAL AERONAUTICS AND SPACE ADMINISTRATION/NASA OPPORTUNITIES FOR VISIONARY ACADEMICS (NASA/NOVA) FACULTY PROFESSIONAL DEVELOPMENT MODEL

In response to the needs identified in undergraduate STEM teaching in the mid-1990s, guidelines were developed for a national professional development model for higher education faculty that incorporated research based processes noted in the literature (Loucks-Horsley, Hewson, Love, & Stiles, 1998; Peterman, 1993; Sunal et al., 2001; Sunal & Bland, 2004; Weimer & Lenze, 1994; Zollman, 1997, 2004). The implications of previous research on faculty development in higher education within the supporting framework of cognitive apprenticeship, provided the guidelines for a more effective model of professional development. These guidelines were used to develop, implement, and sustain, through National Aeronautics and Space Administration (NASA) funding, a large, nationally disseminated model for STEM faculty professional development, 1995–2006. The model incorporated specific conditions cited in the research literature as necessary for successful undergraduate course reform, implementation, and institutionalization:

1. interaction of faculty between colleges (e.g., arts and sciences, education, and engineering)
2. participation in a collaborative team representing differing expertise
3. maintenance of a positive college and department climate relating to the reform effort's goals
4. maintenance of administrator presence and support in the change process
5. initiation by beginning with the reform goals to be accomplished rather than with personnel or contextual barriers
6. collaboration via interactions building on effective interpersonal skills and trust
7. planning for incremental, rather than initial massive, change to take place over several years through the sustaining phase of the reform
8. ongoing and consistent monitoring and support system for implementation of the reform activities using action research
9. sustaining through collaboration in a network of faculty within and outside of the institution (Sunal et al., 2001).

The original authors of the model were researchers experienced in reforming undergraduate STEM courses and successful in completing externally funded related projects in previous years, 1991–1995. The model's development was the result of STEM researchers' focus on reform at three universities: University of Alabama (Michael Freeman, Dennis Sunal, and Kevin Whitaker), Fayetteville State University (Leo Edward and Ronald Johnston), and the University of Idaho (Michael Odell). The researchers proposed and developed a program that became the National Aeronautics and Space Administration's pre-college preparation program, NASA Opportunities for Visionary Academics (NASA/NOVA). Beginning in 1995, NASA/NOVA annually invited the participation of undergraduate faculty concerned with how to prepare undergraduates including preservice teachers. Through NASA/NOVA entry-level reform, STEM courses were newly developed or modified and enhanced for all majors using existing courses by collaborative teams of faculty in the sciences, mathematics, engineering, and education. The NASA/NOVA program's goals were

1. disseminating a national preservice model based on national standards and benchmarks for mathematics, science, and technology
2. utilizing the research and development activities from NASA's strategic enterprises (Earth science, aerospace technology, human exploration and development of space, and space science) in STEM instruction
3. building collaboration across faculties in education and science, mathematics, and engineering to develop innovative approaches to teacher preparation for enhanced student learning
4. sustaining the change process through continued professional development and collaboration with networking and mentoring
5. stimulating and conducting action research and monitoring visits.

To address these goals, the NASA/NOVA program began national dissemination of the model presenting a set of professional development activities for as many as 35 science, engineering, technology, mathematics, and education faculty teams per year. Using the NASA mission, facilities, and resources, NASA/NOVA provided these faculty teams with enhanced knowledge and skills for implementing change in their university STEM courses.

Four annual requests solicited proposals for participation in the NASA/NOVA program from faculty at colleges and universities nationwide. Participation in NASA/NOVA included opportunities for enhanced knowledge and skills developed through workshops, demonstration of exemplary models of curriculum and instruction, grant funding to institute course reform, collaboration, and action research to investigate the effects of components of the reform process, along with a number of other complementary professional development activities within and between higher education institutions and NASA resources. The NASA/NOVA professional development model was delivered in three phases: (1) *planning and preparation,* involving training, collaboration, and action planning for addressing baseline needs in faculty skills and knowledge enhancement; (2) *development and implementation,* involving initial course change, action research, mentoring, and sharing of expertise; and (3) *continuing development and long-term sustaining activity,* involving action research, networking, monitoring including site visits, and dissemination. The NASA/NOVA professional development (PD) model included:

1. a collaborative team approach with STEM and education faculty and administrators in a systemic initiative (collaboration) *Phases 1, 2, 3*
2. intensive professional development (30 hours) addressing

    a. higher education concerns reflected in the national science standards (learning environment, course structure, pedagogical content knowledge [PCK]) (AAAS, 1993; NRC, 1996; Siebert & McIntosh, 2001)
    b. "best" practices from research in STEM curriculum, pedagogy, assessment, collaborative learning, and working with student diversity in higher education (learning environment, PCK) (Backer, 2002; Christopher & Atwood, 2004; Francis, Adams, & Noonan, 1998; Krinsky, Anderson, & Kidane, 1996; Project Kaleidoscope, 2005; Scharmann, Stalheim-Smith, & James, 2004; Slater et al., 2003; Sunal & Bland, 2004; Swanson & Bilderback, 1998; Wycoff, 2000)
    c. demonstration course models from research on STEM learning action research planning and methodology (PCK, collaboration) (Raubenheimer, 2004)
    d. "best" practices in research and methods for the use of technology to facilitate STEM learning (learning environment, PCK) (Odell, Badger, Kennedy, Ewers, & Klett, 2004)
    e. strategies for creating course change in the department context (curriculum and instruction)
    f. strategies for enhancing grant writing skills (collaboration) *Phase 1*

3. development of a proposal for course change that applied the NASA/NOVA professional development model in order to obtain funding and resources, which was reviewed with feedback provided (collaboration) *Phase 1*

4. development of standards-based reform undergraduate STEM courses in a range of institutions from Bachelor's degree granting through research universities (learning environment, course structure, PCK, collaboration) *Phases 1, 2*
5. resources including financial support to implement reform STEM courses on a long-term basis (collaboration) *Phases 1, 2, 3*
6. continuous mentoring and monitoring of progress including evaluation site visits during development and implementation (collaboration) *Phases 1, 2, 3*
7. action research conducted by in-house faculty teams that examined student and faculty development (PCK, collaboration) and the sharing of results annually at a national NASA/NOVA conference. *Phases 2, 3*
8. continuous long-term professional development activities based on best practices research over multiple years (PCK, collaboration) Phase *1, 2, 3*
9. collaboration and sharing of expertise and practices between faculty within an institution and among different institutions (PCK, collaboration) *Phase 3*

## NASA/NOVA Phase 1 Professional Development Activities

During each year of the project, NASA/NOVA personnel conducted a series of national workshops for interdisciplinary teams of college and university faculty. A total of 23 Phase 1 faculty professional development workshops were conducted (see Figure 1.1). Each institution's STEM faculty team consisted of three or more members, including at least one member from the education department with the other members from STEM Departments. A number of teams included an administrator. Approximately 8–10 teams attended each initial NASA/NOVA Phase 1 faculty development workshop.

Faculty experienced continuous professional development within a national network of institutions. Teams from 240 institutions involving over 550 faculty from 44 states, Washington DC, Puerto Rico, and the U.S. Virgin Islands attended NASA/NOVA workshops during the 12-year period. Proposals were funded to support a reform effort from 103 institutions' faculty teams. The funded teams involved 354 faculty involved in developing 185 undergraduate STEM courses serving over 15,000 students each year (see Figure 1.2). Once funded, teams joined the NASA/NOVA network and were encouraged to continue during the life of the 12-year project.

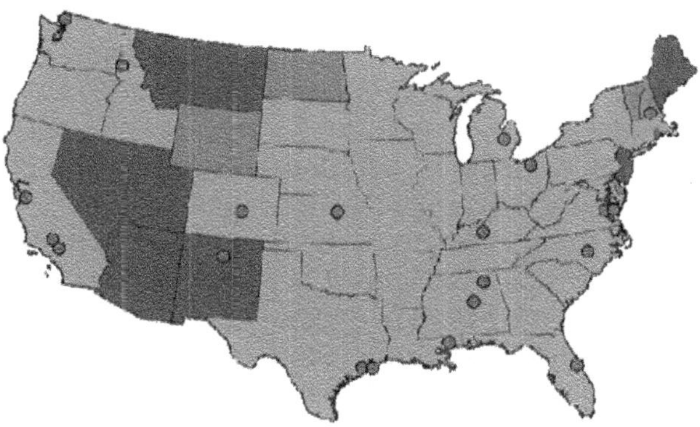

Figure 1.1. NASA/NOVA National University Phase I Workshop Sties, 1995–2006.

Figure 1.2. NASA/NOVA network of institutions involved in STEM course reform.

Phase 1 workshops focused on the implementation of STEM standards in higher education, use of innovative inquiry pedagogy with active and interactive learning, use of prior knowledge as a starting point for lessons, authentic assessment, experiences with examples of successful STEM course models, use of educational technology for facilitating learning, strategies for integrating STEM content, utilization of research on development activities from NASA's strategic enterprises, use of action research in course development as continued professional development, and emphasis on collaboration and diversity. The workshop's focus was on enhancing learning for all student majors in undergraduate STEM courses with special attention to developing a new paradigm for educating preservice teachers. Participants were mentored during and after the workshop. Faculty teams received mentoring concerning grant proposal preparation, proposal revision(s), implementation of reform courses in departments, and institutional barriers. For those institutions that became part of the NASA/NOVA network, additional mentoring took place through site visits by the NASA/NOVA management team.

A listing of key action elements of the NASA/NOVA professional development model is provided in Figure 1.3. Upon successful completion of the initial three-day workshop, a participating team was eligible to submit a proposal for a competitive implementation planning grant of up to $34,000, with at least a 1-to-1 cost-share match. An important element following the Phase 1 activities was the annual leadership development conference (LDC) at which all the NASA/NOVA network institutions were brought together to receive further professional development, present their projects and research results, and collaborate in new network activities. All NASA/NOVA activities took place on university campuses or at one of the NASA research centers such as the Kennedy Space Center and the Marshall Space Flight Center.

## NASA/NOVA Phase 2 Professional Development Activities

A second funding opportunity for NASA/NOVA institutions, the Phase 2 grants, was competitively available for NASA/NOVA network teams, enabling them to propose further development and dissemination of their NASA/NOVA reformed courses. Funded institutions developed methods and techniques for replicating the NASA/NOVA change model among additional institutions not directly reached by the NASA/NOVA workshops. Emphasis was placed on fostering partnerships with community colleges and other educational organizations. The research aspect of the Phase 2 grants was designed to assess the onsite impact of the NASA/NOVA model. Successful research projects used metrics to examine student engagement, student achievement and attitudes, innovation sustainability, the extent to which support and collaboration had been maintained, and the overall impact the project had on students, faculty, and administrators. The funded projects also were expected to produce publishable reports and share files of materials developed, videos, and websites for dissemination.

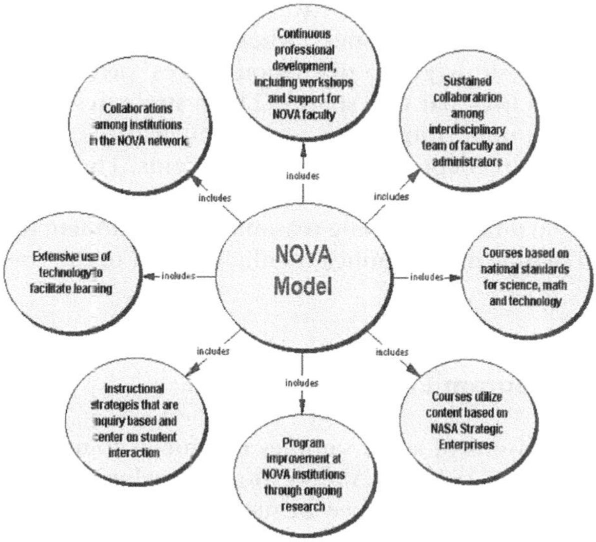


**Figure 1.3.** Key action elements of the NASA/NOVA professional development model.

## NASA/NOVA Phase 3 Professional Development Activities

The NASA/NOVA program further connected local network activities with NASA's strategic plan and field centers via a funding program in Phase 3. The overall goal of this program was to create and maintain a state-of-the-art U.S. STEM research and development presence infusing cutting-edge scientific data into innovative higher education courses. To ensure the widest possible participation by NASA/NOVA network institutions in this program, participants integrated technology to facilitate learning as well as developing blended and fully online interactive learning activities on websites designed for their NASA/NOVA reformed courses. Technology focused on the use of web integration and NASA web products in education. The NASA/NOVA technology initiative encouraged the introduction of technologies into higher education courses and sought to effect change in the manner in which content courses were taught to undergraduate students, including preservice teachers.

Meaningful integration of technology was an integral part of NASA/NOVA and of web-based learning. The electronic design of an environment for learning is quite different from a design for giving out information. Incorporating social media technology in a meaningful way utilized a number of strategies including electronic asynchronous interactions in the form of "whole class idea containers" where the student's view is presented to the class, small study group discussion spaces, private journal spaces, and synchronous live chat web spaces. These strategies aimed at providing students with opportunities to examine their ideas, make comments, and reflect in varying degrees of safe environments. The use of technology promoted integration of content knowledge with reflective assignments, while at the same time maintaining reasonable commitment of faculty time, with the goal of offering a significant enhancement to the course learning structure.

## NASA/NOVA Program Evaluation

Initial comparisons of small samples of courses found positive results indicating that the NASA/NOVA professional development model met the specific conditions identified above for successful course reform (Bland-Day, 1999; Staples, 2002; Sunal et al., 2001; Sunal, MacKinnon, Raubenheimer, & Gardner, 2004). These preliminary studies found increased undergraduate student achievement, long-term change in efficacy in science teaching, positive attitudes toward science, and more effective use of research-based science teaching practice among graduated elementary teachers. Long-term impacts increased as preservice teachers gained additional experiences in coursework and through their initial years of classroom teaching. The NASA/NOVA model design fostered the formation of diverse partnerships and networks that provided for institution and faculty collaboration on a wide range of research and educational endeavors including reforms compatible with new teacher certification. Faculty and students could exchange and access exemplary reform course models, materials, and activities, as well as updated STEM research and information. Numerous external proposals and awarded grants were fostered, extending the reforms developed under the original NASA/NOVA activities.

The NASA/NOVA model defined undergraduate standards-based reform STEM courses as consisting of multiple levels of collaboration and multiple contexts for communication (Sunal et al., 2004). The following NASA/NOVA reform course characteristics resulted from this definition:

- emphasis on facilitating all students' learning of STEM
- use of pedagogy engaging students' prior knowledge
- use of structured inquiry pedagogy with active and extended student participation as a regular part of the instruction
- refocusing of the role of the instructor who works to become a reflective practitioner using action research

- use of integrated multiple learning formats, not only separated lecture and lab
- refocusing of STEM content on a few key ideas covered in depth
- use of interdisciplinary approaches making connections in course content
- use of student group reflection and learning activities focused on interactive and collaborative learning through shared responsibility
- emphasis on evidence-based learning strategies, using relevant and real data reflecting the way science is done
- use of diverse technology in most course activities to facilitate learning
- focusing on performance assessment forming the greater part of course assessment (DeBoer, 2004; Heppert & French, 2004; Mason & Gilbert, 2004).

Other studies reporting results of faculty development and course reform projects supported the NASA/NOVA model reforms (Gabel, 2004; Jordan, Elmore, & Sundberg, 2004; McCormick, & MacKinnon, 2004; Sunal et al., 2001; Sunal et al., 2004; Waggoner, Schaffner, Keller, & McArthur, 2004). Evaluation and research results reported by these studies supported the key elements of the NASA/NOVA model.

## NATIONAL STUDY OF EDUCATION IN UNDERGRADUATE SCIENCE: RESEARCH MODEL

The goal of the National Study of Education in Undergraduate Science (NSEUS) was to determine the feasibility of creating reforms in undergraduate science in order to provide an alternative to existing teaching and learning in undergraduate courses. The major elements of this study included (1) a *literature review* of research on previous undergraduate science reform; (2) a *national survey* of faculty and courses at institutions involved in a professional development program aimed at undergraduate reform program; (3) a *pilot study* determining the feasibility, or proof of concept, of the procedures and instruments for gathering data; and (4) a *national study* of a sample from a population of institutions involved in a long-term professional development program and reform of one or more undergraduate courses with the intention of improving student outcomes. In this chapter we address findings from the first three elements. Chapters 2 and 3 address the method and findings related to the fourth element.

The research model for the NSEUS study links faculty and student variables. To identify possible linkages the study investigated the beliefs, perceptions, experiences, and actions related to the professional knowledge of the course instructor. These faculty variables included pedagogical content knowledge and ability in planning, implementing, and conducting meaningful learning in an undergraduate science course. Student variables investigated were their daily experiences of a course's learning environment, including content structure, and organization, and the instructor's pedagogical content knowledge and beliefs about student learning. These variables are experienced by students and used to construct their perceptions of, and values concerning, the overall character and nature of the course's science content. As a result of these cumulative experiences, student short-term learning outcomes including perceptions, beliefs, and achievement can be predicted in a general sense. Long-term outcomes may be predicted from short-term outcomes. These long-term outcomes can evolve from short-term predispositions established during a course, for example, toward openness to considering the extent of evidence available to support a hypothesis or the obverse, or a lack of willingness to search for additional evidence. Long-term outcomes may include perceptual understandings such as a conception of the nature of science as evidence based, or the conception of the nature of science as a set of final form statements that can simply be memorized and assumed to have support. Another long-term outcome may be teacher pedagogical content knowledge (PCK) that is demonstrated in specific ways as a means to represent knowledge for learning by students.

The overall sequence of events begins with existing undergraduate science courses (see Figure 1.4). Faculty teaching these courses participated in a multiyear professional development program, NASA/NOVA, and reformed courses by instituting changes in the learning environment and course structure. Faculty also worked to foster and develop their own instructor PCK. Collaboration in a team fostered these reform efforts. It was hypothesized that the sum total of faculty engagement in processes supporting reforms may result in altering their beliefs about teaching and student learning. Several years later, the reformed course was investigated during NSEUS with short- and long-term data collected on faculty and undergraduate students in these courses and with comparison undergraduate science courses at the same institution.

A variety of factors can play a role in this model and may lead to alternative hypotheses. These factors include the history and tradition of the teaching occurring in the science departments at an institution, student backgrounds and motivations in courses (e.g., student's academic major), and the nature and prevalence of teaching and learning science in the region's elementary schools. To address these factors, and because such factors are context-specific, comparison courses were selected from the same institution, in most cases from a similar or the same science department. As a means of further addressing such factors, NSEUS utilized random and stratified selection of participant courses and students from the population of institutions.

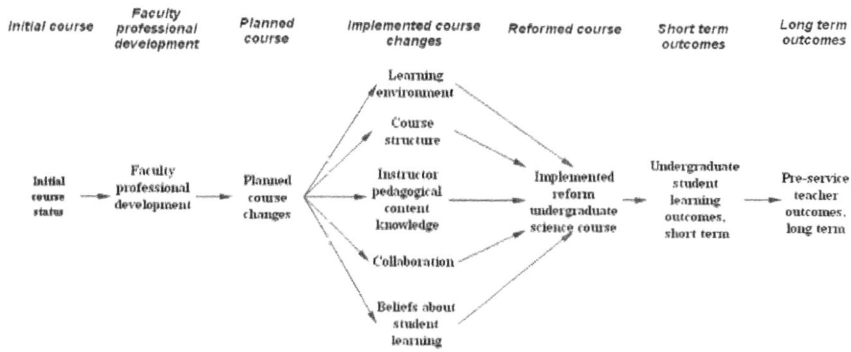


**Figure 1.4.**–Undergraduate science course impact research model.

The NSEUS study population involved 103 institutions at which undergraduate reform courses were taught as a result of participation in the NASA/NOVA professional development. From this population, 20 higher education institutions were selected for the sample. At these institutions, one NASA/NOVA reform course (experimental course) was matched with one undergraduate science course not associated with the NASA/NOVA program (comparison course), totaling 35 courses serving over 2,500 students. Data were collected from course faculty, current students, and past student participants who had graduated and were elementary classroom teachers. Additional data were collected on the learning environment found with the undergraduate courses at the college or university and in the K–6 classrooms of inservice teachers who were graduates of the selected courses.

One overall research question and four sub-questions link the NSEUS Undergraduate Science Course Impact Research Model outlined in Figure 1.4.

> **Problem:** Do efforts in changing undergraduate science courses through professional development that are based on the key issues in national reports of education make a difference in student learning outcomes?

**Central Research Question:** How do undergraduate entry-level science courses, differing in level of reform, affect student learning outcomes?

## Sub-questions

- Does faculty professional development change undergraduate science faculty teaching practice?
- How do science teaching/learning course characteristics differ between courses?
- How do these differences relate to the learning outcomes of undergraduate students?
- How do the differing levels of course reform relate to the short-term learning outcomes of undergraduate students and long term outcomes for graduated inservice K–6 teachers in their own school classrooms?

To investigate the problem of the study, its central question and sub-questions, NSEUS was conceptualized as developing from several primary study components sequenced so that earlier parts informed later parts. The eight sequenced NSEUS study components were:

1. review of the NASA/NOVA faculty development program, 1995–2006
2. review and synthesis of the research literature on reform in undergraduate science
3. development of the National Study of Education in Undergraduate Science research model
4. survey of the population of institutions involved in a program (NASA/NOVA) reforming undergraduate science, 2006–2007
5. pilot study of a small sample of courses selected from the population of institutions conducted to determine the efficacy of the procedure and instruments planned for gathering data in a large scale national study, 2006–2007 (described elsewhere, see Chapter 2)
6. plan and procedure for conducting a national study with a sample of institutions involved in a program of reforms in education in undergraduate science, 2007–2012 (described elsewhere, see Chapter 3)
7. overview of the results, conclusions, and lessons learned from the national study of a sample of institutions involved in a program of reforms in education in undergraduate science, 2007–2012 (described elsewhere, see Chapter 3)

8. sponsorship of two national dissemination research conferences and publications on undergraduate science teaching—June, 2011 and May, 2012 (described elsewhere, see NSEUS website at http://nseus.org)

This chapter focuses on discussion of the items 1–4, while Chapter 2 discusses item 5, and Chapter 3 presents items 6–8.

## THE RESEARCH LITERATURE ON REFORM IN UNDERGRADUATE SCIENCE

An extensive review was conducted of numerous sources on research and reform in undergraduate science teaching published from 1999–2008. Two major strands relating to undergraduate course reform were determined: a faculty strand and a student achievement strand. Within the strands, specific themes were identified. The research literature review indicated that reform efforts are needed in undergraduate science courses serving preservice elementary teacher candidates; undergraduate science faculty have benefited from professional development on teaching; and support exists for specific types of reform. The original literature review was an initial means of determining criteria with which to examine and identify the level of implementation of reform in a course. The literature review also identified data collection procedures and instruments that might hold promise in the national study. Following the review, 23 studies meeting stricter standards of research criteria were extensively analyzed and synthesized in a process guided by the research question, "What criteria can be used to identify the level of implementation of reform in an undergraduate science course?"

The synthesis further identified criteria for determining if a characteristic found in a course can be considered as a reform component. These criteria were (1) universality: all undergraduate reform science courses include this component, (2) uniqueness: change or enhancement without which student outcomes would not result if it were not present, and (3) validity: relevant and valid in that the presence of this variable is measurable in short- and long-term student science performance. The detailed research paper can be read at the NSEUS web site (nseus.org) as Background Research Paper No. 2: *What Do We Know About Undergraduate Science Course Reform? Synthesizing Themes* (Sunal & Sunal, 2008).

Factors causing change were considered as a component of the criteria. These factors are of two kinds: individual and situational (Henderson &

Dancy, 2007). *Individual factors* are those that impact moving from traditional to reformed practice. Individual factors are specific to an instructor and include prerequisite knowledge, skills, and beliefs. Other individual factors may exist. *Situational factors* include student resistance to change; time, structure, and expectations for content coverage; lack of instructor time; departmental norms; lack of resources; promotion and tenure and merit pay concerns; lack of a core of supporters such as a team of colleagues; and the grading system. We further assumed that other situational factors may exist.

Using the literature review as the basis, criteria were developed to identify the level of implementation of reform in an undergraduate STEM course including the following:

- emphasizing facilitation of all students' learning of science
- using pedagogy that engages students' prior knowledge
- using structured inquiry pedagogy with active and extended student participation as a regular part of instruction
- refocusing of the role of the instructor who works to become a reflective practitioner
- using integrated multiple learning formats
- refocusing of science content on a few key ideas covered in depth
- using interdisciplinary approaches in course content
- faculty interacting between colleges regarding teaching and learning
- using student group reflection and learning activities focused on interactive and collaborative learning
- emphasizing evidence-based learning, using relevant and real data
- using diverse technology to facilitate learning
- focusing on performance assessment
- faculty participating in a collaborative team
- positive climate relating to the reform effort's goals
- administrator presence and providing support
- focusing on goals to be accomplished rather than barriers
- planning for incremental rather than initial large change
- ongoing and consistent monitoring of the reform activities

In addition to the review and synthesis of the research literature to identify criteria necessary for an undergraduate science course to claim that it is a reform course, we also explored the question in a national survey of the NSEUS institutions.

## NSEUS SURVEY OF POPULATION OF INSTITUTIONS INVOLVED IN A PROGRAM REFORMING UNDERGRADUATE SCIENCE

An important step in the process of implementing a national study was conducting a national survey of a population of institutions whose faculty participated in a large professional development program that led to reforms in undergraduate entry-level STEM courses. The NASA/NOVA provided such a population. It was the largest program at that time involved in professional development of undergraduate science faculty (Sunal & Sunal, 2008; Sunal, Sunal, Sundberg, Mason, Lardy, & Matloob-Haghanikar, 2008; Sunal, Sunal, Mason, & Zollman, 2008). The NSEUS national survey enabled description of the population of institutions, faculty, and courses. We used the description to determine variables related to reform activities, the implementation process, and the learning environment including the content structure and organization (curriculum) and pedagogy used in the courses. This survey was intended to describe the sustained impact of reforms conducted in undergraduate STEM courses over the previous 12 years. The survey focus was on factors associated with institutionalized courses created as a result of the NASA/NOVA program. We collected survey data from the population from 2005 through spring of 2007. The major categories of survey findings are identified in the sections that follow.

### The NASA/NOVA Population of Institutions

The national survey found the population was represented by a diverse set of 103 institutions (see Figure 1.5) ranging from tribal colleges to doctoral/research universities—extensive (R-I) using the Carnegie (1994) classification. Approximately 28% of the institutions were primarily Hispanic, Historically Black, and Native American serving institutions. Most institutions (52%) were master's degree granting, with 47% MA-I and 5% MA-II institutions. The representation of bachelor's degree granting institutions (21%) included BA-GEN (14%), BA-LA (6%) and BA-SPECI (1%) classifications. Doctoral granting universities (26%) were represented by DR-EXT (16%) and DR-INT (10%). Community colleges in the AA classification represented 1% of the population.

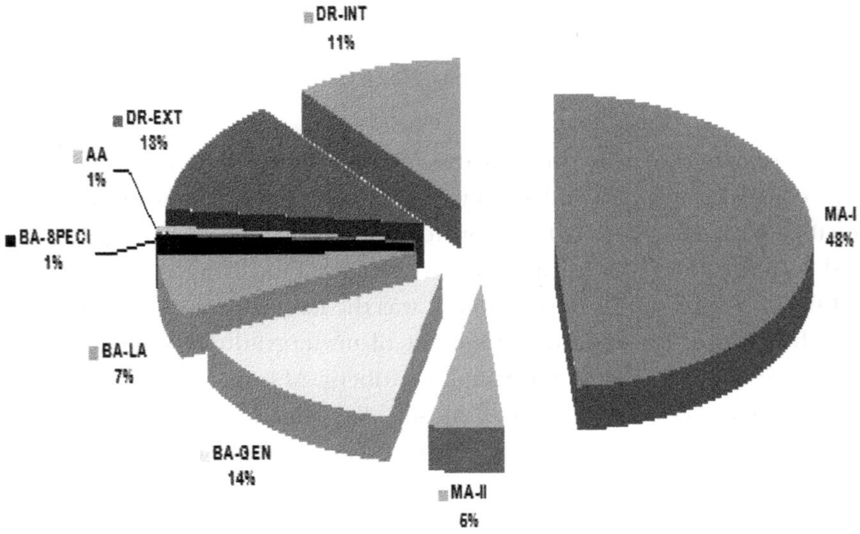


**Figure 1.5.** Carnegie classification of NSEUS institution population.

## NASA/NOVA Reformed Undergraduate STEM Courses at the Time of the Survey

Within the national population of institutions involved in NASA/NOVA, 72 (70%) of the 103 institutions were still offering the implemented courses at the time of the survey (see Table 1.1). The reforms were sustained in the courses. Courses were no longer being offered at 31 institutions (30%) involving 39 courses, 21% of the 185 total courses originally funded. The earliest and latest implemented reforms and course offerings occurred between 3 and 12 years before the national survey. The average course implementation occurred seven years before the survey was completed. An important finding here was that the reform courses, once implemented, were sustained by factors in the context for which they had been planned.

At many institutions, NASA/NOVA reformed courses influenced other faculty teaching various courses at the institution, as instructional practices were changed. Additional courses ($n$ = 118) were created at institutions as a direct result of the impact made by the original NASA/NOVA team reform process on campus (see Table 1.1). Faculty creating these additional courses had not received NASA/NOVA professional development or grant funding. The NASA/NOVA trained team instead acted as facilitators with peer faculty who expressed interest in the results of the original reforms.

Physics faculty at an Ohio institution, for example, created a reformed introductory Physics 101 course after seeing the results of student experiences with the introductory Biology 101 NASA/NOVA course. We refer to these courses as NASA/NOVA-like courses (or NASA/NOVA course clones), with some developed in the same science department and others in different departments. The 118 NASA/NOVA-like courses represent an impact 64% greater than the original 185 courses developed in the professional development program at the time of the survey in 2007. Of the 103 institutions, 49 (48%) reported at least one NASA/NOVA-like course with many developing more than one of these courses (see Figure 1.6).

**Table 1.1. Institutions Offering Reform Courses Created Under the NASA/NOVA Program**

| *Course Status* | *Institutions* | *Number of Reform Courses* | *Institutions Offering NASA/NOVA-Like Courses* not *Developed Under NASA/NOVA Program but Influenced by NASA/NOVA Campus Activity (Number of Courses)* |
|---|---|---|---|
| Reform courses still offered | 72 | 146 | 41 (104) |
| Reform courses no longer offered | 31 | 39 | 8 (14) |
| Total | 103 | 185 | 49 (118) |

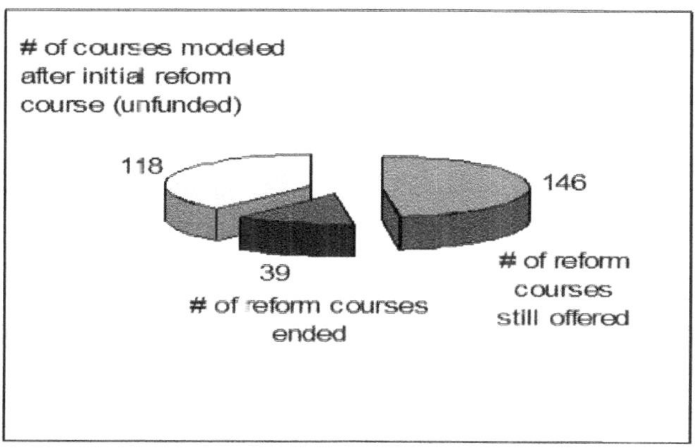


**Figure 1.6.**–Reform courses at study population of higher education institutions.

## Faculty Collaborative Teams at Institutions Involved in Course Reform

Faculty from across the campus at each institution made up NASA/NOVA collaborative teams. The teams included faculty from STEM departments, with two to three most common, and at least one education faculty member. The teams collaborated in setting goals, planning, and implementing courses strong in content and pedagogical methods. These teams encompassed 354 faculty members, with 37 new faculty members added over time following the original implementation of the course to replace members leaving at 24 of the original institutions. Sustained offering of the NASA/NOVA reform course over many years was strongly related to the presence of a permanent faculty team monitoring and, at times, directly involved in the teaching of the STEM content course. Most of the institutions no longer offering the original NASA/NOVA courses reported that the collaborative teams disbanded and were no longer involved with or monitoring the course.

Although collaborative faculty teams were a significant factor in developing and sustaining reform, teams were not stable at a majority of the NASA/NOVA institutions. Only 11 (10%) of the 103 institution teams, remained intact by the time of the survey in 2007 (see Table 1.1). To be considered intact, the team contained the same members from its inception until 2007. Nine (82%) of the institutions with intact teams were still offering the NASA/NOVA reform courses at the time of the survey although two others were not. Of the remainder with active teams having one or more of the original members ($n$ = 61), a subtotal of 51 (84%) were still offering the original reform course, while 10 were not.

Teams were completely replaced or abandoned at 31 institutions (31%). Twelve institutions at which reform courses were still being offered had completely replaced all faculty members of their original teams, while 19 institutions whose team members gradually left or whose team disbanded had stopped offering the NASA/NOVA reform course. In each of the institutions where the teams were completely replaced, the sustained NASA/NOVA courses were being taught by new team faculty members added over time. Statements in the surveys report that the replacement members were trained and mentored by the outgoing and remaining team members in using the NASA/NOVA model.

The rationale given for the team member replacement, reduction in team size, or the disbanding of the team was varied. The reasons included new responsibilities for the team's faculty, members had left the institution, members were now deceased, the need for the NASA/NOVA course had decreased, or the institutional budget had decreased affecting the course offering. Among those institutions with completely replaced teams where

NASA/NOVA reform courses were no longer offered, 15 (80%) of the total of 19 teams had lost faculty because they all left the university or were deceased (see Table 1.2).

**Table 1.2. Collaborative Teams at Each /NASA/NOVA Institution in 2007**

| *Course Status* | *Team Intact* | *Teams With Two or More Continuing Members* | *One Team Member Continues* | *Entire Team Left* | *Institutions Replacing Team Members (Members Added)* |
|---|---|---|---|---|---|
| Reform courses still offered at 72 institutions | 9 | 22 | 29 | 12 | 21 (34) |
| Reform courses not being offered at 31 institutions | 2 | 3 | 7 | 19 | 3 (3) |
| Total Teams at 103 institutions | 11 | 25 | 36 | 31 | 24 (37) |

Once developed, reforms created in the NASA/NOVA professional development program had continued in the large majority of cases. The single-most important factor in the sustaining course reform was identified as the continuous involvement of a collaborative faculty team. When a team stopped functioning, this was reported as mainly due to external factors and not that the faculty had lost interest in continuing the reform. Teams did change membership in some cases with full replacement; however, the existence of the team was the primary factor in sustaining course reform.

## Characteristics of NASA/NOVA Reformed and Comparison Courses

Common characteristics of the 185 reform courses developed by NASA/NOVA faculty were reported in the national survey. Reform courses were implemented as single discipline area courses (e.g., biology or physics) and also as interdisciplinary courses (e.g., aerospace science, natural science). All of the entry-level reform courses were available to non-majors, with elementary education majors forming one student component. Every course was one that could be used a means of fulfilling the science or mathematics requirement in an elementary teacher education certification program at the respective institutions. The NASA/NOVA reform courses differed from each other not only in STEM content but also in the pedagogy implemented. Many of the reformed courses reported in the survey described a majority of the students' time in class sessions as involving them in integrated classroom and lab learning activities during about two-thirds of the class time per week (see Table 1.3). This strategy was compatible with an inquiry approach using a "just-in-time" learning technique. The other one-third of class time involved supportive interactive discussions, use of technology, and lecture (see Figure 1.7). Although there was diversity among the courses, several consistently used reform characteristics were identified by course instructors and team members who developed the courses, to include:

- starting classes with materials or data students can concretely see or touch
- querying students about their prior knowledge of the current concept and/or ascertaining that prior knowledge through students' interactions with materials
- planning for and allowing students to become aware of their alternative conceptions (misconceptions)
- actively involving students during most class sessions
- using simulations and role playing, as in demonstrating the citric acid cycle
- using collaborative and cooperative student learning groups during class activities
- overtly commenting to initiate student reflection on their learning as individuals and in groups
- emphasizing student construction of knowledge through inquiry oriented learning
- deemphasizing memorization of science concepts and algorithms
- applying continuous alternative assessment, rather than using only a few traditional exams.

**Table 1.3. Instructional Methods Reported Used in NASA/NOVA Reform Courses**

| *Instructional Method* | *Average % of Time Per Week* |
|---|---|
| Lecture | 15% |
| Traditional lab | 03% |
| Discussion/Interaction with student groups | 10% |
| Inquiry-based integrated lab | 68% |
| Integrated use of technology | 04% |

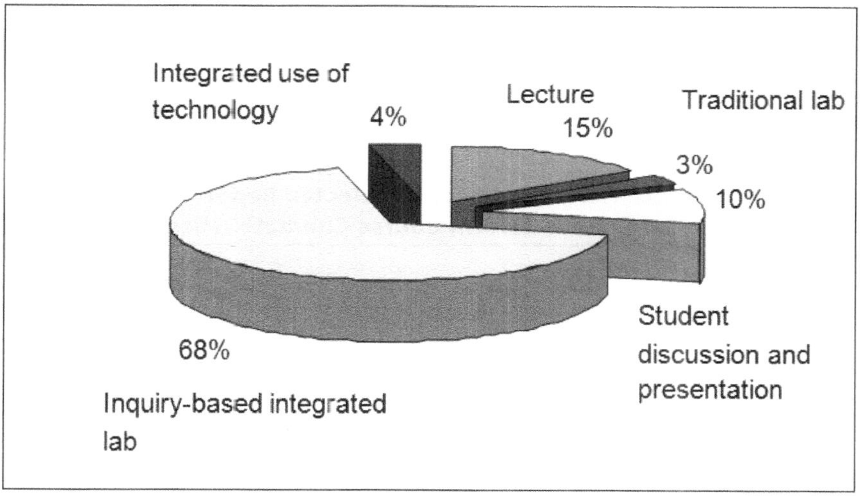


**Figure 1.7.** Pedagogy in surveyed population of reform courses

Instructors were asked to describe the rationale they used to identify the major characteristics incorporated in the NASA/NOVA course. All instructors reported a focus on implementing pedagogy that helped their students construct an understanding of the nature of science and of the key concepts basic to the STEM topics studied. Since elementary education majors formed one of the student groups in the courses and sometimes the entire student body, instructors also had a focus on helping students transfer conceptual learning to everyday life and future careers. Laboratory activities and discussions, for example, at times considered which aspects of key concepts should be taught to students at various elementary grades. Such an approach was reported to be informative to all students in the course as they considered the complexities and abstractness of a concept.

The identified reform NASA/NOVA course characteristics align with the national science standards whose aim is to make STEM concepts more meaningful for all students.

Overall characteristics of the NASA/NOVA reform courses surveyed included frequent offerings averaging twice a year, generally in fall and spring (see Table 1.4). The courses had a high average enrollment of minority students (25%), were generally four credit hours rather than three credit hours, and included laboratory work as a regular part of the course, not as an add-on that may be taken at another time. Lab and lecture were commonly reported as including the same content during the same week of the semester or instructors reported that the lab and lecture portions of the course were integrated in the same room and at the time. The active NASA/NOVA courses at the time of the survey in 2007 enrolled about 10,000 students annually. The population of science courses surveyed represented a sample of the total number of courses and students enrolled in undergraduate science courses in the United States during 2007.

**Table 1.4. Frequency of Selected Reported NASA/NOVA Reform Course Characteristics**

| *Characteristic* | *Average* |
|---|---|
| Years course has been offered at institution | 8 |
| Number of times offered in a year | 1.8 |
| Course enrollment | 39 |
| Minority enrollment | 25% |
| Credit hours | 4 (range = 3–5) |
| Number of sections per semester | 1.5 (range = 1–6) |
| Number of student enrolling in reform courses per year | ~10,000 |

The national survey data included information on 12 STEM comparison courses that had not been involved in the NASA/NOVA course reform process but were at the same institutions and in the same or similar departments. The survey information from these courses indicated that lecture was utilized 68% of the time during weekly class periods, laboratory work utilized 10% of the time, discussion and interaction of student groups, 10%, integrated lab, 6%, and integrated use of technology, 6% (see Figure 1.8). Based on the reported survey data, there was a difference in the instructional process occurring in NASA/NOVA reform courses when compared to similar non-NASA/NOVA courses in the same or other departments at these institutions.

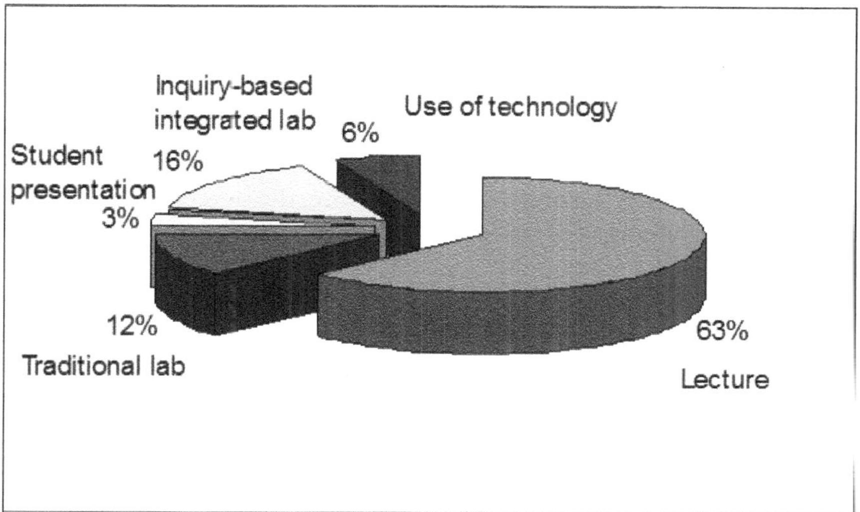


**Figure 1.8.** Pedagogy in comparison (non-reform) undergraduate courses.

## National Survey Summary

The NASA/NOVA institutions surveyed represented a range of Carnegie types from Research 1 to bachelor's degree granting, and from large in size to small, with a representation of primarily Hispanic, Historically Black, and Native American serving institutions. The learning environment in reform courses at these institutions shared overall common course features: (1) involving all students in an inquiry/investigative approach to learning science, (2) including fully integrated inquiry/investigative activities that involved the majority of a week's class time, (3) using collaborative and cooperative learning groups during course activities, and (4) using continuous alternative assessment, rather than only a few traditional exams.

After 12 years, 72% of the population of institutions continued to offer a total of 146 reform courses In addition to the reform courses created as a result of the NASA/NOVA professional development program, additional courses (n = 118) were created at these institutions as a direct result of the impact made by the original reform process on campus. These value-added reform courses represented an impact 64% greater than the original 185 courses developed in the professional development program over a 12-year time period. Once developed, reforms created in the faculty professional development program continued in the large majority of cases. The single-most important factor in the sustained offering of these innovative "reform" courses was the continuous functioning of a collaborative faculty team. When compared to a small sample of similar non-NASA/NOVA courses, one may ask, "Which of the two types of courses more likely represented undergraduate STEM learning for all students?" If we are to make a change in the way STEM disciplines are viewed by students, we have to change the way STEM content is being experienced by students. In traditional courses, where the majority of the time is spent in lecture, the stereotype that STEM concepts are irrelevant will continue to be perpetuated.

## THE NEXT STEP

Having reviewed and synthesized the research literature, hypothesized a possible research model based on the synthesis, and surveyed a national population of institutions involved in a common professional development experience, NSEUS embarked on an intensive national study of undergraduate entry-level science courses at a sample of the institutions. Chapters 2 and 3 describe the study, its findings, and its conclusions and implications.

## ACKNOWLEDGMENT

This material is based upon work supported by the National Science Foundation under Grant No. 0554594. Any opinions, findings, and conclusions or recommendations expressed in this material are those of the author(s) and do not necessarily reflect the views of the National Science Foundation.

## REFERENCES

American Association for the Advancement of Science. (1993). *Benchmarks for scientific literacy, Project 2061*. New York, NY: Oxford University Press

Backer, P. (2002). *Mathematics and science teacher education program (MASTEP) final report*. Retrieved from http://www.mastep.sjsu.edu/final_report.htm

Bland-Day, J. (1999). *Barriers associated with faculty creating change in teaching at higher education institutions*. Unpublished doctoral dissertation, University of Alabama, Tuscaloosa, Alabama. *Dissertation Abstracts International*.

Carnegie Foundation for the Advancement of Teaching. (1994). *Carnegie classification data file*. Retrieved from http://classifications.carnegiefoundation.org/resources/

Christopher, J. E., & Atwood, R. K. (2004). Interdisciplinary curriculum planning in a college course In D. Sunal, E. Wright, & J. Bland-Day (Eds.), *Research in science education: Reform in undergraduate science teaching for the 21st century* (pp. 123–136). Greenwich, CT: Information Age Publishing.

DeBoer, G. E. (2004). A science research organization perspective on reform in teaching undergraduate science. In D. Sunal, E. Wright, & J. Bland-Day (Eds.), *Research in science education: Reform in undergraduate science teaching for the 21st century* (pp. 317–328). Greenwich, CT: Information Age Publishing.

Francis, G., Adams, J., & Noonan, E. (1998). Do they stay fixed? Results of a longitudinal study of students 3 1/2 years after taking reformed physics courses at MSU-Bozeman. *The Physics Teacher, 36,* 488.

Gabel, D. L. (2004). A model for reform in teaching chemistry: With a focus on prospective elementary teachers. In D. Sunal, E. Wright, & J. Bland-Day (Eds.), *Research in science education: Reform in undergraduate science teaching for the 21st century* (pp. 425–444). Greenwich, CT: Information Age Publishing.

Henderson, C., & Dancy, M. (2007). Barriers to the use of research-based instructional strategies: The influence of both individual and situational characteristics. *Physical Review Special Topics: Physics Education Research, 3*(2), 020102.

Heppert, J. A., & French, A. (2004). Scientists' perspective on reform in teaching undergraduate science: Serving pre-service science teachers in the context of a heterogeneous student population. In D. Sunal, E. Wright, & J. Bland-Day (Eds.), *Research in science education: Reform in undergraduate science teaching for the 21st century* (pp. 329–350). Greenwich, CT: Information Age Publishing.

Jordan, W., Elmore, B., & Sundberg, C. W. (2004). A model for reform in teaching in engineering and technology: With a focus on prospective elementary teachers. In D. Sunal, E. Wright, & J. Bland-Day (Eds.), *Research in science education: Reform in undergraduate science teaching for the 21st century* (pp. 511–524). Greenwich, CT: Information Age Publishing.

Krinsky, E., Anderson, K., & Kidane, G (1996). *Los Angeles collaborative for teacher excellence (LACTE) evaluation report.* Retrieved from http://oerl.sri.com/reports/te/report1/report1.html

Lederman N. G., & Gess-Newsome, J. (1999). *Examining pedagogical content knowledge*. Boston, MA: Kluwer.

Loucks-Horsley, S., Hewson, P., Love, N., & Stiles, K. (1998). *Designing professional development for teaches of science and mathematics.* Thousand Oaks, CA: Corwin.

Mason, C., & Gilbert, S. (2004). A science education research organization's perspective on reform in teaching undergraduate science. In D. Sunal, E. Wright, & J. Bland-Day (Eds.), *Research in science education: Reform in undergraduate science teaching for the 21st century* (pp. 401–416). Greenwich, CT: Information Age Publishing.

McCormick, B., & MacKinnon, C. (2004). A model for reform in teaching in the biological sciences: Changing the culture of an introductory biology course. In D. Sunal, E. Wright, & J. Bland-Day (Eds.), *Research in science education: Reform in undergraduate science teaching for the 21st century* (pp. 393–408). Greenwich, CT: Information Age Publishing.

National Academy of Sciences. (2007). *Rising above the gathering storm: Energizing and employing America for a brighter economic future.* Washington, DC: National Academies Press.

National Commission on Excellence in Education. (1983). *A nation at risk.* Washington, DC: U.S. Department of Education, the Commission.

National Research Council. (1996). *National science education standards.* Washington, DC: National Academy Press.

National Research Council. (1997). *Science teaching reconsidered: A handbook.* Washington, DC: National Academy Press.

National Research Council. (2001). *Educating teachers of science, mathematics, and technology: New practices for the new millennium.* Washington, DC: National Academy Press.

National Research Council. (2003). *Evaluating and improving undergraduate teaching in science, engineering, and technology.* Washington, DC: National Academy Press.

National Research Council. (2012). *A framework for K–12 science education: Practices, crosscutting concepts, and core ideas,* Committee on Conceptual Framework for the New K–12 Science Education Standards. Washington, DC: National Academies Press. Retrieved from http://www.nap.edu/catalog.php?record_id=13165

National Science Foundation. (1996). *Shaping the future: Strategies for revitalizing undergraduate education; a conference (Washington, DC, July 11–13, 1996).* Retrieved from http://www.nsf.gov/publications/pub_summ.jsp?ods_key=nsf9873

National Study of Education in Undergraduate Science (NSEUS). Retrieved from http://nseus.org

Odell, M., Badger, S., Kennedy, T., Ewers, T., & Klett, M. (2004). Integrating information technology into undergraduate science. In D. Sunal, E. Wright, & J. Bland-Day (Eds.), *Research in science education: Reform in undergraduate science teaching for the 21st century* (pp. 167–180). Greenwich, CT: Information Age Publishing.

Peterman, F. (1993). Staff development and the process of changing: A teacher's emerging constructivist beliefs about learning and teaching. In K. Tobin (Ed.), *The practice of constructivism in science education* (pp. 227–245). Hillsdale, NJ: Erlbaum.

Project Kaleidoscope. (2005). *Project Kaleidoscope Volume IV: What works, what matters, what lasts?* Retrieved from http://www.pkal.org

Raubenheimer, C. D. (2004). Professional development of university science faculty through action research. In D. Sunal, E. Wright & J. Bland-Day (Eds.). *Research in science education: Reform in undergraduate science teaching for the 21st century.* (pp. 199-224). Greenwich, CT: Information Age Publishing.

Scharmann, L., Stalheim-Smith, A., & James, M. (2004). Assessment in college science courses. In D. Sunal, E. Wright, & J. Bland-Day (Eds.), *Research in science education: Reform in undergraduate science teaching for the 21st century* (pp. 137–152). Greenwich, CT: Information Age Publishing.

Siebert, E., & McIntosh, W. (2001). *College pathways to the science education standards.* Arlington, VA: NSTA Press.

Slater, C., Sireci, S., Berger, J., Brown, M., O'Meara, K., Hintze, J., .... DeCecco, L. (2003, August). *STEMTEC II Evaluation report for year 1—fall 2002/spring 2003.* Retrieved from http://k12s.phast.umass.edu/~stemtec/

Smith, D. C. (1999). Changing our teaching: The role of pedagogical content knowledge in elementary science. In J. Gess-Newsome & N. G. Lederman (Eds.), *Examining pedagogical content knowledge* (pp. 163–198). Boston, MA: Kluwer.

Staples, K. (2002). A university student's perspective on reform in teaching undergraduate science. In D. Sunal, E. Wright, & J. Bland-Day (Eds.), *Research in science education: Reform in undergraduate science teaching for the 21st century* (pp. 351–370). Greenwich, CT: Information Age Publishing.

Sunal, D., Bland, J., Sunal, C., Whitaker, K., Freeman, M., Edwards, L., Johnson, R., & Odell, M. (2001). Teaching science in higher education: Faculty professional development and barriers to change. *School Science & Mathematics, 101*(5), 246–257.

Sunal, D., & Bland, J. (2004). Improving undergraduate science teaching through educational research. In D. Sunal, E. Wright, & J. Bland-Day (Eds.), *Research in science education: Reform in undergraduate science teaching for the 21st century* (pp. 1–12). Greenwich, CT: Information Age Publishing.

Sunal, D., MacKinnon, C., Raubenheimer, C. D., & Gardner, F. (2004). A case study of a national undergraduate science reform effort. In D. Sunal, E. Wright, & J. Bland-Day (Eds.), *Research in science education: Reform in undergraduate science teaching for the 21st century* (pp. 225–240). Greenwich, CT: Information Age Publishing.

Sunal, D., & Sunal, C. (2008). Research Brief No. 1: What levels of institutions in a national population are reforming undergraduate science courses? Retrieved from http://nesus.org

Sunal, D., Sunal, C., Sundberg, C., Mason, C., Lardy, C., & Matloob-Haghanikar, M. (2008). Research brief No. 3: What characteristics are found in reformed and non-reformed undergraduate science courses? Retrieved from http://nesus.org

Sunal, D., Sunal, C., Mason, C., & Zollman, D. (2008). Research Brief No. 5: What are the characteristics of a faculty collaborative team at institutions involved in undergraduate science course reform? Retrieved from http://nesus.org

Sundberg, M. D., & Monaca, G. J. (1994). Creating effective investigative laboratories for undergraduates. *Bioscience, 44*(10), 698–704.

Swanson, E., & Bilderback, B. (1998). *Montana STEP Project: Systemic teacher preparation.* Retrieved from http://www.montana.edu/%7Ewwwse/

U.S. Department of Education. (2001, August 21). *No child left behind.* Retrieved from http://www.ed.gov/offices/OESE/esea/nclb/titlepage.html

Waggoner, C., Schaffner, M., Keller, K. L., & McArthur, J. (2004). A model for reform in teaching in the biological sciences: Infrastructure for inquiry in an introductory biology laboratory. In D. Sunal, E. Wright, & J. Bland-Day (Eds.), *Research in science education: Reform in undergraduate science teaching for the 21st century* (pp. 409–424). Greenwich, CT: Information Age Publishing.

Weimer, M., & Lenze, L. F. (1994). Instructional interventions: A review of the literature on efforts to improve instruction. In K. Feldman & M. B. Paulsen (Eds.), *Teaching and learning in the college classroom* (pp. 653–682). Needham Heights, MA: Ginn Press.

Wycoff, S. (2000). *ACEPT: Arizona collaborative for excellence in the preparation of teachers' evaluation reports, 1995–2000*. Retrieved from http://acept.la.asu.edu

Zollman, D. (1997). Learning cycle physics. In E. F. Redish & J. S. Rigden (Eds.), *The changing role of physics departments in modern universities* (pp. 1137–1149). College Park, MD: American Institute of Physics.

Zollman, D. (2004). A model for reform in teaching physics: Large enrollment physics classes. In D. Sunal & E. Wright (Eds.), *Research in science education: Reform in undergraduate science teaching for the 21st century* (pp. 445–458). Greenwich, CT: Information Age Publishing.

CHAPTER 2

# NATIONAL STUDY OF EDUCATION IN UNDERGRADUATE SCIENCE

## Research Design

**Dennis W. Sunal, Cynthia Szymanski Sunal, Donna Turner, Erika M. Steele, Cheryl L. Mason, Corinne Lardy, Dean Zollman, Mojgan Matloob-Haghanikar, and Sytil Murphy**

**ABSTRACT**

Colleges and universities are increasingly being held more accountable for meaningful student learning, sustaining students in their science, technology, engineering, and mathematics (STEM) majors, and increasing student graduation rates. There is a need to assure students, parents, and critics that undergraduate science faculty can transform science content knowledge in their courses representing it in a way that promotes student learning using effective research-based strategies. It is important to investigate the impact of instructor professional development in teaching on student outcomes and develop procedures for studying these efforts at undergraduate science course reform. The National Study of Education in Undergraduate Science (NSEUS), supported by the National Science Foundation, was designed to address these critical needs. The NSEUS study focused on the need

*Research Based Undergraduate Science Teaching*, pp. 35–65

for teaching undergraduate science effectively to diverse undergraduate majors. The study also focused on the subsequent long-term development of pedagogical content knowledge of elementary preservice teachers of science. The study investigated basic questions asked by faculty regarding how to measure and develop more effective entry level undergraduate science courses to sustain students in their science coursework and lead them to graduation in their chosen majors. We describe here the process of determining the NSEUS research design, population, sample, instruments, and extensive data collection procedure. Other aspects of the complete study are presented in Chapters 1 and 3.

## NATIONAL STUDY OF EDUCATION IN UNDERGRADUATE SCIENCE: RESEARCH DESIGN

---

E-mail Date: Friday, 29 June

From: katy29@bioteach.edu

To: john@univ.edu

Subject: "what to do?"

Hi, John:

Thanks for sending your Biology 101 course syllabus via email. It looks like you have good ideas for a freshman course. Modifying my introductory biology BSC 101 course for the fall semester is proving to be difficult for me. As a gateway course 55% of the students regularly drop out or fail the course. The course has on average 250 students broken down into several discussion groups and labs each week. My department administration has given me the charge to correct this situation. Let me explain.

First, I feel the primary responsibility that I have is teaching my students to understand basic biology concepts for more advanced courses. My attempt at changing the course, as well as teaching it at the same time, is very challenging. The department administration, however, does not seem to see it that way. They think each course only has to have a good lecture sprinkled with some technology. The problem is seen as not engaging students to listen to the lecture and that these changes will create the reform necessary to reduce the dropout rate. My students tell me that the lecture seems to be the problem. Students seem to want to be a part of the material presentation and have more control of their own learning. I, honestly, do not know what to try next.

Second, I have tried some things. I found that the students enjoyed movie clips and the demonstrations, but they do not seem to learn the material any better. I always ended up having to explain more material than if I would have just lectured. Many of the students never made the connection. I attempted several things to make BSC 101 more interesting and relevant to the students. I tried clickers. The students were able to answer the questions during the lecture, but the further away from the lecture, fewer students could answer the questions. I do the p-Glo activity with the students even though they are not learning what I want them to learn. I do the activity because students really like it. I am not convinced that teaching students innovatively really works, or maybe I just cannot do it.

Third, I am having trouble getting my colleagues in my department to see a need to put effort into a different way of teaching biology. Our lecture courses have high attrition among all students, even some of our brightest, most of whom would be excellent in a b career. I have informally asked many of my former advisees why they were changing their majors, and a majority said the classes were exercises in memorization techniques. They were very anxious and felt intimidated by the content. Most felt like they got little out of the courses and were changing their majors "to more relevant" ones.

John, I am beginning to question my choice about leaving government research for academia. I never knew there were so many factors that go into creating an innovative course where most students actually can learn. Please give me some insight about how you would create effective change in this course. I need some guidance from someone who has been there.

Frustrated!

Katy Ann

---

Higher education increasingly faces calls to be accountable for more meaningful student learning; sustain students in their science, technology, engineering, and mathematics (STEM) majors; and increase student graduation rates. Katy Ann makes a personal appeal, as a representative example, for help in transforming undergraduate science teaching. Effective undergraduate teaching is a complex process requiring specialized knowledge and skills to do it well. There is a need to assure students, their parents, and critics that undergraduate science faculty, as instructors, can transform science content knowledge representing it in a way that promotes student learning using research-based strategies (Loughran, Mulhall, & Berry, 2004; Sorensen, Evans, & Andersen, 2009). How can faculty gather research-based evidence that would help the many Katy Anns in higher education with her dilemma?

To address this national need it is important, first, to investigate current efforts underway in undergraduate science course reform, and second, to study the impact of such actions on student outcomes in those courses. Funded by the National Science Foundation, the National Study of Education in Undergraduate Science (NSEUS) investigated the impact of entry-level science course reform on undergraduate students. Data were collected on undergraduate science teaching in classrooms of different types of courses, and included both those identified as reformed and as comparisons. The study focused on the short-term impact of a student sample with diverse undergraduate majors. Long-term impact in the study was specifically focused on a student group composed of K–6 elementary preservice teachers who experienced these reformed science courses during their undergraduate programs and were inservice elementary teachers after graduation.

## BACKGROUND

Four assumptions connected to the *National Science Education Standards* (National Research Council [NRC], 1996, 2012; Siebert & McIntosh, 2001) were addressed in NSEUS. The assumptions were (1) what students learn is greatly influenced by how they are taught, (2) teachers' actions are deeply influenced by their perceptions of science as an enterprise and as a subject to be taught and learned, (3) student understanding is actively constructed through individual and social processes, and (4) actions of teachers when teaching science are deeply influenced by their understanding of, and relationships with, students in their classrooms.

Science teaching requires specialized knowledge refined by faculty over time, and developed through extensive instructional experience (Loughran et al., 2004). We should expect to see differences among faculty instructors in undergraduate science classrooms based on their beliefs about teaching and the institutional context in which they find themselves (Sorensen et al., 2009). For those undergraduate science courses involved in reforms applying national report guidelines such as the *National Science Education Standards,* the knowledge of teaching science, as opposed to a person's knowledge of science, has a greater impact on the quality of teaching and the learning of science by students (Lederman & Gess-Newsome, 1999;

Magnusson, Krajcik, & Borko, 1999; Mason, 1999; Morine-Dershimer, & Kent, 1999).

The research literature describes the undergraduate classroom learning environment as having an effect on student approaches toward learning, which impacts students' learning outcomes (Diseth, Pallesen, Brunborg, & Larsen, 2007; Kreber, 2003). When students have a more positive perception of their learning environment, they are more likely to take a deep approach to learning, attempting to use evidence to make connections between the concepts being presented as opposed to memorizing the concept statements for recitation later on a test.

## RESEARCH DESIGN

We began the NSEUS study by examining faculty beliefs and goals, instructional actions, the learning environment provided to students, and the classroom context experienced in courses that had undergone reform at a national sample of higher education institutions. Data were also collected from comparison courses not involved in a planned reform process. We compared several factors in these courses to the learning outcomes of students. Also considered was how differences within and between courses affected current students' short-term and long-term outcomes. The use of both quantitative and qualitative methods using concurrent comparison strengthened the mixed-method research design. The research design allowed triangulation of data from the literature, the faculty and student interviews, classroom observations, and other faculty and student data (see Figures 2.1 and 2.2).

## RESEARCH SAMPLE

The study sample included 20 higher education institutions at which extensive data sets were collected. The diverse participating institutions were located throughout the U.S. The final sample represented the original population of institutions closely. Pre- and posttesting along with institutional visits were completed. In the sample, 35 faculty teaching entry-level reformed and comparison undergraduate science courses were examined closely.

### Reform Course Faculty Sample

The participating institutions, and one or more of their science courses, were involved in the National Aeronautics and Space Administration (NASA) program, NASA Opportunities for Visionary Academics (NASA/

NOVA), faculty professional development program initiated in 1995 (NOVA, n.d.). This multifaceted program was designed to foster reform in higher education through the development and modification of entry-level, undergraduate science courses. Our study's population included teams of faculty from a diverse national group of 103 institutions that had undergone reform over a 12-year period in one or more undergraduate science courses by the time the NSEUS study began.

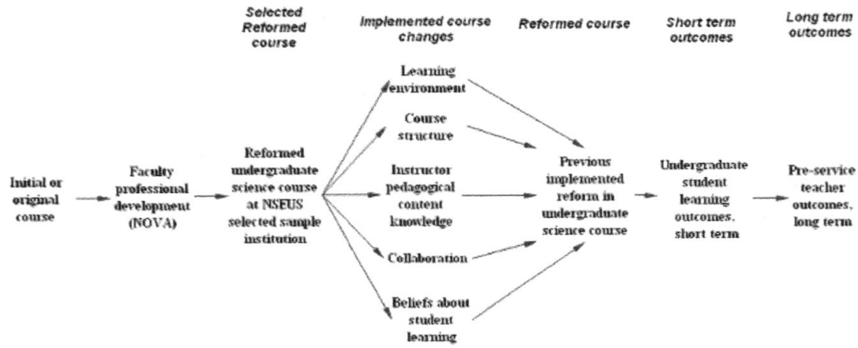


**Figure 2.1.** NSEUS reform undergraduate science course research model.

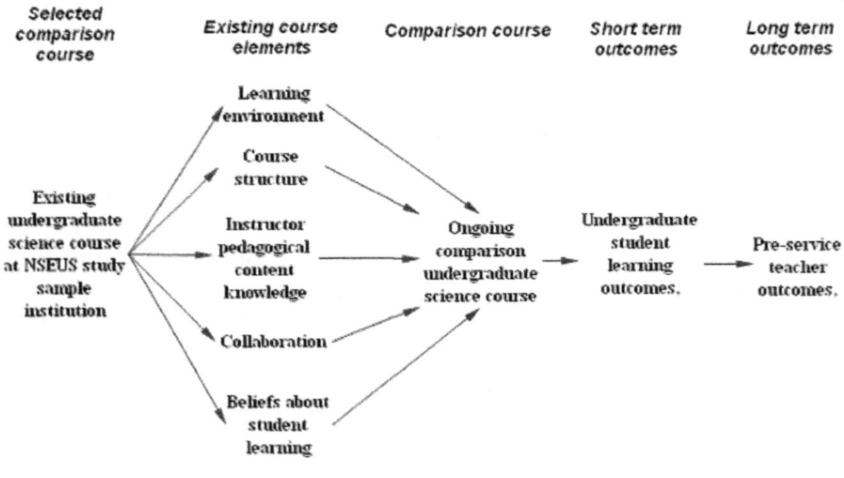


**Figure 2.2.** NSEUS comparison undergraduate science course research model

From this national population, faculty at institutions who had participated in the NASA/NOVA professional development project and who taught undergraduate science were selected to participate in the NSEUS study. A stratified random sample of "reform" courses at 30 institutions was selected based on seven criteria: (1) the NASA/NOVA course was still being taught, (2) the course focused on the science disciplines (single or integrated), (3) the course was an entry-level science course, (4) the course instructor was involved in the NASA/NOVA professional development program or related training activities, (5) the course was required or an elective in the program of study for elementary education majors, (6) institutions sampled represented the population in diversity of geographical distribution and in character (Carnegie Classification level and type), and (7) the reformed course contained elements of the original NOVA model. Faculty were called and emailed to solicit their participation.

Data collection from the beginning initial survey (see Chapter 1) to the final course visitation semester for the selected institutions was completed over a four-and-a-half-year period. Sample attrition, resulting in a final sample size of 20, occurred for several documented reasons. First, four NASA/NOVA courses ceased to be offered due to death of the course instructor or instructor transfer to another institution (see selection criterion 1). Second, the change in curriculum focus of courses at two of the institutions limited the NASA/NOVA course to secondary science education majors only (see selection criterion 5). Third, the course curriculum changed at one institution making the NASA/NOVA course a capstone course to be taken during the senior year (see selection criterion 3). Fourth, a faculty instructor at one institution decided not to complete the study partway through the data collection period. Fifth, posttest data were not completely collected at one institution. Sixth, the curriculum, instruction, and structure of the course were changed at one institution so that its reform character ceased to exist (see selection criterion 7). (Note: the course character change at this institution resulted from budgetary deficiencies and included the creation of large lecture classes and dropping of labs and other features of the original course, so the original NASA/NOVA team members indicated that the reformed course had ceased to exist.)

## Comparison Course Faculty Sample

The comparison sample courses were located at the same institutions as the reformed course sample. The comparison courses were taught by faculty who had not received professional development or funding for course development through the NASA/NOVA Program. Thirty comparison course faculty instructors were planned. All but one of the selection criteria used with the "reformed" sample courses were similar to those used with the comparison courses. Criterion 7 was inapplicable and so deleted. The comparison course selection criteria remaining were (1) the course was currently being taught at a NASA/NOVA selected institution, (2) the course focused on the science disciplines (single or integrated), (3) the course was an entry-level science course, (4) the course instructor was *not* involved in the NASA/NOVA professional development program or related training activities, (5) the course was required or an elective in the program of study for elementary education majors, and (6) institutions sampled represented the population in diversity in geographical distribution and in character (Carnegie Classification level and type). Comparison faculty reported professional development experiences and attempts at reform of their undergraduate science courses; however, these efforts were not related to the original NASA/NOVA program of professional development nor influenced by the existing NASA/NOVA course at the same institution. Comparison sample attrition resulted in a final sample size of 15. Documented reasons for attrition were that their NASA/NOVA institutions were not a part of the final sample (see selection criterion 1), a comparison faculty instructor at one institution decided not to complete the study partway through the data collection period, and posttest data failed to be completely collected at four institutions.

## DATA COLLECTION INSTRUMENTS

The research model included four sets of data. *Course instructor* information was derived from surveys and interviews of faculty. These data were designed to construct inventories describing expectations, knowledge, and beliefs about teaching and student learning (course curriculum, beliefs about student learning, teaching efficacy, pedagogical content knowledge, and faculty collaboration). *Course* data included observational ratings and narrative descriptions of events in the undergraduate science course associated with the instructor's teaching in the course class sessions (classroom learning environment and pedagogical content knowledge). *Undergraduate student learning outcomes—short-term* data included students' ratings, surveys, and narrative description of events through focus groups of their reformed undergraduate science course (classroom learning environment, attitudes, and knowledge achievement). *Undergraduate elementary education student outcomes—long-term* data encompassed observational ratings, narrative description, and surveys of graduated preservice teachers from the sample courses related to science lesson events in their elementary classrooms (classroom learning environment, science curriculum, and pedagogical content knowledge) (see Figures 2.1 and 2.2). The data were analyzed to determine differences between groups and relationships between the instructor and students' course outcomes as a result of the course experiences. As an example, course instructor and elementary teacher pedagogical content knowledge (PCK) were determined from several data sources including responses to background and demographic survey questions, individual structured interview responses, and classroom observation field notes supported by reports of teachers' expressed thoughts and actions in debriefing interviews.

A description of the instrumentation is presented below. Several NSELS-trained observers participated during each site visit to assure coverage and observer reliability. Since personal bias is a factor in any study due to the influence of observers' experiences and beliefs (Janesick, 2003), training sessions, a pilot visitation trial, and purposeful reliability checks on instrument use and data analysis were implemented to avoid this bias.

## Constructivist Learning Environment Survey (CLES)

The *Constructivist Learning Environment Survey* (CLES) was designed to monitor the development of interactive strategies in the classroom (Taylor, Fraser, & Fisher, 1997). The CLES was intended to document teachers' understanding of their students' perceptions of the extent to which the classroom learning environment enabled them to reflect on their prior knowledge, develop as autonomous learners, and negotiate their understandings with other students. The instrument is divided into five key dimensions of a critical constructivist, standards-based science learning environment.

1. *Personal Relevance Scale*—measures how relevant students feel the course content is to their lives outside of the classroom
2. *Shared Control Scale*—measures students' perceptions of their control over classroom learning
3. *Critical Voice Scale*—measures students' perceptions of their ability to question the teacher's pedagogy
4. *Student Negotiation Scale*—measures students' perceptions of their ability to share their ideas with other students in the classroom
5. *Uncertainty Scale*—measures students' perceptions of the level of inquiry based science knowledge in the classroom

Categories of responses in the instrument are (1) learning about the world, (2) learning about science, (3) learning to speak out, (4) learning to learn, and (5) learning to communicate. Five items are provided for each of the five dimensions with possible responses of: almost always, often, sometimes, seldom, and almost never.

Two versions of the instrument were given to the students in the classrooms sampled during the semester. The first version involved participants in identifying the kind of classroom learning environment they *prefer* and was given at the beginning of the course. The second version asked participants to identify the classroom experience they had as they *perceived* that experience, and it was given at the end of the semester. In this study, the CLES was used to examine students' preferences (expectations) and perceptions (perceptual understanding) of the learning environment in the sample of reform and comparison undergraduate science classes. The instrument was delivered online outside of the regular class time. After completing consent forms in class, students were instructed to respond to an email providing the Internet URL where the CLES could be completed. Students who had not completed the instrument after a brief time were reminded by email on a periodic basis over a few days. Students' data were stored electronically and could be downloaded for analysis.

The CLES instrument has been found to have high reliability on small and large scales (Johnson & McClure, 2004) and in higher education (Wright & Sunal, 2004). Results on the CLES have been shown to be very close to what researchers would expect from classroom observation. The alpha coefficient for reliability in the study was 0.92 for an individual student and 0.98 for the class mean.

## Thinking About Science Survey Instrument (TSSI)

The *Thinking About Science Survey Instrument* (TSSI) uses a quantitative Likert scale format to assess students' understandings of nature of science and the sociocultural resistance to and support for science in nine areas of significant cultural concern (Cobern, 2000):

1. Knowledge is derived through scientific inquiry.
2. Scientific progress is vital to the economy.
3. Science is a positive for the protection of the environment and resource development.
4. Science should influence public policy and be publicly supported but not publicly controlled,
5. Science is a positive force for public health.
6. Science is neutral with regard to religion and morality.
7. Emotions and aesthetics are part of science.
8. Science is open to people regardless of race and gender.
9. Science is a goal for all students.

Table 2.1 provides a description of each of the TSSI categories and the number of items falling under each category of the TSSI.

**Table 2.1. Description of the Sub-scales on Thinking About Science Survey Instrument**

| # | *Scale* | *Sub-Scale Description* | *Number of Items* |
|---|---|---|---|
| 1 | Epistemology | "Science is a superior, exemplary form of knowledge that produces highly reliable and objective knowledge about the real world." | 9 |
| 2 | Science and the Economy | "Modern industrial commercial and information-based economies depend on scientific developments for increasing production, wealth, and general public welfare." | 10 |
| 3 | Science and the Environment | "Science is necessary for the discovery, development, conservation, and protection of natural resources and the environment in general." | 4 |
| 4 | Public Policy and Science | "Science acts in the public interest. Science thus should be supported by public funds; however, the science community is more than capable of policing scientific activity." | 10 |
| 5 | Science and Public Health | "The conquering of diseases and physical affliction and the great advances in public health are made possible by science and will not continue without science." | 4 |
| 6 | Science, Religion, & Morality | "People make moral choices about the use of scientific findings but science itself is morally neutral. Science is also neutral with regards to religion. The importance of science, however, is such that science must be protected from the intrusive activities of some religions." | 7 |
| 7 | Science, Emotions, & Aesthetics | "Scientists are often passionate about their work but the work of science best proceeds on the basis of objective reason and empiricism. There is a beauty to science. Indeed 'elegance' is often required of scientific ideas. " | 4 |
| 8 | Science, Race, & Gender | "Science is an 'equal opportunity employer.' Race, gender and other personal factors are irrelevant in science. This is the ubiquitous claim of the science community." | 4 |
| 9 | Science for All | "The importance of science is such that it should be taught at all levels of schooling. Every citizen should have attained at least a minimal level of science literacy." | 8 |

*Source:* Scale name and description are quoted from Cobern (2000).

Analyses of TSSI data followed the author's use of categories and weightings for the items (Cobern, 2000). Each of the nine TSSI factors included 4 to 10 items from the instrument to measure objections to science and defenses for science. Responses are given in a Likert-type scale ranging from 1 to 5: "Strongly Agree" at 5 points, "Agree" at 4 points, "Uncertain" at 3 points, "Disagree" at 2 points, and "Strongly Disagree" at 1 point. Each category produces a score that measures a model science position if the student disagrees/disaffirms (score of 10–25), is neutral (score of 26–35), or agrees/affirms (score of 36–50) with commonly presented images of science in the literature and in science courses. Instrument reliability was established with an alpha reliability coefficient of 0.78. The category item alpha coefficient was 0.79 (Cobern, 2000).

## Draw-A-Scientist Test (DAST)

The *Draw-A-Scientist Test* (DAST) was developed by Chambers (1983) to examine people's beliefs about what a scientist is. Students draw a picture of what they think a scientist looks like. We provided students with a sheet of white paper and the instructions: "Draw a scientist" or "Draw what you think a scientist looks like." If a student asked for clarification, he or she was instructed to draw an "image of a scientist" or "what you think a scientist looks like," that "stick figures were okay," and/or that "artistic talent is not being evaluated." Students were told that the drawings were data for a research study and were not going to be graded. The students were asked to indicate their gender and major.

In response to concerns that students were drawing societal stereotypes they thought they *should* draw, rather than what they actually thought, Symington and Spurling (1990) added the prompt: "Do a drawing that tells what you know about scientists and their work." The students next were asked to write a few sentences under their drawings to explain why they drew what they did. In our study, DAST was administered pre and post to undergraduate students during the reform and comparison science course using the prompts reported by Chambers (1983) and Symington and Spurling (1990). Students were given approximately 10–12 minutes to draw the images. The scoring rubric was adapted from Mason, Kahle, and Gardner's work (1991) and used a 14 indicator standard. The analyses were checked for high inter-rater reliability. We trained raters using the 14 indicator standard and developed correlation ratings above 0.95 while rating the same class sample sets of DAST drawings.

## Science Teaching Efficacy and Beliefs Instrument (STEBI-A)

The *Science Teaching Efficacy and Beliefs Instrument* (STEBI) is a survey used to measure two components of teacher efficacy beliefs, the "extent to which teachers believe they have the capability to positively affect student achievement" (Riggs & Enochs, 1990, p. 626): (1) a teacher's self-efficacy (level of confidence in own teaching abilities) and (2) a teacher's outcome expectancy (belief that student learning can be influenced by effective teaching).

The STEBI-A is a version designed specifically to relate to teaching science in a classroom lesson (Riggs & Enochs, 1990). Each question on STEBI-A is specific to one of the two components, but the questions are mixed together. Teachers respond to each item by circling "Strongly Agree" (score = 5), "Agree" (4), "Uncertain" (3), "Disagree" (2), or "Strongly Disagree" (1). The STEBI-A was administered in this study to course instructors and inservice teachers about midterm in the semester (see Table 2.2).

**Table 2.2. Sample Responses Requested or Faculty Instructors on STEBBLA-A**

| | | |
|---|---|---|
| 1. | When an undergraduate student does better than usual in science, it is often because the instructor exerted a little extra effort. | SA A UN D SD |
| 2. | I am continually finding better ways to teach science. | SA A UN D SD |
| 3. | Even when I try very hard, I don't teach science as well as other instructors. | SA A UN D SD |
| 4. | When the science grades of students improve, it is most often due to their instructor having found a more effective teaching approach. | SA A UN D SD |

## Reformed Teaching Observation Protocol (RTOP)

The *Reformed Teaching Observation Protocol* (RTOP) is a classroom observation protocol designed to measure quantitative characterization of the degree to which a science classroom is "reformed" (Sawada & Piburn, 2000). The RTOP draws guidelines from documents published by the National Research Council, American Association for the Advancement of Science, Project 2061, and the National Council for the Teaching of Mathematics. For this instrument, the characteristics of reformed teaching practices are based on national standards for science education. The RTOP observer uses extensive note taking followed by rating a list of classroom characteristics on a scale of 0–4 ("never occurred" to "very descriptive"). The RTOP is reported to have high inter-rater reliability (Sawada & Piburn, 2000). In our study, observers trained in its use developed correlation ratings above 0.85 while observing the same reform and comparison classroom settings. The RTOP analyses followed the authors' use of categories and weightings for the items (Piburn et al., 2000).

Interviews and RTOP rating observations were conducted with both reformed and comparison undergraduate science course instructors. The RTOP is comprised of three sections to be rated by the observer: (1) lesson design and implementation, (2) content (propositional knowledge and procedural knowledge), and (3) classroom culture (communicative interactions and student/teacher interactions) (see Table 2.4). Each section contains specific items to be rated by the observer (see Table 2.3). Specific criteria additionally are used when scoring items under each section.

**Table 2.4. Reformed Teaching Observation Protocol (RTOP) Rating Categories**

| |
|---|
| **1. Background Information** |
| **2. Contextual Background and Activities (Narrative)** |
| **3. Lesson Design and Implementation (5 rating statements)** |
| Ratings focus on students' prior knowledge and the pre-conceptions, engagement in a learning community, whether exploration precedes formal presentation, alternative modes of investigation or of problem solving, and ideas originating with students. |
| **4. Content (10 rating statements)** |
| Ratings focus on propositional knowledge (content, coherence, abstraction, and connections) and procedural knowledge (representing, predicting, thought provoking, reflecting, challenging). |
| **5. Classroom Culture (10 rating statements)** |
| Ratings focus on communicative interactions (communication through a variety of means, teacher questions, student talk, student questions, climate of respect) and student/teacher relationships (active participation, alternative strategies, teacher patience, teacher as resource person, and teacher as listener). |
| **Additional comments** (Optional) |

## Content Representation (CoRe) and Professional and Pedagogical experience Repertoire (PaP-eR)

Pedagogical content knowledge (PCK) in science refers to combining pedagogical and science content knowledge to best communicate and represent science so that it is comprehensible to others—in this case, to students. A teacher's PCK is specific to the particular science concept taught. Two means of capturing and portraying science PCK linked to particular science content and teaching practice were used: observing individual science lessons and interviewing both the undergraduate course instructors and elementary inservice teachers.

**Table 2.3. Sample Observations and Analyses From the Reformed Teaching Observation Protocol**

| LESSON DESIGN AND MPLEMENTATION | | | | | | |
|---|---|---|---|---|---|---|
| | | Never Occurred | | | | Very Descriptive |
| 1) | The instructional strategies activities respected students' prior knowledge and the preconceptions inherent therein. | 0 | 1 | 2 | 3 | 4 |
| 2) | The lesson was designed to engage students as members of a learning community. | 0 | 1 | 2 | 3 | 4 |
| 3) | In this lesson, student exploration preceded formal presentation. | 0 | 1 | 2 | 3 | 4 |
| 4) | This lesson encouraged students to seek and value alternative modes of investigation or of problem solving. | 0 | 1 | 2 | 3 | 4 |
| 5) | The focus and direction of the lesson was often determined by ideas originating with students. | 0 | 1 | 2 | 3 | 4 |

Observation, interview, and analysis protocols for PCK were developed by van Driel, Verloop, and de Vos (1998) and later by Loughran et al. (2004) in the *Content Representation* (CoRe) and *Pedagogical and Professional Experience Repertoires* (PaP-ers) instruments. According to Loughran et al., the purpose of the CoRe is to discuss science teachers' understanding of particular aspects of PCK related to a specific science lesson (e.g., main idea of the class session/lesson, knowledge of alternative conceptions, insightful ways of testing for understanding, known points of confusion, effective sequencing, and important approaches to the framing of ideas). The purpose of PaP-eR is to expand the participant's description of the proposed lesson by observing how such knowledge might inform effective classroom practice (Loughran et al., 2004). A PaP-eR is a narrative account of a teacher's PCK highlighting a particular piece, or aspect, of science content to be taught. Specifically, a PaP-eR is designed to purposefully unpack a teacher's thinking about a particular aspect of PCK in that given content and thus is largely based around classroom practice (Loughran et al., 2004). The PaP-eR is intended to represent the teacher's reasoning, that is, the thinking and actions of a successful science teacher in teaching specific aspects of science content (Loughran et al., 2004). The PaP-eR, thus, develops through the interaction of the prompts, questions, issues, and difficulties influencing the particular approach to teaching that content to which the PaP-eR is tied and reflects the richness of the teacher's understanding of science teaching and learning in that field (Loughran et al., 2004). Accordingly, the more specific a PaP-eR is, the more it informs the relationship between teaching and learning. Every action of undergraduate science instructors specifically designed to help students learn science concepts meaningfully is a part of PCK.

Both instruments, CoRe and PaP-er, were used together to capture and portray the pedagogical content knowledge (PCK) of both undergraduate science course instructors and K–6 inservice teachers. See Table 2.5 for questions to be answered while using CoRe and PaP-eR. These questions attempt to access teachers' knowledge about how to teach particular content in science. These data were posted on a chart identified as Content Representations (CoRe) of teachers' thoughts. Pedagogical and Professional Experience Repertoires (PaP-er) are detailed descriptions from classroom observations and of interviews with individual teachers as a result of discussions pertaining to the CoRe.

The procedures for the use of each instrument, modified from the original (Loughran et al., 2004) for our study, are described below.

### *CoRe*

1. In an interview, the instructor or teacher is provided with questions and prompts related to general and specific aspects in the planned lesson for a science concept or topic to be taught. A narrative is developed that represents a first draft of an instructor's/teacher's PCK for this science concept. The prompts include:

   a. What will be the main ideas of this class session or lesson?
   b. What do you intend the students to learn about these ideas?

**Table 2.5. Co-Re and PaP-eR Questions**

| *Interview Questions* | *Important Science Ideas/Concepts* |
|---|---|
| What will be the main ideas of this course session/science lesson? | |
| What do you intend the students to learn about these ideas? | |
| Why is it important for students to know this? | |
| What do you anticipate will be some difficulties and/or limitations connected with teaching this idea? | |
| What knowledge about students' thinking influences your teaching of this idea? | |
| What are other factors that influence your teaching of this idea? | |
| a) Describe how you will teach the main ideas in this lesson. | |
| b) Why will you be using this procedure to teach these main ideas? | |
| What are specific ways you will use to determine students' understanding or confusion around the idea(s)? | |

c. Why is it important for students to know this?
d. What do you anticipate will be some difficulties and/or limitations connected with teaching this idea?
e. What knowledge about students' thinking influences your teaching of this idea?
f. What other factors influence your teaching of this idea?
g. Describe how you will teach the main ideas in this lesson.
h. Why will you be using this procedure to teach these main ideas?
i. What are specific ways you will use to determine students' understanding or confusion around this idea?

2. Each CoRe statement is developed, refined, and then validated during interviews.
3. The process provides a way to enable instructors and teachers to flesh out explanations of their own PCK for a specific science concept they may not normally think about.

### *PaP-eR*

1. The specific science lessons discussed using CoRe are observed with extensive notes are taken. In some cases, these lessons are audio- or videotaped.

2. Specific instances, or examples, of how PCK was applied in the observed lesson described in CoRe are noted in a matrix for each CoRe prompt.
3. Additional actions and interactions, observed or otherwise recorded in the science lesson taught, that expand the instructor's or teacher's PCK repertoire are added to the matrix.
4. Each PaP-eR matrix is developed, refined, then validated during interviews, finally completing a description of an instructor's or teacher's PCK related to the science concept or topic taught.

Four PCK rubrics were used to assess the results of the CoRe and PaP-er narratives: (1) teachers' content knowledge of the science concept(s) involved in the lesson, (2) teachers' knowledge or understanding of student thinking related to the science concept(s) involved in the lesson, (3) teachers' knowledge of how to represent the teaching of science concept(s) involved in the lesson, and (4) teachers' professional development in teaching the science concept(s) involved in the lesson (Sunal, Sunal, Ogletree, Steele, Mason, & Zollman, 2008). The ratings included use of a continual progression of faculty or teacher responses and actions from Expert, Competent, Emergent, to Novice.

Faculty course instructor interviews and field notes were taken during observation of undergraduate reform and comparison science class sessions taught over a weeklong period. Interviews and classroom science lesson observations also were conducted with graduated elementary, now inservice, teachers who participated either in the reformed or the comparison undergraduate science courses while in their preservice program. The inservice observations of science lessons were conducted in the teacher's elementary classroom.

### *Science Content Achievement* Instrument (SCA)

Designed to assess science concept achievement, the *Science Content Achievement* instrument (SCA) was developed in conjunction with each of the reform and comparison course instructors and completed by undergraduate science students in each course. The approach to assessing meaningful understanding described by Nieswandt and Bellomo (2007) was used in the *Nieswandt Model Standardized Test*. The questions included three levels of related concepts: descriptive, hypothetical, and theoretical. Students needed to combine and link different levels of concepts and apply them in a complex system in scenarios or contexts different from that studied in the course. For example, questions developed in a paper by Nieswandt and Bellomo (2007) for biology included the following:

a. Seals remain underwater without breathing for nearly 45 minutes as they hunt for fish but their ancestors could stay underwater only a couple of minutes. Use the concepts of mutations and genetic drift to explain how the ability to not breathe for long periods of time has evolved.

OR

b. White hair in animals (such as in polar bears) is due to the absence of a chemical pigment. Biologists think that ancestors of polar bears had dark fur. Use the theory of natural selection (e.g., directional selection) to explain how polar bears evolved to have white fur.

The *Nieswandt Model Standardized Test* was modified to meet NSEUS study goals. Science content was assessed as a posttest only, short essay achievement test item near the end of the undergraduate science course. In order to compare student learning outcomes across institutions, sample courses, and course science content, each SCA administration contained different science concepts. The content was unique to each course. The SCA question(s) were developed to be a measure of students' science content and critical thinking achievement in the course. The question(s) represented an application and conceptual understanding of what students have learned when given a real-life novel problem situation.

The question(s) assessed a key course concept developed in each course that related strongly to one of the instructor's main content objectives. Each course instructor first was asked to identify a key science course goal. Then, each instructor was asked to select a science concept related to the course goal that was "very important for the students to learn during your course." Using the identified concept, one or two open-ended questions were developed. We edited the question(s), which then were validated by the course instructor as being a critical and relevant assessment measure of achievement in the course. Usually, several drafts were developed before a final form emerged. The SCA science content question(s) were designed using the following characteristics:

1. Require higher order thinking beyond memorization. The questions should be rich in detail and require logic and reasoning connecting the concepts to be evaluated. This means that students will be comparing and contrasting two or more concepts, discussing relationships and differences while explaining the phenomenon involved.

2. Include a related phenomenon in which similar rules and concepts hold. Ask a question about the phenomenon in a way that the explanation employs the course concepts to be evaluated along with their logical connections.
3. Assess why students used the particular concepts and certain connections, and what led them to determine the relationships among concepts.
4. Assess the appropriate use of a concept and a realistic setting reflecting students' understanding of the meaning of that concept.
5. Evaluate whether or not students know when, why, or how to use the rules and/or principles related to the concepts. The caveat is that, sometimes, students know the concept or rule, but while applying the concepts they choose the wrong aspect of the rule or principle.
6. Lead students through a thinking process to first restructure, and next synthesize, their scientific conceptual knowledge, then apply that knowledge to explain the new context.
7. Utilize qualitative rather than quantitative features of concepts.
8. Encourage students to provide explanations from which you can infer whether or not students can identify the related concepts. The targeted concept(s) are not to be obviously identified within the questions.
9. Provide some scaffolding to guide students through these procedures (Zollman, Matloob-Haghanikar, & Murphy, 2012).

Criterion-based rubric measures focusing on levels of concepts developed and linked were used to compare science achievement across courses and disciplines.

The *Science Content Achievement* instrument questions created were to be open-ended, requiring students to recognize and generalize facts and concepts learned in the course and apply them to a new context. The question(s) were analyzed using hierarchies of knowledge, taking them from a low level of recall to a synthesis of ideas. Using Anderson's revision of Bloom's taxonomy (as cited in Zollman et al., 2012) each question was examined for the type of knowledge students were required to use. The knowledge was divided into three knowledge dimensions: factual, conceptual, or application.

a. Factual knowledge is knowledge of the basic vocabulary, terminology, symbols, and so forth students must know to be acquainted with a discipline or solve problems in it. Evidence for factual knowledge included basic elements, single entities, and separate pieces of information mentioned in the student essays, or addressed symbolically, or inferred from other statements

b. Conceptual knowledge/conceptual schema represents the relationships between the basic facts within a body of knowledge about a subject that enables us to form our understanding of that subject. To find evidence of this type of knowledge, essay data were analyzed to see whether the appropriate concepts were employed, the definitions of concepts were clear, and correct meanings were attributed to the concepts. Essays were evaluated for understanding of relations between the concepts and concept links as well as whether attributed properties and specific features were associated with appropriate categories or classes.

c. Application is the ability of students to use the information they recall in a new context, which is found on the cognitive scale of Anderson's revision of Bloom's taxonomy. To find evidence for this type of knowledge, the essays were analyzed to see whether students had the skills, knew the techniques and algorithms, or knew rules and steps of applying a principle. Procedural knowledge manifests itself in different ways, such as executing an algorithm, performing a series of mathematical manipulations as in solving an equation in mathematics or science, or doing an experiment. In essence, procedural knowledge is the knowledge of how to execute the prescribed steps of solving particular problems, or perform a process or series of steps required for verifying a principle.

The cognitive scale examined the types of skills students used to answer the question, such as, are they simply recalling information to state facts or concepts? Or, do they have the ability to use the information to apply, infer, or explain the information in novel situations? (see Table 2.6).

### *Example of an SCA Rubric Analysis*

As an example, we present the analysis of a SCA question asked on an exam in an astronomy course for elementary education majors at a small Midwestern university. This question is based on one developed for an inquiry-based physical science course (McDermott et al., 1996) in which 50 students completed this question on the final exam.

**Question:** You look outside and see a first-quarter moon. Suppose that an astronaut were on the moon looking at Earth. Make a sketch of the Earth as seen by the astronaut. How will the illuminated portion of the Earth appear different three days later?

**Table 2.6. Science Content Assessment Rubric: Cognitive Processes Dimension**

| Cognitive Process | Description |
|---|---|
| **Understand /Compare and Contrast** | Evidence was sought by comparing those aspects and features that were fundamental for justifying cause and effect changes, or comparing variables that provide plausible evidence for why and how and what changed that caused the effect. |
| **Understand/ Infer** | Determining whether the written response showed evidence that the student recognized the patterns that connected a series causally. |
| **Understand/ Explain** | Evidence was sought for a cohesive and convergent argument that described the situation, explained the outcome, and was well supported by showing why and how things are happening. |
| **Apply** | Evidence was sought for the association of facts, concepts and procedures that were reconstructed in connection with the features of a new scenario to present a plausible answer. |

Two responses representative of the types of reasoning that show evidence or no evidence for the traits of our rubric are discussed. The students submitted the drawing below each response as the part of the response.

**Response 1.** *The astronaut would see a 3rd quarter, waning moon. The moon will have moved slightly more in its evolution, making earth see the moon as slightly more than 1st quarter. In contrast, the earth would appear less full to the astronaut on the moon.*

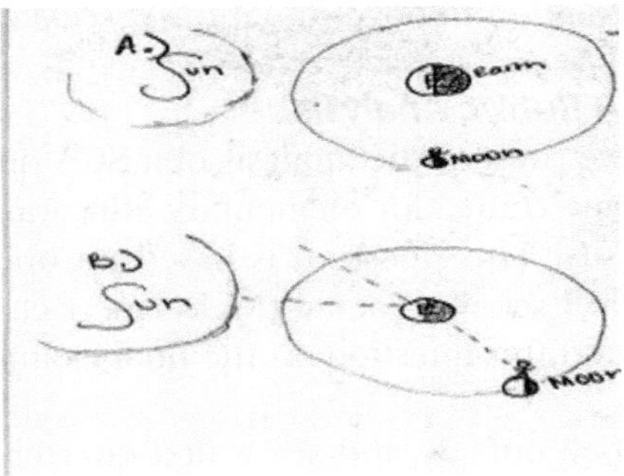

**Response 2.** *The earth-illuminated portion would decrease the same, it would be a waning gibbous instead of a third quarter. It would be even a waning crescent almost a full earth, depending on the rotation.*

Below is a sample analysis of the evidence of several of the response characteristics:

> *Understand/Compare and Contrast*: The first response compared the moon's position and moon and earth's sunlit portions in the diagram showing sun–earth–moon model. The second response just compared the appearance of the sunlit portions of earth that are analogous to the moon. Although the first student's response shows more in-depth reasoning than the second, in both cases we see evidence for the cognitive process of compare.
>
> *Understand/Infer:* The first response showed evidence for cause-and-effect connections, including the changes in the location of the observer in two situations on the sun–earth–moon geometrical model that, in effect, cause a change in the appearance of the sunlit portions of the earth and the moon. The second response, however, includes a series of disconnected concepts without any plausible connection between the described events.
>
> *Understand/Explain:* The first student described personal understanding of the situation and predicted the outcome with a series of explanations that describe why and how the predicted outcome is true. The second student, however, did not provide any additional explanation to support the predicted outcome.
>
> *Apply:* The first student reconstructed facts, concepts, and procedural skills from the perspective of the observer who is located on the moon. The second student did not consider the observer on the moon. Therefore, the discussion of the question was not set in the context of a new situation.

### *Interviews and Focus Groups*

Faculty course instructors and inservice teachers who were reform or comparison course graduates were interviewed individually. Faculty interviews focused on their experiences related to planning, developing, and teaching the undergraduate science course. Inservice elementary school teacher interviews focused on the purpose and rationale for teaching the observed K–6 science lesson and how it related to other lessons recently taught in science during the science unit and to the elementary school setting and science curriculum.

Current undergraduate students in the sample courses were interviewed in focus groups of four or five students. These interviews centered on students' understandings of, opinions of, and perceptions about science and science in courses at the college or university; the specific science course setting in which they were enrolled; and views of science teaching (see Figure 2.3). All interviews took place during site visits. The interviews usually occurred immediately after their course session was observed by the NSEUS research observers.

**Sample Undergraduate Student Focus Group Questions**

**Science Courses taken at the University:**

Have you taken other university level science courses (e.g. biology)? If so, identify them.

When I say the word "science" what is your reaction? Describe your' interest in this course and in science in general.

Have your attitudes towards science changed as a result of a college science course(s)? If so, in what way(s)? If not, why do you think this is so?

What has been the most important university science course (to date) that will prepare you for your career? Why do you think so?

How has your understanding of science changed as a result of this course?

What specific activity or assignment enabled you to change your belief about an issue in science or a science concept that you held prior to taking this course?

Which instructional strategy did you experience as most beneficial to your learning science in this science course?

**Science lesson just experienced/observed:**

How typical is this lesson for this class? If this is not typical describe the typical class session in this course.

What were the main ideas of this class session or lesson?

Why is it important for students to know this?

What confusion did you experience in learning this idea?

Describe how you would like to see an instructor teaching the main ideas in this lesson. Why should this strategy be used to teach these main ideas?

How did the instructor determine students' understanding of, or confusion about, this idea?

*Figure 2.3.* *Continues on next page.*

**For education majors only:**

Have your ideas of how you will teach science changed as a result of taking this course/ studying science this term?

Do you think that you can become an effective teacher of science? Why or why not?

What do you feel is the best way to teach science in elementary school? Why?

What science content areas do you feel most prepared to teach?

**Figure 2.3.** Sample undergraduate student focus group interview questions.

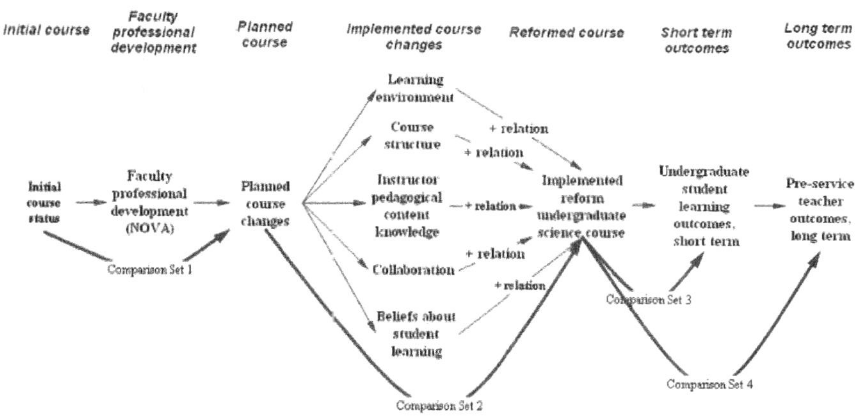


**Figure 2.4.** Generalized NSEUS Research Model indicating four comparisons completed during data analysis.

## DATA COLLECTION PROCEDURE

Multiple quantitative and qualitative data sources were used (see Table 2.7). The research procedures and instruments from reviewed studies were used to inform and develop the study data collection instruments and procedures. The full population survey was conducted during Year 1 of the study (see Chapter 1). A pilot study was also conducted during Year 1. During the full study, Years 2–5, data were collected from all undergraduate students in a course section at three times during the semester for each of the site visits.

### Data Collected at the Beginning of the Course

Data were collected from faculty instructors using an online questionnaire requesting descriptive data on the course, the course curriculum, and activities to take place over the semester. All approvals and arrangements for the site visit were scheduled in advance along with completion of institutional review board (IRB) consent forms at the beginning of the semester. Undergraduate student data were collected during the first two weeks of each semester and included online and in-class surveys of students' perceptions of their course's learning environment, attitudes towards science, and perceptions of what it means to be a scientist. Online data collection instruments included the CLES *preferred* version and TSSI. The DAST was administered in class by the course instructor.

**Table 2.7. Variables, Data Collection, and Analysis**

| *Research Question* | *Variables Measured* | *Instruments* | *Analysis* |
|---|---|---|---|
| 1 *Does faculty professional development change undergraduate science faculty teaching practice?* | Reform faculty curricular priorities, lesson planning, instruction, & pedagogical decision making (PCK) | CLES (Instructor), RTOP, CoRe & PaP-er, faculty interviews, content analysis of course materials and artifacts | Qualitative and quantitative comparisons will be made between reform variables predicted by the undergraduate science course impact research model |
| 2 *How do science teaching/ learning course characteristics differ between courses?* | Reform and comparison faculty curricular & instructional priorities, classroom learning environment, course structure, PCK, collaboration, undergraduate student learning outcomes | CLES (instructor & student), RTOP, CoRe & PaP-er, faculty & undergraduate student interviews, content analysis of course materials and artifacts, | Qualitative methods, multivariate analysis of variance, and effect sizes will be used to compare reform and comparison courses based on the implementation of reforms and student learning outcomes |
| 3 *How do these differences relate to the learning outcomes of undergraduate students?* | Reform faculty curricular & instructional priorities, classroom learning environment, course structure, PCK, collaboration, student learning outcomes | CLES (instructor & student), RTOP, CoRe & PaP-er, faculty & student interviews, content analysis of course materials and artifacts, | Qualitative comparisons will be used as well as regressing student achievement onto the levels of reform science course characteristics |

*(Table continues on next page)*

**Table 2.7. (Continued)**

| *Research Question* | *Variables Measured* | *Instruments* | *Analysis* |
|---|---|---|---|
| 4 *How do the differing levels of course reform relate to the short term learning outcomes of undergraduate students and long term outcomes for graduated inservice K-6 teachers in their own school classrooms?* | Reform and comparison faculty curricular & instructional priorities, classroom learning environment, course structure, PCK, collaboration, graduated K-6 teacher outcomes in elementary school classroom learning environments, undergraduate student achievement | CLES (instructor & student), RTOP (instructor), CoRe & PaP-er, faculty interviews, content analysis of course and K - 6 science materials and artifacts | Qualitative methods, multivariate analysis of variance & effect sizes compare undergraduate course characteristics & teaching characteristics of graduates of reform courses and comparison courses |

## Data Collected at the Middle of the Course

During the middle of the semester, a week-long site visit to each institution was scheduled by two to four of the NSEUS-trained observers. The typical week involved arriving Sunday afternoon and meeting the institution's liaison faculty member. During that meeting, the schedule of activities would be discussed, generally revising some items and formalizing the rest. A tour of the campus and facilities occurred then or on Monday morning to locate all class venues.

A personal semi-structured interview was conducted with every instructor whose classroom was visited, including both reformed and comparison courses. The interview occurred at a planned time and was also part of informal discussions during the week. To determine instructor PCK, CoRe charts (Loughran et al., 2004) from interviews were created for each of the faculty instructors.

All class activities occurring were observed during the weeklong visit. Class observations generally involved the main class instructional periods designated as "lecture," or a class with a combined lecture and laboratory for both the reformed and comparison courses. All other class learning activities also were observed, including laboratory and discussion periods and other class meetings and activities scheduled during the week. Observers used the RTOP (Sawada & Piburn, 2000) instrument to describe and rate the instructional and interactive learning events in the classroom during each of these class learning activities. The main class instructional period observations lasted from 50 to 180 minutes each for the undergraduate courses.

Course artifacts were collected including, for example, syllabi, samples of lab sheets and classroom lesson handouts, other course handouts, and quizzes. Notes from the interviews, course artifacts, and observation narrative field notes from the RTOP were used to create PCK profile matrices of instructor pedagogical content knowledge using *PaP-eR* (Loughran et al., 2004). Five or six undergraduate students enrolled in each of the reformed and comparison courses were invited, following one of the visited class sessions, to participate in separate focus group interviews using a semi-structured interview format.

At least two trained observers observed classroom instruction. The RTOPs were individually scored immediately after the classroom observation and then discussed to resolve any discrepancies between researcher ratings. The lectures, laboratories, and/or discussion sessions were combined to give an average RTOP score for each faculty member/course.

### *Data Collected in the Elementary Classroom*

The typical elementary school visit involved arriving in the morning or afternoon, registration and badging at the school office, and meeting the school principal and the sample teacher. During that meeting, the schedule of activities was discussed before the teacher interview and visiting the elementary school classroom occurred. To determine teacher PCK, CoRe charts (Loughran et al., 2004) from the teacher interviews were created for each of the elementary teachers.

The on-site elementary school observations of the main class instructional period lasted from 30–60 minutes and typically consisted of a transition to the science lesson, pre-lesson activities, science lesson with hands-on activities or investigation or lab, post-lesson discussion, and transition to next classroom subject activity. Notes from the interviews, course artifacts, and observation narrative field notes from the RTOP were used to create PCK profile matrices of teacher pedagogical content knowledge using *PaP-eR* (Loughran et al., 2004).

Multiple observers visited each elementary school. The RTOP ratings were individually scored after the classroom observation of the science lessons and then discussed to resolve any discrepancies between researcher ratings. The class observations, artifacts, and teacher interviews were involved in creating a PCK determination using CoRe and PaP-er for each elementary teacher.

## Data Collected at the End of the Course

At the semester's end, post-test data were collected from the students. The undergraduate students completed the *perceived CLES* and the *TSSI* online. During the last two weeks of the course, in-class assessments included the DAST as well as the SCA essay question(s). The in class assessments were administered by the course instructor.

## PILOT STUDY

A pilot study was planned and implemented to determine the feasibility, or proof of concept, of the NSEUS procedures and instruments for gathering data in the full national study in Years 2–5. Prior to beginning the pilot study, NSEUS observers trained on the use of the instruments and completed reliability checks on the observation ratings and description notes for sample classes. The pilot study was conducted at one institution with a sample of two comparison and reform courses and the elementary classroom science lessons of six elementary teachers who had completed one of the undergraduate science courses. The purpose of the pilot study was to train the full NSEUS team of observers from three institutions on the data collection protocol described above in the sections titled Data Collected at the Beginning, Middle and End of the Course. All NSEUS team members were from the University of Alabama, Kansas State University, and San Diego State University. Several aspects of the protocol were refined as a result of the pilot study including scheduling of times for visits to the elementary classrooms, use of multiple observers for a single observation or interview rather than one, extending the full visit to one complete week, and observing all undergraduate student learning experiences during the time of the visit to account for variations in students' experiences. The pilot study determined that the efficacy of the planed study procedure and instruments was workable and appropriate for the planned gathering data in a large scale phase of the NSEUS national study.

## SUMMARY

NSEUS embarked on an intensive national study of the impact of undergraduate science courses by defining the problem and research questions, reviewing and synthesizing the research literature, and surveying a national population of institutions where faculty instructors were involved in a common professional development experience (see Chapter 1). Each of these steps can help faculty gather evidence that informs undergraduate science teaching and would help the many Katy Anns in higher education with her dilemma. The next steps of the NSEUS study, as described in this chapter, included determining the research design, population, sample, instruments and extensive data collection procedure. Chapter 3 describes the full study results, presenting the findings, conclusions, and implications.

## ACKNOWLEDGMENT

This material is based upon work supported by the National Science Foundation under Grant No. 0554594. Any opinions, findings, and conclusions or recommendations expressed in this material are those of the authors and do not necessarily reflect the views of the National Science Foundation.

## REFERENCES

Chambers, D. W. (1983). Stereotypic images of the scientist: The draw-a-scientist test. *Science Education, 67,* 255–265.

Cobern, W. W. (2000). *The thinking about science survey instrument (TSSI) –SLCSP 151.* Kalamazoo, MI: Scientific Literacy and Cultural Studies Project.

Diseth, Å., Pallesen, S., Brunborg, G., & Larsen, S. (2007). Academic achievement among first semester psychology students: The role of course experience, effort, motives, and learning strategies. *Higher Education. 59,* 335.

Janesick, V. (2003). *"Stretching" exercises for qualitative researchers* (2nd ed.). Thousand Oaks, CA: Sage.

Johnson, B., & McClure, R. (2004). Validity and reliability of a shortened, revised version of the constructivist learning environment survey (CLES). *Learning Environments Research,* 7(1), 65–80.

Kreber, C. (2003). The relationship between students' course perception and their approaches to studying in undergraduate science courses: A Canadian experience. *Higher Education Research and Development, 22,* 57–75.

Lederman N. G., & Gess-Newsome, J. (1999). *Examining pedagogical content knowledge.* Boston, MA: Kluwer.

Loughran, J., Mulhall, P., & Berry, A. (2004). In search of pedagogical content knowledge in science: Developing ways of articulating and documenting professional practice. *Journal of Research in Science Teaching, 41*(4), 370–391.

Magnusson, S., Krajcik, J., & Borko, H. (1999). Nature, sources, and development of pedagogical content knowledge for science teaching. In J. Gess-Newsome & N. C. Lederman (Eds.), *Examining pedagogical content knowledge* (pp. 95–132). Dordrecht, The Netherlands: Kluwer.

Mason, C. L. (1999). The TRIAD approach: A consensus for science teaching and learning. In J. Gess-Newsome & N. J. Lederman (Eds.), *Pedagogical content knowledge: Its role and usefulness in science teacher education* (pp. 277–299). Amsterdam, The Netherlands: Kluwer Academic Publishing

Mason, C. L., Kahle, J. B., & Gardner, A. L. (1991). Draw-a-scientist test: Future implications. *School Science and Mathematics, 91,* 193–198.

McDermott, L. C., (1996). *Physics by inquiry.* Hoboken, NJ: John Wiley.

Morine-Dershimer, G., & Kent, T. (1999). The complex nature and sources of teachers' pedagogical knowledge. In J. Gess-Newsome & N. C. Lederman (Eds.), *Examining pedagogical content knowledge* (pp. 21–50). Dordrecht, The Netherlands: Kluwer.

National Research Council. (1996). *National science education standards.* Washington, DC: National Academy Press.

National Research Council. (2012). *A framework for K–12 science education: Practices, crosscutting concepts, and core idea.* Committee on Conceptual Framework for the New K-12 Science Education Standards. Washington DC: National Academies Press. Retrieved from http://www.nap.edu/catalog.php?record_id=13165

Nieswandt, M., & Bellomo, K. (2007). *Written extended-response questions as assessment tools for meaningful understanding of evolutionary theory.* Toronto, ON: Ontario Institute for Studies in Education of the University of Toronto.

NOVA. (n.d.) *NASA Opportunities for visionary academics.* Retrieved from http://www.education.nasa.gov/divisions/higher/programs/high_ed_faculty.html

Piburn, M., Sawada, D., Turley, J., Falconer, K., Benford, R., Bloom, I., ... Juddson, E. (2002). *Reformed teaching observation protocol (RTOP): Reference manual* (ACEPT Technical Report No. IN00-3). Tempe, AZ: Arizona Collaborative for Excellence in the Preparation of Teachers. (Eric Document Reproduction Services, ED447 205.)

Riggs, I., & Enochs, L. (1990). Toward the development of an elementary teacher's science teaching efficacy belief instrument. *Science Education, 74*(6), 625–637.

Sawada, D., & Piburn, M. (2000). *Reformed teaching observation protocol (RTOP).* (ACEPT Technical Report No. IN00-1). Tempe, AZ: Arizona Collaborative for Excellence in the Preparation of Teachers.

Siebert, E., & McIntosh, W. (2001). *College pathways to the science education standards.* Arlington, VA: NSTA Press.

Sorensen, H., Evans, R. H., & Andersen, A. M. (2009, March). *Content knowledge and efficacy.* Paper presented at the Annual International Conference of the National Association for Research in Science Teaching, Garden Grove, CA.

Sunal, D., Sunal, C., Ogletree, G., Steele, E., Mason, C., & Zollman, D. (2008) Research Brief No. 4: What characteristics are found among institutions still offering and no longer offering reformed undergraduate science courses? Retrieved from http://nesus.org

Symington, D., & Spurling, H. (1990). The draw-a-scientist test: Interpreting the data. *Research in Science and Technological Education, 8,* 75–77.

Taylor, P., Fraser B., & Fisher, D. (1997). Monitoring constructivist classroom learning environments. *International Journal of Educational Research, 27*(4), 393–301. Retrieved from http://surveylearning.moodle.com/cles/?PHPSESSID=6620efe47ced32ae6b57c4744487345f

van Driel, J. H., Verloop, N., & de Vos, W. (1998). Developing science teachers' pedagogical content knowledge. *Journal of Research in Science Teaching, 34*(8), 673–695.

Wright, E., & Sunal, D. (2004). Reform in undergraduate science classrooms. In D. Sunal & E. Wright, E. (Eds.), *Research in science education: Reform in undergraduate science teaching for the 21st century* (pp. 33–52). Greenwich, CT: Information Age.

Zollman, D., Matloob-Haghanikar, M., & Murphy, S. (2012). Background Research Paper No. 34: Student's reasoning and the level of interactivity in science content courses. Retrieved from http://education.ua.edu/centers/ortd/background-research-papers/

CHAPTER 3

# NATIONAL STUDY OF EDUCATION IN UNDERGRADUATE SCIENCE

## Lessons Learned

**Dennis W. Sunal, John Dantzler, Cynthia Szymanski Sunal, Donna Turner, Erika M. Steele, Cheryl L. Mason, and Dean Zollman**

### ABSTRACT

The National Study of Education in Undergraduate Science (NSEUS) was a multiyear (2006–2012) National Science Foundation supported research project. We focused on critical needs in teaching undergraduate science to diverse majors. A specific long-term emphasis in this study was on the development of pedagogical content knowledge of preservice elementary teachers of science. The research question guiding the investigation was, "How do undergraduate entry-level science courses, differing in level of reform, affect student outcomes?" Entry-level undergraduate science courses were analyzed in a faculty professional development impact design model involving a national sample of reformed and comparison undergraduate science courses from a population of 103 universities. The NSEUS study investigated several sub-questions related to outcomes of faculty professional development. The results discussed centered around four sub-questions. The NSEUS Research

*Research Based Undergraduate Science Teaching*, pp. 67–120

Model provides evidence for an undergraduate reform process extending from initial faculty professional development to increased student learning outcomes. What we learned was that the reform process is not a short-term endeavor. It is a process that is long-term and for most, rarely ends. The broader impacts of this study advance the professional development, course planning, and teaching of undergraduate science faculty and the development of more effective entry-level courses. Other aspects of the complete study are presented in Chapters 1 and 2.

## NATIONAL STUDY OF EDUCATION IN UNDERGRADUATE SCIENCE: LESSONS LEARNED?

Entry-level science undergraduate courses frequently are structured to be gateway courses for science, technology, engineering, and mathematics (STEM) careers and professional degrees. Gateway courses influence the science literacy of all their students in important ways. For example elementary teachers' ability to implement science standards-based reform in their own classrooms can be affected by the gateway science courses they take. Changing entry-level undergraduate science courses is perceived to be an easy task, although the faculty who teach the courses usually are the less experienced instructors in a department and rarely have had professional development focused on teaching.

Research studies have suggested that many issues that higher education faculty must confront before innovations can be implemented and sustained successfully in undergraduate coursework (Sunal, MacKinnon, Raubenheimer, & Gardner, 2004). To develop expertise in teaching undergraduate science, faculty instructors require deep knowledge of their science discipline in addition to knowledge of and professional skill in research supported best teaching practices, effective course components to implement for science course reform, procedures to implement successful innovation, and methods for evaluating the effectiveness of course change, such as action research.

The National Study of Education in Undergraduate Science (NSEUS) was designed, first, to establish criteria for identifying varying levels of reform in undergraduate science courses. Second, the study examined the impact of reform on the learning outcomes among all undergraduate students. As part of the second element, the pedagogical content knowledge of preservice teachers that translates into more effective science teaching in the school classroom was specifically examined. The major components of the NSEUS investigation included (1) a literature review of research on previous undergraduate science reform, (2) a national survey of faculty and courses at institutions involved in a professional development program aimed at a program of undergraduate science reform, (3) a pilot study determining the feasibility or proof of concept of the procedures and instruments for gathering data, and (4) a national research study of faculty, students, and undergraduate science courses including a large population of higher education institutions. We describe results from component 4 in this chapter. Other aspects of the complete study are presented in Chapters 1 and 2.

## NATIONAL STUDY OF EDUCATION IN UNDERGRADUATE SCIENCE

The goal of NSEUS was to determine the feasibility of using faculty professional development experiences to create reforms in undergraduate science courses providing an alternative to existing traditional undergraduate courses. The completion of the preliminary work of faculty professional development with the original population of 103 institutions guided development of a process to investigate undergraduate course reform.

Among the population of institutions, one or more of their science courses had been involved in the National Aeronautics and Space Administration (NASA) program, NASA Opportunities for Visionary Academics (NASA/NOVA), a faculty professional development program begun in 1995 (NOVA, n.d.). See Chapter 1 for a description of the NASA/NOVA program, which was designed to foster reform in higher education through faculty professional development and the modification of entry-level undergraduate science courses. The NASA/NOVA program included teams of faculty from a diverse national group of 103 institutions, who conducted and sustained reform over a multiyear period in one or more undergraduate science courses.

The NSEUS investigation selected a research sample from the national NASA/NOVA population of institutions. Sample selection used a two-stage stratified random process. Institutions were stratified by Carnegie type and by the sustained offering of the reform science courses. The characteristics of the institutional sample related closely to the population of institutions from which they were selected. Two undergraduate science courses then were drawn from each of the selected sample institutions. See Chapter 2 for a description of the selection process for the institutions and courses. One course sample included NASA/NOVA reformed courses where faculty experienced a long-term professional development program, referred to here as reform courses. The second course sample included comparison courses where faculty instructors were not involved in this same long-term professional development program, referred to here as comparison courses. Research model designs for both samples hypothesizing course effects on short- and long-term student learning outcomes are shown in Figures 3.1

and 3.2. Data collection from the national sample of 20 institutions was concluded in 2012. Other aspects of the study's sample procedure and its instruments are discussed in Chapter 2.

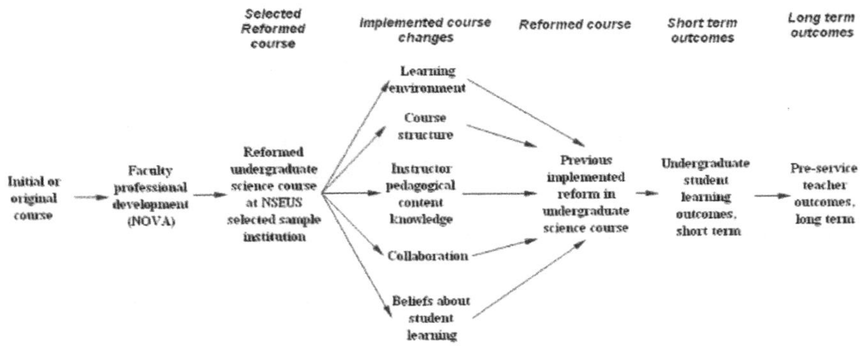


**Figure 3.1.** NSEUS reform undergraduate science course research model.

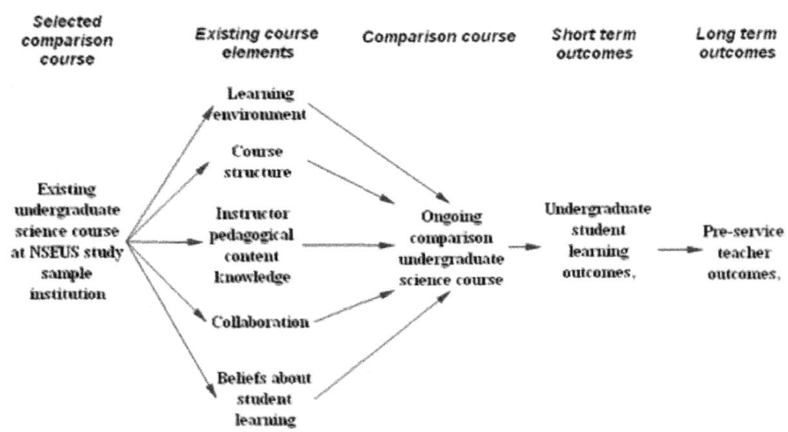


**Figure 3.2.** NSEUS comparison undergraduate science course research model.

The research question guiding the NSEUS study was "*How do undergraduate entry-level science courses, differing in level of reform, affect student outcomes?*" The four study sub-questions were:

1. Does faculty professional development change undergraduate science faculty teaching practice?

2. How do science teaching and learning course characteristics differ between courses before and after a faculty professional development program?
3. How do course differences relate to the short term learning outcomes of undergraduate students?
4. How do the differing levels of course reform relate to the long term learning outcomes for graduated inservice K–6 teachers teaching science in their own school classrooms?

## RESULTS

The NSEUS study involved comparative and relational analyses that investigated, first, similarity between two types of courses. These two types were reform and comparison undergraduate science courses at the same institution. Second, NSEUS investigated similarity among all undergraduate courses demonstrating differing levels of reform from among higher education institutions in the national sample. The relationship between NSEUS research questions, variables measured, instruments used, and analysis conducted with the study instruments is summarized in Table 3.1. Quantitative and qualitative procedures in a mixed-method research design were used for analyzing data in different stages of the research model to increase validity. Comparative analyses examined differences between the courses Relational analyses examined relationships between the characteristics of the courses and level of student outcomes.

**Table 3.1. Variables, Data Collection, and Analyses**

| *Research Question* | *Variables Measured* | *Instruments* | *Analysis* |
|---|---|---|---|
| 1. *Does faculty professional development change undergraduate science faculty teaching practice?*<br>**Comparison 1** | Reform faculty curricular priorities, lesson planning, instruction, & pedagogical decision making (PCK) | CLES (Instructor). RTOP. STEBI. CoRe & PaP-er, faculty interviews, content analysis of course materials and artifacts | Qualitative and quantitative comparisons will be made between reform variables predicted by the undergraduate science course impact research model |

*(Table continues on next page)*

**Table 3.1. (Continued)**

| *Research Question* | *Variables Measured* | *Instruments* | *Analysis* |
|---|---|---|---|
| 2. *How do science teaching/ learning course characteristics differ between courses?* **Comparison 2 & Comparison 3** | Reform vs. comparison faculty curricular & instructional priorities, classroom learning environment, course structure, PCK, collaboration, and undergraduate student learning outcomes | CLES (instructor & student), RTOP, CoRe & PaP-er, faculty & undergraduate student interviews, content analysis of course materials and artifacts | Qualitative methods, multivariate analysis of variance, and effect sizes will be used to compare reform and comparison courses based on the implementation of reforms and student learning outcomes |
| 3. *How do these differences relate to the learning outcomes of undergraduate students?* **Comparison 3** | Reform vs. comparison faculty curricular & instructional priorities, classroom learning environment, course structure, PCK, collaboration, and undergraduate student learning outcomes | CLES (instructor & student), TSSI, STEBI, RTOP, CoRe & PaP-er, faculty & student interviews, content analysis of course artifacts | Qualitative comparisons were used as well as regressing student achievement onto the levels of reform science and comparison course characteristics |
| 4. *How do the differing levels of course reform relate to the short term learning outcomes of undergraduate students and long term outcomes for graduated inservice K–6 teachers in teaching science in their own school classrooms?* **Comparison 4** | All faculty curricular & instructional priorities, classroom learning environment, course structure, PCK, collaboration, undergraduate student learning outcomes, and graduated K–6 teacher outcomes in elementary school classroom | CLES (instructor & student), RTOP (instructor and teacher), CoRe & PaP-er, faculty interviews, content analysis of course, and K–6 science artifacts | Qualitative methods, multivariate analysis of variance compare undergraduate course characteristics & teaching characteristics of graduates of all courses |

Analysis of the theoretical research model (see Tables 3.1 and 3.2) involved four major comparisons along with other confirming analyses. Qualitative analyses and results were used to confirm and triangulate quantitative results. The four major comparisons made up the linear model linking faculty professional development and development of reform courses to short- and long-term student outcomes (see Figure 3.3). The results for each of the four major comparisons are provided in the sections that follow.

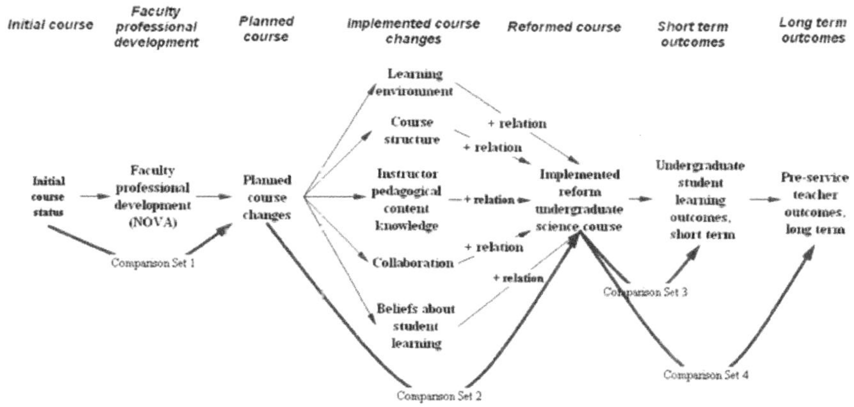


**Figure 3.3.** Generalized NSEUS Research Model indicating four comparisons completed during data analysis.

**Table 3.2. RTOP Means and Standard Deviations for the Reform and Matched Comparison Courses**

| *Course* | *N* | *Mean* | *Standard Deviation* |
|---|---|---|---|
| Reform | 20 | 70.81 | 16.38 |
| Comparison | 15 | 49.29 | 26.20 |

## Comparison 1

**Research Question 1:** *Does faculty professional development change undergraduate science faculty teaching practice?*

Comparison 1 involved a qualitative analysis of the initial baseline course data prior to NASA/NOVA faculty professional development began for instructors. Comparison 1 examined the baseline course learning environment, instructional strategies, and curriculum of the original sample course at each institution. These baseline results were compared to the later planned reformed course following the professional development experience of the 20 reform faculty (see Figure 3.4).

The analyses used secondary data from the archives of the NASA/NOVA professional development program. Each participating institution was required to describe in detail its baseline targeted undergraduate entry-level science course before entering the program. The pre- data were included in extensive application materials prepared to participate in the NASA/NOVA professional development workshop. Following the professional development experience, faculty teams from each institution developed an extensive proposal for a new design for the course using their professional development experiences. The proposals were reviewed by a program committee for possible funding, which ranged between $5,000 and $34,000.

*Initial course* *Faculty professional development* *Planned course*

**Figure 3.4** Comparison Set 1.

Qualitative analysis of the documents describing the courses before and after professional development aggregated the data into categories in order to identify themes. Common themes were captured from analyses of individual institutional data sets and from the comparison of data sets. Overarching themes were then identified. The overarching themes were used to qualitatively analyze course descriptions, prior to and after professional development for each of the sample institutions. The overarching themes were identified as classroom learning environment, course structure, pedagogical content knowledge, collaboration, and beliefs about teaching and student learning. Faculty who participated in the NASA/NOVA project expressed a desire to change the way in which they taught STEM discipline knowledge to their undergraduate students.

## Narrative Description of Findings

### *Pre-NASA/NOVA Professional Development*

Prior to participation in the NASA/NOVA professional development program, the identified characteristics of the science courses offered by faculty were very traditional in format. Most were teacher-centered. Some of these courses were lecture courses with a high failure rate. The courses did little to dispel students' fear of learning or, for education majors, fear of teaching science. Faculty instructors described their courses in ways that indicated lack of relevance for students, science being presented as a set of facts, and students taking on passive roles. Many of the courses involved students in a separated lab and lecture addressing curriculum that was disconnected, without common prerequisites, and without similar content objectives. These characteristics were features reported in the research literature as turning students off from science (Seymour, 1995).

The emphasis of the baseline courses was on broad-based science content. In these courses, an extensive range of topics was covered to ensure students were given enough information to have an understanding of science. In designing the courses, the content was not selected for its relevance to the students nor was the students' prior knowledge considered. The topics covered were selected for traditional and historical reasons, or because the topics were covered in the course textbook. The use of traditional memorization-type assessments was commonly described. The traditional assessment discouraged students from learning the course material beyond a low level. There was little student–student or student–teacher interaction in the classroom environment.

### *Post-NASA/NOVA Professional Development*

Following the professional development experience, the faculty from each study sample institution had submitted a funding proposal that planned reform of an undergraduate science course. Each of these proposals was accepted and funded by the NASA/NOVA program. All of the courses were collaboratively designed by an interdisciplinary faculty team of three or more members representing science faculty, education faculty, and in a number of cases, college administrators. The collaboration involved all members who participated in the NASA/NOVA professional development experience, so they had similar expectations and goals for course reform

Significant change was proposed for the existing courses. Along with the NASA/NOVA professional development experiences, resources were used to upgrade college level instruction: for example, the *National Science Education Standards* (NSES) (National Research Council, 1996), *Shaping the Future* (National Science Foundation, 1996), and *College Pathways to the Science Education Standards* (Siebert & McIntosh, 2001). These resources were cited in many proposals for approaches to teaching, how students should actively learn, and be able to work collaboratively in an undergraduate science course. The various resources tended to center beliefs about teaching and learning that faculty used in development of their goals for the reform course.

Within the context of the NASA/NOVA professional development, pedagogical content knowledge encompassed the instructor's ability to selectively choose instructional strategies for specific science concepts that would best facilitate students' understanding of science content. Many of the proposals emphasized inquiry with about 1/3 doing so through use of the learning cycle (Karplus, 1979) during instruction. The typical learning cycle in a reform course integrated lab and lecture to connect aspects science concepts being taught. The learning cycles developed considered students' prior knowledge, engaged students in relevant and interesting activities, allowed students to construct their own understanding through their own actions, applied students' knowledge to novel situations.

The reform courses focused on providing students with the opportunity to experience the social importance of science and learning science. In the reform course's classroom learning environment students were more often provided the opportunity to collaborate with other students and the instructor in order to construct their own understanding of the science concepts. The roles of the student and the teacher were changed in the proposed reform courses. The instructor was no longer the holder of all knowledge and the teaching methods were no longer fully teacher centered. The courses were to be more student-centered where the teacher served as a facilitator to help guide learning through inquiry, problem solving, or other active learning experiences. The students were to be generally engaged by actively participating in the building of knowledge through student centered learning activities

The science knowledge presented in the reform course structure was broadened to facts, concepts, problem solving, and investigation. Inquiry skills were emphasized in the proposals submitted. The science knowledge proposed was to be presented in a more interdisciplinary way. The specific discipline science concepts were connected to STEM content as well as to the importance of science in the everyday lives of citizens. The proposed courses covered fewer science concepts with greater depth. Many proposals developed plans to begin to move away from using traditional tests to assess students' meaningful understanding using alternatives such as more frequent testing, portfolios for part of the assessment, and performance based testing.

## Summary Findings for Comparison Set 1

Pre-post faculty professional development course description findings were as follows

1. The analysis of NASA/NOVA funding proposals found five overarching elements or themes in course descriptions that differed

from pre (baseline) to post (following professional development) course descriptions.

a. Classroom learning environment
b. Course structure (curriculum and sequence of learning activities)
c. Pedagogical content knowledge
d. Collaboration
e. Beliefs about teaching and student learning.

2. The summary findings for Comparison Set 1: Description of post-course designs, following professional development as compared to pre-course designs included

a. New common characteristics in the courses generally were related to a college level interpretation of the national science standards.
b. Inquiry-based instructional methods and learning goals focusing on increased active learning dominated post course descriptions.
c. Greater use of instructional techniques focusing on pedagogy engages students' prior knowledge.
d. Collaborative interdisciplinary faculty teams function in a supportive way to develop innovative course elements.
e. Rationale provided for course change is related to increasing emphasis on facilitating all students' learning.
f. Course curriculum generally emphasized a greater integration of STEM knowledge over one specific discipline in each course.
g. Course descriptions described refocusing the role of the instructor, who works to become a reflective practitioner using action research.
h. Course design incorporates greater use of integrated multiple learning formats as opposed to separated lecture, discussion, and lab sessions.
i. Greater planned use of classroom activities incorporates interactive and collaborative learning.
j. Greater use of diverse technology is planned in many course activities.
k. Greater emphasis is placed on using relevant applications and real data reflecting current science.
l. Performance assessment is used as the greater part of course assessment.

## Comparison 2

> **Research Question 2:** *How do science teaching and learning course characteristics differ between courses before and after a faculty professional development program?*

In order to investigate Research Question 2, Comparison 2 related the planned course characteristics described in the NASA/NOVA proposals to the characteristics of the later implemented reform undergraduate science courses observed during extended NSEUS site visits. This analysis was long-term, examining the sustainability of implemented reform over a 4–12 year period following the initiation of the original plan for the course as described in the funding proposal (see Figure 3.5). The analyses occurred within the group of 20 sample reform courses. Quantitative and qualitative data were collected from the earlier planned course proposals and the later implemented reform courses. The data set was analyzed to determine differences in characteristics.

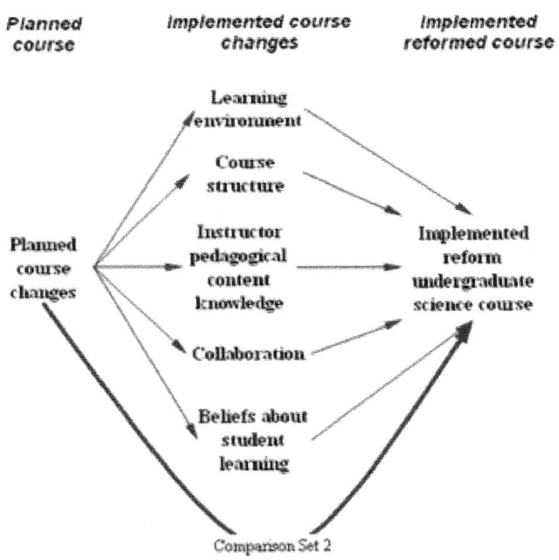


**Figure 3.5.** Comparison Set 2.

## Narrative Description of Findings

Five overarching themes emerged from qualitative analysis of data from the planned reform in undergraduate science courses reviewed in the submitted proposals following NASA/NOVA professional development experiences and the resulting implemented courses (see Figure 3.6). Collecting observational and interview data from the observed courses supported the existence of each of the themes: classroom learning environment, course structure, pedagogical content knowledge, collaboration, and beliefs about teaching and student learning

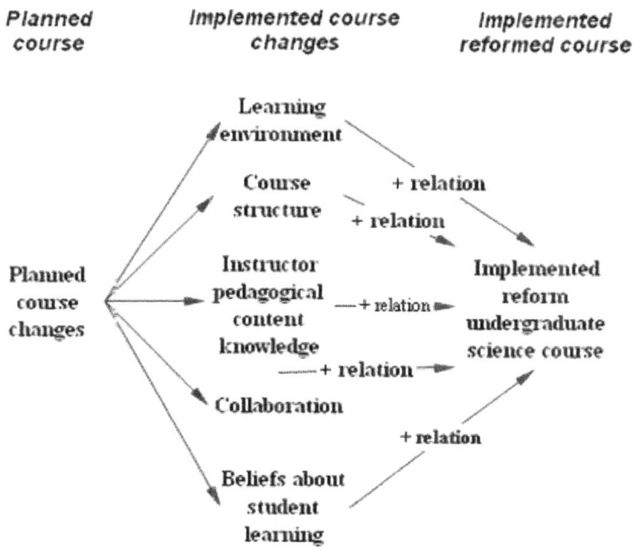


**Figure 3.6.** Themes representing differences in characteristics between the planned and implemented reform courses and the matched comparison courses.

### *Themes 1 and 2: Classroom Learning Environment and Course Structure*

Extensive quantitative and qualitative data were collected in sampled courses from the 20 institutions. The site observations occurred during the fall and spring semesters from 2007 through 2012. The sampled reform courses continued to have an observable level of reform implemented into the course. Many of the courses had fully integrated laboratory and lecture format. In the proposals, most of the teams indicated they intended to move away from teacher-centered instruction to more student-centered instruction. This movement was observed later during site visits where the emphasis in courses was on science inquiry skills over rote memorization of science content. Many instructors continued to use aspects of the learning cycle to create an environment conducive to student learning. In the reformed courses, the classroom learning environment was more likely to engage student interest allowing students to construct their own knowledge by applying critical problem-solving skills. Students were given opportunities to examine their understanding of the course content, share and compare their understanding with other students in the classroom, develop a more scientific understanding, and learn to apply their knowledge to solve problems in novel situations. There was a higher level of interaction between students and with the instructors. Although the majority of the reform instructors indicated traditional assessments were used to evaluate students, they mentioned some formal and informal alternative assessments. Several instructors created new assessments by using activities students performed in course lessons as a form of informal assessment. Portfolios, journals, and independent science projects were mentioned as alternative forms of assessments not used previously. This result shows implementation of the intentions of the teams writing the proposal to use alternative assessments to monitor the development of students' use of science inquiry skills.

A few of the activities observed in the reformed courses had reverted back to the traditional course pattern in use before the NASA/NOVA professional development experience. Collaboration and professional development were important aspects for maintaining the reforms in the learning environment. There were two sample cases where observations of the reform course indicated the course had few characteristics of the planned reformed course. In the first case, the majority of the original team had left the institution. The instructor was the only faculty member in the department interested in maintaining the course as a reformed course. The instructor indicated she was interested in maintaining the course the way it was written in the proposal, but forces outside of her control made it difficult. In the second case, most of the team members had left the institution, but it was unclear as to whether or not the remaining team member collaborated on the course. The instructor of this course was a new member of the team who did not experience the original NASA/NOVA professional development activities.

### *Theme 3: Pedagogical Content Knowledge*

Within the context of the NASA/NOVA professional development, pedagogical content knowledge encompassed the instructor's ability to selectively choose instructional strategies for specific science concepts that would best facilitate student understanding of science content. Instructors with pedagogical content knowledge (PCK) for specific science concepts display knowledge of ways to meaningfully represent the concept, plan and schedule the daily class curriculum for effective learning, organize the classroom context to increase interaction and student engagement, and use pedagogy to engage students in deeper learning. Instructors exhibiting

PCK for specific science concepts also plan for and structure classroom activities that elicit students' prior knowledge, using discrepant experiences that help students recognize their misconceptions, building a complete lesson sequence to accommodate students' learning. Prior to the NASA/NOVA professional development project, the majority of the instructors stated that the pre-NASA/NOVA course was a lecture-dominated, teacher-centered course. Their statements suggested that most science concepts were essentially taught using a similar didactic format.

Qualitative analysis of the proposals as well as of later interviews and classroom observations of the implemented course found that reform instructors exhibited enhanced pedagogical content knowledge for the specific science concepts taught during the classroom observations. First, instructors were reflective about their teaching. They took into account the limitations specific activities or lessons had and made adjustments accordingly. Instructors noted their students' misconceptions and the learning difficulties associated with key science concepts. Second, several instructors designed lessons based on students' prior knowledge and misconceptions. Many instructors used the learning cycle to teach science concepts because it allowed them to draw out students' prior knowledge and misconceptions and then engage students in applying new knowledge in novel contexts. One instructor made the following statement regarding the strategy used to teach a specific science concept:

> I teach this course differently ... [this course] is inquiry based.... Elicitation addresses prior knowledge and misconceptions.... Exploration is the discovery portion ... and Explanation—students give presentations. (Instructor comment)

Lastly, instructors used alternative assessments to gauge students' understanding of science content. This included frequent quizzes with immediate feedback using clickers and exams requiring student synthesis of ideas in completing performance tasks.

### *Theme 4: Collaboration*

Within the context of the NASA/NOVA professional development project, collaboration involved building communication among faculty including scientists, science educators, and science department administrators to create reform in entry-level undergraduate science courses. Years later, the majority of the instructors who participated in the NASA/NOVA project were still actively involved in ongoing collaboration with science educators and reflected on the quality of their collaboration (Sunal, Sunal, Ogletree, Steele, Mason, & Zollman, 2008). Collaboration was the key variable making possible the initiation and sustaining of the reformed science course over time. Analysis of faculty instructor interviews found that collaboration was not without resistance. Several instructors from various institutions suggested that improvement still was needed in collaborative efforts. Comments from some of the reformed course instructors included: "Faculty members were questioning the need for a smaller [*reform*] course for non-majors, why can't they just take… [*regular large lecture science content courses*]" (faculty comment), "Nobody at the university [administration] cares about what I do in classes with non-majors.… Anybody who does really good teaching in undergraduate science is probably a rebel" (faculty comment), and "Acceptance by my chair and the department!…There's a lack of respect here in the sciences for the education department ('Those people who play with construction paper')…I still fight that battle of acceptance" (faculty science educator comment).

### *Theme 5: Faculty Beliefs about Teaching and Learning*

Within the context of the NASA/NOVA professional development program, "beliefs" encompassed the instructor's disposition regarding science teaching and learning. Qualitative analysis of faculty interviews indicated that instructors' initial beliefs about science teaching and learning revolved heavily around knowledge of content. Although many instructors wanted to change the way in which science was traditionally taught, their initial suggested reforms maintained an emphasis on knowledge of content and a teacher-centered instructional focus in courses. Some instructors initially believed that science must be relevant and connect to other disciplines as well as emphasize investigation promoting critical thinking skills. Qualitative analyses of interviews and classroom observations found that instructors who participated in the extended NASA/NOVA program were able to plan and implement a course embodying their newly stated course goals based on changed beliefs about science teaching, being more student-focused and including more active student learning. Many instructors were able to modify their initial beliefs to incorporate a more constructivist approach to science teaching and learning. The NASA/NOVA courses observed were more "constructivist" in nature with emphasis placed on the nature of science and scientific inquiry as course objectives. The notion that less coverage results in more depth was stated throughout the faculty interviews. Instructors modified and expanded their beliefs about science teaching and learning to include instruction that was inquiry-oriented for courses that included majors and non-majors, use of alternative rather than traditional assessments, regularly engaging in action research, and science teaching and learning that facilitated development of science literacy.

## Summary Findings for Comparison Set 2

1. Significant differences were found before and after faculty experienced a professional development program between the pre- and post-implemented undergraduate science courses in all five course elements as evidenced from multiple data sources:
    a. Classroom learning environment
    b. Course structure (curriculum and sequence of learning activities)
    c. Pedagogical content knowledge
    d. Collaboration
    e. Beliefs about teaching and student learning.
2. Among the sample of *implemented* undergraduate reform courses, several important findings were identified:
    a. Most major planned changes in the reform courses were implemented and sustained long-term.
    b. Common characteristics aligned with the national science education standards (Siebert & McIntosh, 2001) including inquiry-based instructional methods and learning goals dominated reform course descriptions.
    c. Courses, once reformed, typically continued to be offered over the long term. Sustainability of reformed courses ranged from up to 12 years, limited only by the scheduled time of NSEUS study site visit.
    d. Collaborative interdisciplinary faculty teams were long lived.

## Comparison 3

**Research Question 2:** *How do science teaching and learning course characteristics differ between courses before and after a faculty professional development program?*

**Research Question 3:** *How do course differences relate to the short term learning outcomes of undergraduate students?*

Comparison 3 related observed characteristics between courses and involved two separate analyses (see Figure 3.7).

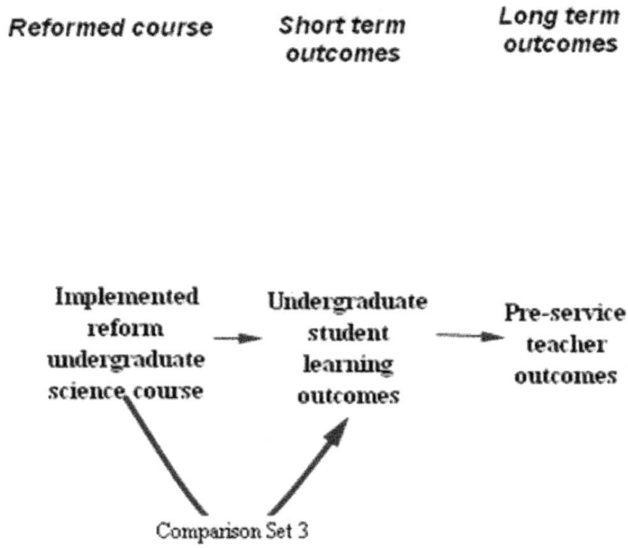


**Figure 3.7.** Comparison Set 3.

**Research Question 2.** The first analysis involved two groups of courses. One group was the NASA/NOVA-implemented reform undergraduate science courses (n=20). A second group of courses was selected and analyzed from each of the same sample of institutions: the matched comparison undergraduate science courses ($n$ = 15). The matched comparison courses involved faculty who had not participated in the NASA/NOVA professional development program but may have participated in other professional development opportunities. Both groups of courses were observed during week-long site visits. Quantitative and qualitative data were collected from both groups of courses relating to the five themes: classroom learning environment, course structure, pedagogical content knowledge, collaboration, and beliefs about teaching and student learning. The data set was analyzed to determine differences in characteristics among reform courses and between reform and comparison courses. Collecting observational and interview data from the observed courses supported the existence of each of the themes based on the analysis of course characteristics in the implemented reform courses, an effect that was not typical in comparison courses. The course characteristics represented by the themes differed between the implemented reform courses and the matched comparison courses.

**Research Question 3.** Comparison 3 involved a second analysis of the relationship between *level* of implementation of reform and the short-term learning outcomes of undergraduate students. This second analysis used data collected on the students and faculty of both sample groups of courses based on level of implemented reform in the courses ($n$ = 35). Both groups, reform and matched comparison courses, demonstrated differences in levels of reform characteristics between. Some matched comparison courses had high levels of reform characteristics due to the randomly stratified course selection process used in the study. So, as members of a single group of 35 courses, the level of reform created a mix of rankings of reform and comparison courses. This second analysis also was based on the course characteristics found in the five overarching themes that emerged from qualitative analysis of data from undergraduate science courses: classroom learning environment, course structure, pedagogical content knowledge, collaboration, and beliefs about teaching and student learning.

## NARRATIVE DESCRIPTION OF FINDINGS

### Themes 1 and 2: Classroom Learning Environment and Course Structure

#### *Instructional Practice*

The study included 35 faculty instructors from 20 institutions. Twenty instructors taught a reformed course and 15 taught a matched comparison undergraduate science course at the same institution. In all courses, instructors were observed multiple times using the *Reformed Teaching Observation Protocol* (RTOP), a classroom observation protocol designed to measure and provide a quantitative characterization of the degree to which a science classroom is reformed (Sawada & Pilburn, 2000). The RTOP is comprised of multiple sections rated by the observer: (1) lesson design and implementation, (2) content (propositional knowledge), (3) content (procedural knowledge), (4) classroom culture (communicative interactions), and (5) classroom culture (student/teacher interactions). See Chapter 2 for an extended description of RTOP. Overall, the ratings of faculty classroom instructors on the RTOP ranged from 17–97 with an average of 63.5. The RTOP ratings in the reform courses ranged from 35–93, while the RTOP ratings from the comparison courses ranged from 17–97. Comparison courses had fewer reform characteristics implemented than were found with the reform courses. The average RTOP rating for the reformed courses was 70.81, whereas the average for the comparison courses was 49.29 (see Table 3.2). Results from quantitative analysis of data found a statistically significant difference in the total RTOP score between faculty who were instructors of reform and matched comparison undergraduate science courses, $t(33, 2.997), p = .005$. The observers' RTOP ratings of faculty were significantly higher in the reformed courses. Analysis of classroom observations on RTOP subscale scores between reformed and comparison courses found statistically significant group effects (Wilks = .028, $F[5, 29] = 3.756$, $p = .010$) when comparing the two faculty groups on the RTOP subscales. Univariate analyses indicated the reform course instructors demonstrated significantly higher ratings on four of the five subscales (see Table 3.3). Significant RTOP subscale differences were found on (1) lesson design and implementation, (3) content (procedural knowledge), (4) classroom culture (communicative interactions), and (5) classroom culture (student/teacher interactions).

**Table 3.2. RTOP Means and Standard Deviations for the Reform and Matched Comparison Courses**

| *Course* | *N* | *Mean* | *Standard Deviation* |
|---|---|---|---|
| Reform | 20 | 70.81 | 16.38 |
| Comparison | 15 | 49.29 | 26.20 |

**Table 3.3. MANOVA Results by RTOP Subscales**

| *RTOP Total and Subscales* | *df* | *Mean Square* | *F* | *Sig.* |
|---|---|---|---|---|
| Lesson Design and Implementation | 1 | 153.242 | 6.242 | .018 |
| Propositional Knowledge | 1 | 26.267 | 2.729 | .108 |
| Procedural Knowledge | 1 | 166.947 | 6.978 | .013 |
| Communicative Interactions | 1 | 289.109 | 13.289 | .001 |
| Student Teacher Relationships | 1 | 223.469 | 8.143 | .007 |
| Total Rating | 33 | | 2.997 | .005 |

Qualitative analyses of instructors' interview data, observational field notes reported on the RTOP instrument, and CoRe and PaP-eR supported findings from the quantitative analysis of data. For example, the faculty interviews were examined to further consider pedagogical practices that provided students with opportunities to engage in learning activities. The following statements are examples of such comments:

> I teach this course differently than I teach my regular physics courses. Physics courses are lecture-based.... Physics [reformed course] is inquiry-based.... Elicitation addresses prior knowledge and misconceptions.... Exploration is the discovery portion.... Explanation—students give presentations.... (Reform course instructor)

> What do you recall about the circulatory system ... taking a look at blood flow ... case study on heart attack; what does heart attack have to do with blood flow; you have to know how something works; take fundamental concepts about circulatory and apply it to this disease.... Physiology—blood pressure; how does it tie it into circulatory system. Take a look at model of heart.... For each topic that I do there is some type of inquiry lesson and case studies along with viewing models. (Reform course instructor)

Interviews with the undergraduate student focus groups from the same course corroborated the statements of the instructors. For example:

> Yeah. I never made the link that those things were part of science. They seemed just mechanical.... I'll remember circuits from working with them.... I know it works, but now I'm starting to get a grasp on how and why it works.... I'm thinking more about the scientific process. In lab we always go through the process. (Undergraduate reform course student in focus group)

Instructors of the comparison course taught in a more traditional manner. Many instructors were more likely to spend the majority of the class time lecturing to their students. The following instructor statements are examples of instructor comments:

> I try to engage students as much as I can. I use Blackboard to post my notes so that they can spend class time concentrating on what I am saying rather than copying down information.... There is a lack of administrative support for faculty to collaborate with one another or to have pedagogical professional development. Rather, we just get some research support. (Comparison course instructor)

> 10% of the time you can make it relevant; the remainder has to be memorized ... use of the board and overheads. .. online simulations and cartoons are very helpful.... I want to do group work but I don't have time.... I mainly straight lecture.... I use the online homework ... students like it now that it is in line with the lectures. (Comparison course instructor)

The key themes arising from qualitative analysis of these data for *matched comparison courses* were (1) instruction was teacher-centered, (2) content focus was more on facts than concepts, (3) emphasis was on student competition, (4) students were passive in learning, (5) lecture dominates instruction, and (6) lecture and laboratory were disconnected. The key themes arising from qualitative analysis of the data for *reformed courses* were (1) reflective practitioner comments, (2) hands-on approach to science teaching through inquiry, (3) science content that is relevant to students and facilitative of student scientific literacy, (4) emphasis on student collaboration, (5) use of strategies that considered students' prior knowledge and misconceptions, and (6) collaboration with colleagues on teaching. Major instructional strategies were described in a survey questionnaire completed before observational visits were made. Reform instructors indicated that the greatest amount of student class time in a week was spent in an inquiry-based integrated laboratory (68%). Comparison instructors indicated that the greatest amount of student class time was spent in lecture (63%). The data collected from both reform and matched comparison course instructors were similar to results from observations made by NSEUS data collectors regarding the average time major course instructional strategies were used weekly (see Figures 3.8 and 3.9). The results were seen as being a positive indicator of an inquiry-oriented classroom learning environment that was experienced by students in the reformed courses.

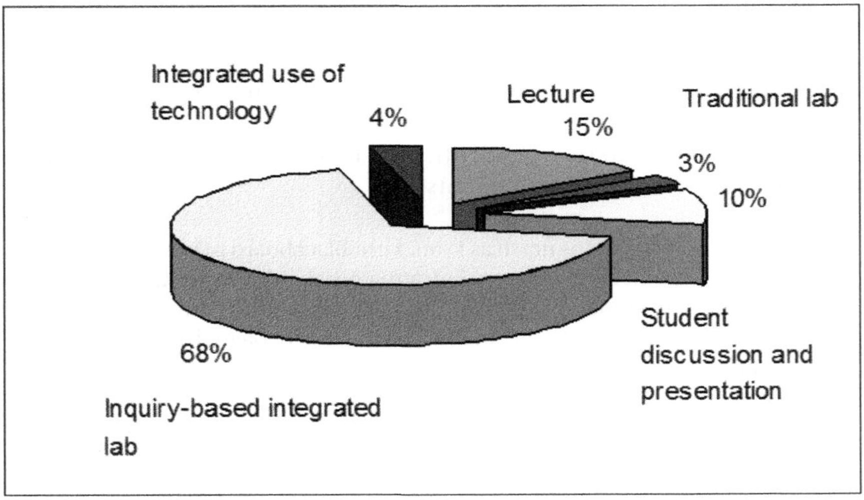


**Figure 3.8.** Pedagogy in study reform courses: Weekly time use.

The quantitative ratings on the RTOP and the qualitative results from the instructor and student focus group interviews converged on the types of teaching and learning that occurred in the courses. Undergraduate science reform courses were significantly different and more student-centered than the matched comparison courses. The differences in ratings using the RTOP instrument and qualitative data analyses opened up the question as to whether the students perceived differences in the level of reform implemented in their undergraduate science courses.

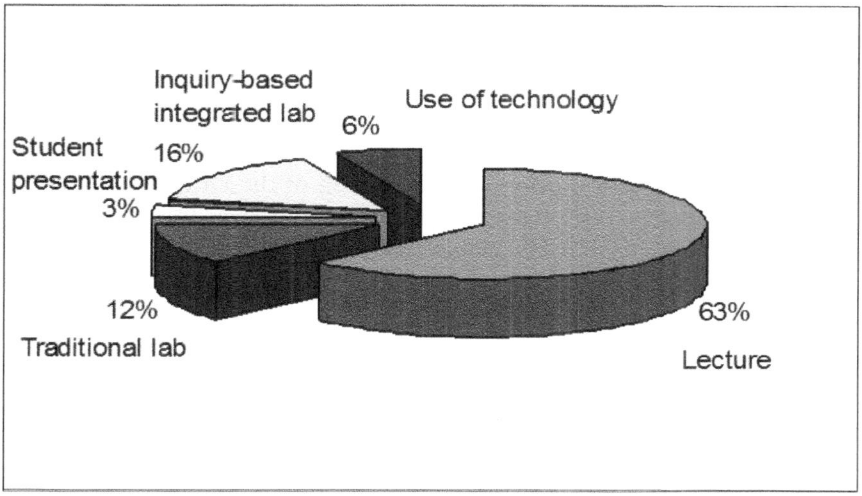


**Figure 3.9.** Pedagogy in study comparison courses: Weekly time use.

## Themes 1 and 2: Classroom Learning Environment and Course Structure

### *Student Engagement*

The *Constructivist Learning Environment Survey* instrument (CLES) was developed to help instructors understand students' perceptions of the extent to which the classroom learning environment enabled them to reflect on their prior knowledge, develop as autonomous learners, and negotiate their understandings with other students. The instrument contains five scales: (1) the *Personal Relevance Scale* measures how relevant students feel the course content is to their lives outside of the classroom; (2) the *Shared Control Scale* measures students' perceptions of their control over classroom learning; (3) the *Critical Voice Scale* measures students' perceptions of their ability to question the teacher's pedagogy; (4) the *Student Negotiation Scale* measures students' perceptions of their ability to share their ideas with other students in the classroom; and (5) the *Uncertainty Scale* measures students' perceptions of the level of inquiry-based science knowledge developed in the classroom. Two versions of the instrument were given to the students during the semester. The first version involved participants in identifying the kind of classroom learning environment they prefer and was given during the beginning of the semester asking for preferences or expectations as a pretest. The second version involved participants in identifying the classroom experience they had as they perceived that experience and was given at the end of the semester asking for perceptions or perceptual understanding of experiences as a posttest. See Chapter 2 for an extended description of CLES. No significant statistical difference was found between the students' scores on the CLES prior to instruction ($F = 1.397; p = 0.238$). The students enrolled in both comparison and reform course types preferred the same level/type of teaching methods. Significant differences were found on the posttest version of the CLES. Overall, students enrolled in the reform courses perceived more constructivist student engagement being used in the classroom ($F = 17.597, p < 0.001$)

## Themes 1 and 2: Classroom Learning Environment and Course Structure

### *Faculty Instructional Practice as Related to Student Engagement in all Sample Courses*

Additional analyses were conducted to determine the source of the students' classroom learning environment perceptions. Which type of faculty instructor made a measurable difference in students' understanding of the classroom environment as being more conducive to learning? Univariate analysis of variance determined that an instructor would have to have a RTOP rating above 71 ($F = 4.5, p = .012$) or below 45 ($F = 5.1, p = .024$) in order for students to perceive the classroom learning environment as different as measured on the CLES posttest. A rating of 71 indicates a high level of reform was observed in the classroom. An instructor would have to be rated at an average of 3 or above on every RTOP item scale to achieve a rating of 71 or above to obtain a classroom rating by students as creating an above average or good learning environment. A rating of 45 or below was considered an average rating on RTOP items of between 1 and 2. Such a rating is indicative of a classroom where very few reformed teaching practices were observed and where students perceived a below average or poor classroom learning environment. The results for upper- and lower-level group differences are shown in Tables 3.4 and 3.5, respectively.

Students in courses where the instructor was rated with a score of 71 or above experienced science instruction that allowed them to share and explore their ideas with others and build their understanding of science concepts through inquiry. The higher an instructor's RTOP observational rating was above 71, the more confident students were that they had learned the course material or that they could learn the material, as reported in CLES and supported in student focus groups. There also was a relationship between instructor RTOP rating and use of higher cognitive skills on end-of-the-semester achievement items on the *Science Content Assessment* (SCA) instrument.

**Table 3.4. Univariate Analysis Determining Relationship of High RTOP Rating and CLES Score: Tests of Between-Subjects Effects**

*Dependent variable: RTOP71*

| *Source* | *Type III Sum of Squares* | *Df* | *Mean Square* | *F* | *Significance* |
|---|---|---|---|---|---|
| Corrected Model | 0.925[a] | 1 | 0.925 | 4.485 | 0.035 |
| Intercept | 47.325 | 1 | 47.325 | 229.364 | 0.000 |
| Post CLES Average | 0.925 | 1 | 0.925 | 4.485 | 0.035 |
| Error | 51.171 | 248 | 0.206 | | |
| Total | 778.000 | 250 | | | |
| Corrected Value | 52.096 | 249 | | | |

a. $R$ Squared = 0.018 (Adjusted $R$ Squared = 0.014)

**Table 3.5. Univariate Analysis Determining Relationship of Low RTOP Rating and CLES Score: Tests of Between-Subjects Effects**

*Dependent variable: RTOP71*

| *Source* | *Type III Sum of Squares* | *Df* | *Mean Square* | *F* | *Significance* |
|---|---|---|---|---|---|
| Corrected Model | 1,229.748[b] | 1 | 1,229.748 | 5.127 | 0.024 |
| Intercept | 157,049.207 | 1 | 157,049.207 | 654.789 | 0.000 |
| Post CLES Average | 1,229.748 | 1 | 1,229.748 | 5.127 | 0.024 |
| Error | 59,482.096 | 248 | 239.847 | | |
| Total | 1,248,617.000 | 250 | | | |
| Corrected Value | 607,11.844 | 249 | | | |

b. $R$ Squared = 0.020 (Adjusted $R$ Squared =0.016

Students in courses with a score of 45 and below were likely to experience science as transmitted mainly through a lecture. They were less likely to be given opportunities to express and explore their ideas for themselves or with others in the course. The students were more likely on the post-CLES and during student focus groups to report the perception that there was little connection between the lecture and the other activities in the course and that the course was not helpful in their learning. The lower an instructor's RTOP rating, the more likely students were to report being dissatisfied with the learning environment.

Students in classes with higher instructor RTOP ratings used higher cognitive skills on end-of-the-semester achievement items on the SCA than students in courses with lower instructor RTOP ratings. Students in classes with higher RTOP ratings were more able to make connections between science facts and concepts, explain their answers, and apply the facts and concepts to a novel situation. Many students in courses with low RTOP ratings were able to answer the questions using factual and conceptual knowledge, but they did not connect, explain, or apply their knowledge to novel situations. The instructor RTOP ratings triangulated with student CLES results and student focus group interviews as well as students' lower use of higher cognitive skills on end of the semester achievement items.

## Themes 1 and 2: Classroom Learning Environment and Course Structure

### *Students' Perceptions Described in Focus Group Interviews*

Focus group interviews were held with students from each of the 35 courses during NSEUS site visits. The interviews occurred at or after the middle of the semester. Two researchers asked questions during the focus group interview and recorded students' responses in field notes. Groups consisted of five to six undergraduate students meeting in a location outside of the classroom. Only students who volunteered and completed the IRB permission form participated. Students' responses from courses varying by level of reform are discussed first. Level of reform was determined using the course RTOP ratings. Three levels were used in analysis of focus group data: (1) RTOP rating of 71 or higher, (2) RTOP rating of 46–70, and (3) RTOP rating of 45 or lower.

Students enrolled in courses or with low RTOP ratings (45 or lower) expressed more dissatisfaction with their courses than students who enrolled in courses with medium (46–70) or high RTOP ratings (71 or above). Students in low RTOP courses described the learning environment as being generally unwelcoming or intimidating. Groups more often described their instructors as judging them negatively if they could not answer questions or ask questions, or if they expressed confusion. The students said they did not feel comfortable speaking in the classroom. A few students mentioned feeling nervous about asking the instructor questions after class. The students more often noted that their instructor explained concepts in one way and usually at a level they could not understand. Students reported that the course material did not relate to their lives or make connections well between the lab and lecture portions of the course. Many students said learning in their courses focused on memorization so that even those receiving a good grade in the course did not perceive themselves as having learned anything useful. Students in courses with a medium level of reform (RTOP ratings of 46–70) reported their courses experiences as positive. These students expressed satisfaction with the hands-on experiences they were provided, but they also stated they would like to have had more explanation and help in connecting the lab and the lectures. The perceptions of satisfaction expressed by students in courses with a medium level of reform were more varied than those found in low or high levels. Students enrolled in courses with high level of reform (RTOP rating of 71 or above) described course experiences as positive. The most positive response was found among students who were non-science majors or reported having anxiety about science before taking the course. Instead of focusing on what the instructor did in the classroom, these students' descriptions of how learning occurred focused on their own actions and the actions of other students. The students also spoke positively of the communication patterns occurring both within and outside of the classroom. They felt more comfortable and confident in expressing their ideas and thoughts with their peers and their instructor. Being allowed to express themselves was viewed as helping them to develop a better understanding of science concepts by being allowed to see the content from multiple perspectives and having to rethink their own ideas and come up with better ways to solve problems

Regardless of the level of reform, students in all courses reported hands-on experiences as important in guiding their learning. Doing science was viewed as allowing them to manipulate the variables so that they could understand the relationships between them. All student groups stated that interacting with peers helped them to learn. Sometimes, seeing another student solve the problems helped give them confidence that they can do it. Other times, interacting with their peers helped them to be more reflective and refine their thinking so they could come to a better understanding of the concept. Student focus groups indicated, in general, that it was important to feel comfortable in expressing themselves in the classroom. Students further indicated the importance of instruction of course concepts to include making connections and being relevant to their personal experiences. They suggested that opportunities to explore course concepts through inquiry made the courses more personally relevant.

As a result of qualitative analysis of the student focus group interviews, four themes emerged as descriptors of the course experience: (1) communication in the classroom, (2) how science is learned, (3) shared control of learning in the classroom, and (4) connections/differences to personal views of science. Within the theme of communication in the classroom, four sub-themes were found: (1) sharing ideas, (2) feeling safe, (3) feedback, and (4) who communicates. The ability to share their ideas, opinions, and thoughts in the classroom was important for students. Within the theme of how science is learned, statements varied depending on the level of reform experienced in a class. Some elements related to science learning were sharing ideas with others; reflecting on, and then communicating, their understanding of science concepts; being able to explore concepts prior to their formal introduction to the concepts; and trying out or applying what they just learned in solving problems. The theme of shared control related to students' reporting they were able to find their own answers and solutions to problems. Related to the views of science themes, relevance and applicability were two important aspects of responses provided.

Qualitative analyses of the undergraduate student focus group interviews supported the observers' course ratings and descriptions of instructors. Important differences existed between courses with varying levels of reform. Students from the focus groups with a high level of reform, for example, reported that experiences in the course enabled them to (1) understand science as a process, (2) conceptualize difficult science concepts through models and investigations, (3) apply science concepts learned in class in novel situations, and (4) reflect on their learning and think about how they could use the knowledge and skills gained in the future.

Important differences also existed between reform and matched comparison courses. Students in courses with low instructor RTOP ratings described the learning environment in the same way as students enrolled in the entire group of matched comparison courses. Students expressed greater satisfaction with the learning experiences in reform courses. Students also perceived the reform course as more helpful in facilitating their understanding of the course concepts.

## Themes 1 and 2: Classroom Learning Environment and Course Structure

### *Student Achievement*

The NSEUS study investigated the quality of students' reasoning and explored the relationship between students' learning of science content and the level of reform in science courses. The sample of undergraduate science courses covered a variety of science disciplines. Accordingly, a direct comparison of students' learning of subject matter was not possible. We concentrated instead on understanding of discipline knowledge by comparing evidence of reasoning within the different content areas that the students had studied. For undergraduates, student reasoning is a primary goal of the (core) science curriculum as well as advanced science literacy. Emphasis on students' reasoning had two important outcomes. First, an assessment tool was designed that was aligned with reform goals in undergraduate science, focusing on evaluating depth of understanding. Second, the NSEUS protocol for developing written content questions provided a common structure for probing and classifying students' level and type of reasoning. In designing the NSEUS protocol, we considered several distinct criteria that were independent of course content and made comparison possible across disciplines. Because of the size of the study, we were not able to interview individual students on understanding of science knowledge and instead relied on written responses to specialized essay questions as the source of our evidence.

To probe students' reasoning, NSEUS developed, with close participation of each sample course instructor, open-ended written content questions with the distinguishing feature of exploring student's higher level thinking. The NSEUS science achievement instrument, *Science Content Assessment* (SCA), was a protocol designed to create similar content items on an end-of-the-semester test matched to each sample course. Questions were phrased to elicit students' reasoning about the relevant science content in short essays and evaluate students' responses through different lenses. These questions exposed students to a new problem scenario that represented a very important course goal and was developed in collaboration with the course instructor. This scenario required students to recognize the concepts they learned in a new context. Further, students were required to use higher cognitive skills on end-of-the-semester achievement items: explore relationships, investigate cause and effect, and develop a hypothesis using their deductive reasoning.

The SCA analysis divided knowledge into three dimensions: factual, conceptual, and application. The analysis further considered the student cognitive processes used to answer the question. The cognitive scale examined the types of skills students used to answer the question such as simply recalling information to state facts or concepts or the ability to use the information to apply, infer, or explain the information in novel situations. (See Chapter 2 for a more detailed description of SCA.) The NSEUS content assessment protocol then used several forms of logistic-regression analysis to build statistical models that described the relationships between our measures of student reasoning and the level of reform observed within the classroom (Zollman, Matloob-Haghanikar, & Murphy, 2012). We conducted a trial logistic regression for a sample of 18 undergraduate courses with about 900 students. We also analyzed the conceptual structure of students' responses based on the conceptual classification scheme and then clustered students' responses into six categories. We derived regression models to estimate the relationship between the sophistication of the categories of conceptual structure and the course instructor RTOP ratings. In the final analysis, the outcome variable with six categories required a more complicated regression model, known as multinomial logistic regression, generalized from binary logistic regression (Zollman, Matloob-Haghanikar, & Murphy, 2012).

We found that higher RTOP overall ratings were correlated with evidence of greater factual and conceptual knowledge in the students' responses, SCA total scores. Regardless of how they improved in reconstructing the knowledge of facts and concepts in connection with the features of question, however, these same students did not improve further to develop higher cognitive processes such as *Infer* and *Explain* to present a plausible answer. As a result, we observed that the usage of higher-order conceptual structures with the science content were less prevalent in higher-rated RTOP courses. Both student groups using higher-order thinking were a small percentage of the total sample of students. At first glance, this result may seem to be a negative statement about inquiry-based classes. That is, students in lower RTOP classes were more likely to use and link higher-level concepts than students in reformed classes, as measured by RTOP. Inquiry-based classes, however, frequently focus much of the instruction on observation, measurement, and reasoning from the students' own observations. Students in lower RTOP classes are more likely to be introduced to higher-level theoretical concepts and thus are more likely than the inquiry students to display higher-level concepts in their answers on written exams. This greater exposure alone does not necessarily relate to meaningful understanding. It is not clear whether these results indicate that the students in the inquiry classes are not able to use the high-level concepts and links or have just not had sufficient practice in doing so. Answering that question is beyond the scope of the current study. With the large amount of collected data, we found that significantly higher total SCA average scores in written responses were related to students' experiences in classes with higher levels of reform. We also saw lower use of higher-level thinking with more abstract concepts in classes with greater levels of reform.

## Theme 3: Pedagogical Content Knowledge

The CoRe and PaP-eR instrument provided a measure of the instructor's science teaching knowledge (pedagogical content knowledge) of the specific discipline concepts taught during the observed classroom lessons. Analysis of the CoRe and PaP-eR qualitative data instrument results for instructors was a two-step process. First, fundamental themes were determined using qualitative analysis of faculty interviews (CoRe) and classroom observational reports (PaP-eR). The themes were used to develop general rubrics considering several categories of planned and observed instructor actions. One was an overall single rubric, *Effective Teaching*, considering an instructor's classroom teaching responses and actions with four levels of rating: (1) novice, (2) emergent, (3) proficient, and (4) advanced. Four additional rubrics, each containing four rating levels, scored subcategories of an instructor's classroom teaching responses and actions: (1) content knowledge, (2) knowledge about how students think and learn, (3) science teaching knowledge, and (4) extent of professional development experienced related to effective teaching of the science concept(s) involved in the lesson.

Quantitative analysis of the rubric data found a statistically significant difference with the *Effective Teaching* rubric, or total score, between instructors of the reform course and comparison courses, $t(33,2.61)$, $p = 0.03$. Instructors of the reform courses total score averaged 3.31 for the CoRe and PaP-eR, while comparison course instructors averaged 2.39. Subcategory rating differences from the CoRe and PaP-eR for reform course instructors indicated a higher level of proficiency in utilizing science content for instruction. The subcategory ratings are related to instructors' higher level of proficiency in facilitating student thinking based on an understanding of students' prior knowledge about the specific science concepts taught. Subcategory ratings also indicated greater proficiency in science teaching knowledge among reform course instructors indicating they selected appropriate teaching strategies to better represent science concepts. These instructors also more actively sought out and participated in professional development workshops geared toward science education reform.

Further qualitative analysis of faculty interviews and CoRe and PaP-eR results for instructors focused on their level of pedagogical content knowledge. Instructors with PCK ratings of *Advanced* shared the following characteristics: (1) strong science content knowledge; (2) orientations congruent with the upgraded national science standards for effective science teaching and learning; (3) knowledge of student learning difficulties and prior knowledge; (4) purposeful selection of activities to teach specific science concepts that would best engender student understanding; (5) participation in ongoing professional development geared toward science education reform; (6) collaboration with scientists, science educators, and inservice teachers in K–12 schools; (7) engaging in reflective practice; and (8) having many years teaching experience. In addition, reform course instructors' intended goals for a lesson were consistent with the enacted goals observed during instruction.

An example of an instructor with PCK ratings of *Proficient* is one who demonstrated evidence of content knowledge regarding biological classification and had taught this particular topic for many years. This proficient instructor did not, however, discuss student learning difficulties associated with this topic based on prior experiences. The instructor said he used this activity because it helped students to understand the basics of classification, but not necessarily deepen their understanding of this topic. An example of an instructor receiving a PCK rating of *Emerging* also had strong content knowledge regarding the concept (e.g., cellular respiration) and many years' teaching experience. The emerging instructor maintained, however, a lecture-based teaching approach rather than adjusting instruction to accommodate student learning difficulties and their prior knowledge. The instructor's belief about science teaching and learning was reflected in his teaching style.

The analysis found that faculty instructors in reformed courses were more likely to be rated as having *Advanced* or *Proficient* PCK regarding the science concept taught during the observed lesson. Instructors who taught the comparison course focused more on content than on pedagogy. The following themes emerged from the qualitative analysis of interviews and CoRe and PaP-eR for instructors in a comparison course: (1) teacher-centered (lecture was the dominant mode of instruction), (2) little student discourse, and (3) separate lecture and laboratory. Regarding pedagogical content knowledge, comparison course instructors generally did not consider student misconceptions or learning difficulties when selecting instructional strategies to teach specific science concepts. These instructors demonstrated strong evidence of content knowledge but did not possess the pedagogical knowledge needed to transform science knowledge to facilitate student understanding. Comparison course instructors rarely attended professional development to improve science instruction and rarely collaborated with other colleagues regarding course design or development.

Qualitative analysis of the interviews from the comparison course student focus groups corroborated the instructional practices observed for their instructors. Students stated that comparison instructors spent the majority of class time lecturing and giving notes. Students also stated they were usually unable to make the connection between the science content discussed during the lecture to the laboratory investigations. Overall, analysis of the data revealed that comparison instructors employed a traditional approach to science teaching and learning.

## Theme 5: Beliefs and Attitudes

### *Student Beliefs About Science*

The *Draw A Scientist Test* (DAST) investigated students' beliefs about scientists and the nature of scientific work. A drawing of a scientist and an accompanying narrative describing the work of a scientist were requested of students at the beginning and end of their courses. The higher the DAST score, the greater the stereotypical view. Overall, there was little difference in the way the students participating in the study chose to depict scientists between the reform and comparison courses and between courses differing in levels of reform. Stereotypical views common in society and the media predominated. There were fewer stereotypical characteristics on the posttest as compared to the pretest for all students.

Prior to instruction, the majority of students drew male scientists. After instruction, on the posttest, when a change in gender occurred in the drawing, it was usually to a gender-neutral scientist. Very few students changed from drawing a male to drawing a female. The increase in the number of gender-neutral scientists in the posttest drawings reflected the belief that anyone can be a scientist as interpreted from the DAST narrative. On the pretest, students drew scientists with several characteristics that served as rubric indicators. The indicators included one or more of the following: lab coat (usually white), eyeglasses, facial hair (beards, mustaches, abnormally long sideburns), symbols of research, symbols of knowledge, signs of technology (products of science), captions (formulae, eureka syndrome), male Caucasian, signs or labeling (such as warning, secrecy, private), pencils or pens in a pocket, and an unkempt appearance. On the posttest, students removed most of the earlier indicators from the drawing. In some cases, however, students increased their scores by adding more indicators. These indicators included symbols of research, signs of technology, and captions reflecting a better understanding of scientific investigations.

On the posttest, most students continued to typically draw a scientist in the lab or without a setting. The scientists were depicted investigating problems connected to concepts in the course on the posttest. In a geology course, for example, the scientist may have been collecting various rocks. The increase in the number of scientists outside of the lab reflected, as described in the narratives, an increased level of beliefs that scientists help society by providing an understanding of the world around us and by solving societal problems. There was an overall decrease in scientists who were depicted as eccentric or sinister in the posttest.

While the type of instruction may not have made a difference in students' beliefs about scientists and their work, the results indicated that the additional undergraduate instruction, as evidenced by the pre- and post-administration of the DAST, changed the way in which students viewed scientists and their work. The students adopted a less stereotypical view of scientists and their work after taking a course in science.

## Theme 4: Beliefs and Attitudes

### *Student Attitudes About Science*

The *Thinking About Science Survey Instrument* (TSSI) examined beliefs and attitudes of the undergraduates towards science and the nature of science (Cobern, 2000). The instrument assessed sociocultural resistance to and support for science. The TSSI data were used to quantitatively document the presence or absence of significant cultural concerns about science. The TSSI instrument uses a five-point scale that is labeled: strongly disagree (1 point), uncertain (3 points), and strongly agree (5 points) to determine the final score. Of the 60 items, 27 were reversed. All of the items were summed. The TSSI was used to determine whether or not science course reform impacts students' attitudes toward science. The scores on the TSSI ranged from 148–268 on the pretest and 104–177 on the posttest. There were no significant differences between the reform and comparison courses on the total score of the TSSI before ($F$ = .007, $p$ = .65) or after instruction ($F$ = .65, $p$ = .43) using Univariate Analysis of Variance. Differences in students' scores based on level of reform in the classroom were not found when the RTOP score was considered high (above 70) or low (45 or below) before or after instruction. The results are outlined in Table 3.6. Statistical differences were not noted when comparing the change in scores between students enrolled in courses even with RTOP ratings of 90 and above ($F$ = .190, $p$ = .664). Differences between the students' pre- and posttest scores were not found to be related to level of reform implemented in the under

When controlling for differences found in the students' pretest subscale scores, significant differences were seen on the TSSI Science, Religion, and Morality Scale ($F$ = 5.84, $p$ = .02). Students in the comparison courses scored significantly higher than students in the reform courses. Comparison students had significantly higher concerns about the interaction of science and religion. These results indicate that inquiry learning may help ease students' distrust in scientific knowledge, especially when it comes to conflicts with religious beliefs.

**Table 3.6. Comparison of Pre- and Posttest Scores on the Thinking About Science Survey Instrument (TSSI) Based on Level of Reform With RTOP Rating**

| *TSSI* | *RTOP Interval* | *N* | *F* | *Sig* |
|---|---|---|---|---|
| **Pre** | RTOP 71 or above | 198 | .53 | .47 |
| | RTOP 70 or below | 200 | | |
| **Post** | RTOP 71 or above | 198 | .11 | .74 |
| | RTOP 70 or below | 198 | | |
| **Pre-** | RTOP 46 or above | 200 | .22 | .64 |
| | RTOP 45 or below | 196 | | |
| **Post-** | RTOP 46 or above | 199 | .77 | .38 |
| | RTOP 45 or below | 196 | | |

Nonparametric statistical analyses revealed differences on two question items on the Science, Religion, and Morality Scale. Students in the comparison courses agreed with these items more than students in the reformed courses. Using the Mann-Whitney U test, the null hypothesis was rejected for the item that read "Religion tends to impede scientific progress" ($p$ = .03) and the item reading "Scientific research is morally neutral" ($p$ = .02).

Based on these results, it can be concluded that reformed science teaching impacts some aspects of students' attitudes and beliefs about the social and cultural values of science differently than the teaching found in comparison courses. There were no differences, however, between the students' total scores based on the type of course in which they were enrolled in or the level of reform implemented in the classroom. It is likely that a single course does not significantly change the students' attitudes and beliefs about scientific knowledge based on the results of the TSSI.

## Theme 4: Beliefs and Attitudes

### *Faculty Beliefs About Teaching and Learning*

Individual interviews were held with instructors from each of the 35 courses studied during NSEUS site visits. These interviews occurred at or just after the middle of the semester. Two researchers asked questions during the interview. Instructor responses from courses varying by level of reform are discussed first. Level of course reform was determined using faculty instructor RTOP ratings: (1) RTOP rating of 71 or higher, (2) RTOP rating of 46–70, and (3) RTOP rating of 45 or lower.

All instructors interviewed stated they wanted the course to be relevant to their students. Other common differences, however, were found in the instructors' interview data and varied according to the level of course reform. These variations were not evident between the reform and comparison groups of instructors. Reform instructors with low RTOP ratings, for example, responded similarly to comparison instructors with low RTOP ratings.

Instructors teaching courses with higher RTOP ratings were more reflective than were those with lower ratings about how they would make science relevant to their students. These instructors reported considering prior knowledge when creating lessons to make science meaningful to their students. They also planned to use a variety of instructional methods to help students learn the content from different types of learning experiences. The instructors in this group made much less use of extensive lectures, relying instead on shorter just-in-time mini lectures and learning activities to help students develop their understanding. Typically, lab and lecture were integrated. Instructors reported planning for the use of student-centered teaching methods over teacher-centered methods. An example of such planning was to set up their classroom beforehand or plan pairing activities so students could interact and collaborate with each other as well as with the instructor. One instructor stated, "Learning is personal and you need to teach to your audience." This statement summarized the approach some of the instructors described when helping students master the content that they were teaching.

Instructors with low RTOP ratings generally stated that giving students the information they needed would help them learn. They commonly suggested that telling students about examples in the media would help connect a science concept to their lives. Instructors used lecture and PowerPoint presentations to give the students the information they needed during course meetings. The instructors in this group reported using teacher-centered methods in planning for classes focusing on what they were going to do when presenting the course information when describing their teaching. These instructors often gave little description of what students did in class except to say that they took notes.

The purpose of teaching was the only area where the reform course instructors differed from the matched comparison course instructors. Comparison instructors were more likely to state the idea that mastering the content was the main goal of the course. Reform instructors emphasized the importance of students' building an understanding of the content, but they also wanted students to develop problem-solving skills, inquiry skills, or the ability to apply scientific knowledge. The level of reform, rather than the reform course vs. comparison course division, was generally associated with how the instructors reported achieving their goals for teaching

## SUMMARY FINDINGS FOR COMPARISON SET 3

**Research Question 2:** *The results of analyses of implemented reform in sample courses (n = 20) versus matched comparison science courses (n = 15) were:*

1. Significant quantitative differences were found between reform and matched comparison instructors' teaching performance as represented in RTOP ratings.
2. Qualitative differences were found between the implemented reform and matched comparison courses in all five course elements as evidenced by using multiple data sources:
   a. Classroom learning environment
   b. Course structure (curriculum and sequence of learning activities)
   c. Pedagogical content knowledge
   d. Collaboration
   e. Beliefs about teaching and student learning
3. Further analyses found that reform and matched comparison instructors differed in their teaching approaches and pedagogical content knowledge (e.g., how to teach physics concepts)
4. Differences were found between reform and comparison course instructors in types of observed class instructional activities using data from RTOP, CoRe and Pa-PeR, and student focus groups
5. In reform courses, a major instructional strategy used was the inquiry-based integrated laboratory (68%), while lecture dominated the comparison course (63%) as found when analyzing data from RTOP, CoRe and Pa-PeR, and student focus groups.
6. Reform course instructors exhibited deeper understanding of how students typically think about science and more often modified their teaching to match students' learning needs as indicated in analysis of data from CoRe and Pa-PeR and from student focus groups.

7. Reform course instructors exhibited greater proficiency in knowledge about science teaching (e.g., use of inquiry methods as an effective means to increase active learning) as found when analyzing data from RTOP, CoRe and Pa-PeR, and student focus groups
8. Collaborative faculty teams performed an important role in developing and sustaining course reform in a majority of the sample reform courses as evidenced by data from faculty interviews.
9. Faculty teaching in reform courses demonstrated significantly greater use of teaching practices aligned with the college level interpretation of national standards for science education as per RTOP and the CoRe and Pa-PeR.
10. Faculty teaching in reformed courses demonstrated a significantly higher level of pedagogical content knowledge relative to matched comparison course faculty as indicated by analyzing data from RTOP and from CoRe and Pa-PeR.
11. Faculty teaching in reformed courses described, in personal interviews, beliefs that were less teacher- and more student-centered.
12. Reform courses were reported and observed to have a significantly greater number of indicators of inquiry, which related to course learning environment, instructional strategies, and curriculum as documented via the RTOP and the CoRe and Pa-PeR.
13. Analyses found that students in reform and matched comparison courses differed in their reaction to faculty development of course characteristics.
14. Students in reformed courses rated the classroom learning environment on CLES more positively than students in the matched comparison courses.
15. Students in focus groups reported a qualitatively different experience and greater amount of learning in the reform course as compared to students' reports of their comparison course experiences.
16. Students demonstrated no difference in their understanding of the nature of science and scientists between the reform and matched comparison courses as indicated on the DAST. (Note: Taking science courses—reformed or comparison—positively impacted the way students viewed scientists and their work.)
17. Students' attitudes towards science and society on TSSI were not significantly different between those in reformed or comparison courses. Little overall change was also noted on the test over the length of the course.
18. Students had higher achievement on the SCA content tests when they were in higher level of reform courses (as measured by the RTOP), demonstrating higher-level thinking.

**Research Question 3:** *Findings from all courses in a single group based on level of implemented reform (N = 35 courses at 20 institutions) are summarized below:*

1. Students experiencing *higher levels of reform* (having had instructors with a significantly higher level of RTOP rating):
   a. Rated their classroom learning environment more positively on CLES and described it more positively in focus group interviews
   b. Reported a qualitatively different experience and greater amount of learning in focus group interviews
   c. Demonstrated attitudes towards science and society on TSSI that were not significantly different from students in the courses with differing levels of reform taught by other instructors
   d. Demonstrated understandings of the nature of science and scientists on DAST that were not significantly different from those students in courses taught by other instructors (Note: Taking science courses—even with different levels of reform—positively impacted the way students view scientists and their work.)
   e. Had higher achievement on the SCA content test demonstrating higher level thinking
   f. Demonstrated lower levels of misconceptions on the SCA about the science concept(s) tested
   g. Demonstrated a complex relationship between their achievement on the SCA based on the level of instructional reform they experienced in the classroom:
      i. Students experiencing higher levels of reform showed a higher but mixed degree of ability to answer the SCA questions using higher order thinking.
      ii. Students' responses on the SCA in courses experiencing lower levels of reform indicated success in gaining factual knowledge but little ability to apply those facts.
2. Faculty instructors showing *higher levels of reform* (having a significantly higher level of RTOP rating)
   a. Demonstrated significantly greater use of teaching practices aligned with the upgraded national science standards as indicated by analysis of data from RTOP and from CoRe and Pa-PeR
   b. Demonstrated a significantly higher level of pedagogical content knowledge as evidence by data from RTOP and from CoRe and Pa-PeR

c. Described, in personal interviews, beliefs that were less teacher- and more student-centered
d. Were observed to have used a significantly greater number of indicators of inquiry in their classroom teaching that related to classroom learning environment, instructional strategies, and curriculum as per data from RTOP and Pa-PeR

## Comparison 4

**Research Question 4:** *How do the differing levels of course reform relate to the long-term learning outcomes for graduated inservice K–6 teachers teaching science in their own school classrooms?*

Comparison 4 involved an analysis of the relationship of the quality of implementation of reform on *long-term* outcomes of undergraduate students—specifically, in their teaching of science as graduated elementary classroom teachers. Quality was determined in two ways that were similar to those described in Comparison 3. First, a comparison was made of data collected on the inservice elementary teachers who had been students in the courses of reform faculty with data collected on the inservice teachers of comparison faculty. Second, a comparison was made using data collected on the inservice teachers who had been students in all of the sample courses based on the level of reform implemented in the undergraduate courses at the time of the study (see Figure 3.10).

# NARRATIVE DESCRIPTION OF FINDINGS

## Instructional Practices of Elementary Inservice Teachers: Quantitative Findings

Two sample groups of inservice elementary education teachers who completed their required undergraduate science courses and graduated from one of the sample institutions were identified. As undergraduate preservice elementary education majors, each sample group had completed either the reform or matched comparison science courses as a part of their teacher certification program. The graduated students were inservice elementary teachers in local elementary schools. The original sample was made up of three students who were to be selected from each reform and matched comparison science course at the 20 NSEUS institutions. Attrition due to schedule conflicts and lack of available science lessons in the elementary school curriculum led to a total of 56 reform teachers and 35 matched comparison course graduates being scheduled and visited. Data collection site visits were made to the elementary classrooms of 91 elementary inservice teachers in grades kindergarten to grade 6.

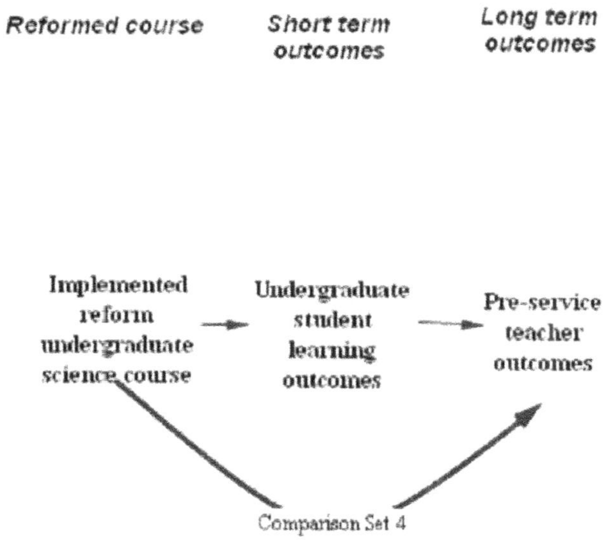


**Figure 3.10.** Comparison Set 4.

Analysis of RTOP ratings found no significant differences between the science pedagogical practices of inservice elementary teachers who experienced the reform course during their undergraduate program and those who did not, $t$ (89, .274), $p$ = .785). Results from the quantitative analysis of classroom observations using RTOP subscale ratings between those teachers who experienced the reformed or comparison courses revealed no statistically significant group effect, *Wilks* = .061, $F$ [5, 85] = 1.514, $p$ = .194 (see Tables 3.7 and 3.8).

## Instructional Practices of Elementary Inservice Teachers: Qualitative Findings

Qualitative analysis of RTOP narratives and teacher interviews, in addition to observers' field notes, indicated that the majority of inservice teachers, whether reform or comparison course graduates, used hands-on

**Table 3.7. RTOP Means and Standard Deviations for the Inservice Elementary Teachers Who Experienced Reform and Matched Comparison Courses**

| *Course* | *N* | *Mean* | *Standard Deviation* |
|---|---|---|---|
| Reform | 56 | 61.13 | 18.34 |
| Comparison | 35 | 60.03 | 18.99 |

**Table 3.8. MANOVA Results for Inservice Teachers by RTOP Subscales**

| *RTOP Total and Subscales* | *df* | *Mean Square* | *F* | *Sig.* |
|---|---|---|---|---|
| Lesson Design and Implementation | 1 | 6.431 | 0.362 | .549 |
| Propositional Knowledge | 1 | 8.798 | 0.712 | .401 |
| Procedural Knowledge | 1 | 10.836 | 0.562 | .456 |
| Communicative Interactions | 1 | 0.102 | 0.006 | .936 |
| Student Teacher Relationships | 1 | 6.915 | 0.351 | .555 |
| Total Rating | 89 | | 0.274 | .785 |

activities during the observed lesson. Importantly, however, confounding variables were identified from several sources as interacting to influence the pedagogical practices of both groups of teachers.

First, many teachers in both groups attended some form of professional development workshop to improve science teaching after completion of their undergraduate science courses and graduation from their teacher preparation program. The ethos of these workshops varied in nature and focus. Some workshops were reported by teachers to focus on use of science notebooks with students during lessons while others focused on setting goals with science standards. Some workshops provided science kits for the inservice teachers to use with their students. These kits typically revolved around a science unit and had some elements of inquiry. The professional development workshops provided comparison course graduates with additional positive science experiences that were not found in the comparison course they had taken as undergraduates. The undergraduate reform courses had provided their students with experiences implementing the national science standards. As comparison teachers participated in professional development opportunities, they reported having experienced and adopted some inquiry-oriented teaching practices similar to those experienced by reform course graduates.

Second, many inservice teachers from both the reform or comparison undergraduate courses reported they were not able to teach science or could not teach science on a consistent daily or weekly basis. Teachers reported that the local school administration emphasized and required a school day prioritizing reading and mathematics instruction, excluding science and other aspects of the full school curriculum. This restriction was reported by more than 50% of the teachers. These restricted inservice teachers, in general, were less able to continue to develop PCK associated with national science standards. Low curriculum priority in the elementary grade levels influenced a number of related effective science teaching variables that were reported by the teachers in interviews: lack of time during each day, regularity of lessons and frequency in teaching science, lack of resources for hands-on and inquiry science, lack of curricular materials that supported teaching hands-on and inquiry science, lack of professional development in science teaching provided by the school system, and just fear that monitoring brings to teachers daily regarding meeting reading standards. As an example, some teachers were not able to teach science in grades K–3 or to teach science only once a week in higher grades, while others taught it three or more times a week. These forms of school organizational priorities restricted the amount of science taught in a given week or in a specific lesson, creating several strong, confounding variables affecting possible differences between reform and comparison elementary teachers.

## Pedagogical Content Knowledge of Inservice Teachers

The findings regarding the pedagogical differences between teachers were enhanced by additional qualitative analyses of inservice elementary teacher's interviews and field notes of observed teaching of science lessons in elementary schools. The qualitative data were used to develop CoRe and PaP-eR descriptions for each teacher. Four rubrics were developed based on the CoRe and PaP-eR descriptions that provided a mean PCK rating for each of the teachers. The PCK rating categories were *advanced, proficient, novice,* and *emerging*. Overall, inservice elementary teachers who experienced the undergraduate science reform course obtained a mean PCK rating of 2.41. Teachers who graduated with experience in comparison courses were found to have an average 2.08 PCK rating. There was no statistically significant quantitative difference between the two groups.

While there was no statistically significant quantitative difference between the two groups, qualitative data from teacher interviews and CoRe and Pap-eR revealed several factors related to both groups of participants' science teaching PCK. Qualitative analyses indicated that differences existed within inservice elementary teachers of both groups. This relationship between PCK ratings for the teachers, however, was complex. Elementary teachers demonstrating advanced or proficient PCK used their knowledge of science content and of pedagogy to inform their science teaching practices. Inservice elementary teachers who had these higher PCK ratings also had the higher overall RTOP ratings. Teacherswho had high RTOP ratings, however, did not necessarily achieve higher PCK ratings regarding the specific science concept that was taught.

The following characteristics emerged from the qualitative analysis of data from teachers with high PCK ratings, advanced or proficient PCK, and high RTOP ratings: (1) demonstrated higher science content knowledge regarding the specific concepts taught during the observed lesson, (2) participated in ongoing professional development geared towards the enhancement of science content and pedagogy, (3) taught in a school culture that emphasized science teaching and learning, (4) considered student learning difficulties and prior knowledge in lesson planning, (5) purposefully used inquiry instructional strategies to present science content to facilitate student learning, (6) used hands-on activities to facilitate student learning, and/or (7) reflected on teaching practice. An elementary inservice teacher, reform group case study sample description is provided in Figure 3.10 and composed from data from the RTOP, interviews, and the CoRe and PaP-eR.

Inservice teachers with PCK ratings of emerging or novice and low RTOP ratings shared these characteristics: (1) demonstrated lower level of knowledge regarding the specific science concepts taught; (2) taught in a school culture that deemphasized science; (3) exhibited a disconnect between their reported teaching beliefs and the enactment of those beliefs in the observed lesson; (4) demonstrated lower levels of participation in professional development related to science teaching; (5) used hands-on activities for their own sake and not for enhancing student understanding—for example, assisted with students' motivation; (6) placed high priority on subject areas, especially mathematics and reading, covered in state testing programs; (7) taught science as a reading lesson; and (8) covered a lot of science in a short time to prepare students for state tests.

While there was no statistically significant quantitative difference among inservice teacher RTOP ratings between the two groups, reform and comparison, narrative analysis demonstrated that a relationship was evident between levels of science teaching PCK and classroom RTOP ratings. Those teachers who had high PCK ratings and high RTOP teaching performance ratings were described as more effective teachers.

### *Inservice Teacher Efficacy*

The impact of undergraduate science courses on science teaching efficacy was also examined. Data were collected form 85 elementary teachers, 35 university faculty, and 190 undergraduate students from across the NSEUS national sample during site visits using the *Science Teaching Efficacy Beliefs Instrument* (STEBI-A). The STEBI-A has two subscales: Personal Science Teaching Efficacy (PSTE) refers to a teacher's belief in his or her own ability to effectively teach science, while Science Teaching Outcome Expectancy (STOE) reflects the extent of a teacher's belief that, if teachers provide appropriate science instruction, then their students will learn. From the data set, eight case studies of inservice elementary teachers also were examined in closer detail.

We examined the relationships among Personal Science Teaching Efficacy (PSTE) beliefs, science teaching practices, and the Science Teaching Outcome Expectancy (STOE) (beliefs about these practices) among the elementary inservice teacher sample. Two comparisons were made of PSTE and STOE: (1) reform and comparison group experiences and STEBI-A total score and subscale scores (PSTE and STOE) and (2) level of reform based on RTOP ratings and STEBI-A scores. Pearson's correlations and independent sample $t$-tests of coded qualitative data and quantitative survey data were conducted in order to compare the beliefs and practices regarding science teaching within and across efficacy scores. In addition, eight case profile teachers were chosen with varying combinations of high and low PSTE and RTOP ratings in order to examine some of the complexities existing between science teaching self-efficacy beliefs and science teaching behaviors in closer detail.

Overall, STEBI, PSTE subscale, and STOE subscale scores were not significantly different for graduated inservice teachers who experienced a reformed undergraduate science course from inservice teachers who experienced the comparison course. While there was no statistically significant quantitative difference between the two groups, qualitative data from interviews with undergraduate students who were elementary preservice majors revealed several factors perceived by participants to influence their science teaching self-efficacy. Factors cited by undergraduate students who were elementary preservice majors as influencing their confidence in science teaching focused primarily on aspects of the undergraduate science courses they had experienced, such as the use of cooperative learning and the integration of lab and lecture. Factors discussed by inservice teachers in interviews were associated with several related to teacher beliefs interacting with self-efficacy and science teaching behavior. They covered a variety of areas and experiences, such as giving students more control over their science learning, using hands-on activities, access to science professional development, quality and type of science curriculum provided to the school, and availability of resources. Most of the factors cited as efficacious to science teaching, however, were not observed in inservice teachers' classroom practices. Interviews and observations demonstrated that self-efficacy levels manifested themselves in different ways with different teachers.

A clear relationship was not evident between science teaching self-efficacy and reformed science teaching. Inservice teachers, for example, stated that hands-on practices were the best way to teach science at the elementary level. Teachers, however, varied greatly in their description of what the term "hands-on" means to them and in turn, the ways that hands-on teaching were manifested in their observed lessons. Teachers' beliefs about the meaning of "hands-on" impacted their beliefs about their own efficacy as teachers of science. Several teachers, for instance, with low RTOP ratings and high STEBI-A scores said in their interviews that they were teaching science in the way it should be taught, which was hands-on. There was no clear understanding of what the terms they were using meant to them. Participants reported that they gained more confidence in their ability to teach science effectively from courses that (a) explicitly connected the science content to the teaching of that content, (b) gave students opportunities to teach the content to others, and (c) sparked students' interest in the content. These results demonstrate the importance of using qualitative data to support quantitative data when studying self-efficacy beliefs of teachers.

## SUMMARY FINDINGS FOR COMPARISON SET 4

Summary findings from the reform courses versus matched comparison courses were as follows (see also Figure 3.10):

1. Reform and comparison course graduates who were teachers of elementary school science did not differ quantitatively in RTOP ratings of their observed performance in teaching science in their elementary classrooms.
2. Reform and comparison course graduates who were teachers of elementary school science did not differ quantitatively in PCK ratings.

Summary findings from all courses based on level of implemented reform ($n$ = 35) at 20 institutions are summarized below:

1. Higher levels of reform experienced as an undergraduate were associated in qualitative analyses with higher levels of pedagogical

content knowledge among inservice teachers. They demonstrated qualitatively greater science PCK in:

a. Depth of science content knowledge of the concepts taught
b. Knowledge of how students think about science
c. Adaptation of their teaching to match students' learning needs
d. Knowledge of science pedagogy

2. Higher levels of reform course experienced as an undergraduate were not related to their observed RTOP performance rating in teaching science in their elementary classrooms as inservice teachers.
3. Higher levels of reform course experienced as an undergraduate were more closely associated with inservice teachers' interview descriptions of teaching and lesson planning behaviors identified in the *National Standards for Science Education* (National Research Council, 1996).
4. Higher levels of reform experienced as an undergraduate were associated with a higher level of perceived science teaching efficacy on the STEBI-A.
5. Science teaching efficacy scores on the STEBI-A were not found to be related to observed classroom practices as indicated in RTOP ratings.

## CONCLUSIONS AND LESSONS LEARNED

The goal of the National Study of Education in Undergraduate Science (NSEUS) was to determine the feasibility of creating reforms in undergraduate science in order to provide an alternative to existing often traditional teaching and learning in undergraduate courses. The overall research question investigated in this national study of undergraduate science courses was, "How do undergraduate entry-level science courses, differing in level of reform, affect student learning outcomes?" The sub-questions were:

1. Does faculty professional development change undergraduate science faculty teaching practice?
2. How do science teaching and learning course characteristics differ between courses before and after a faculty professional development program?
3. How do course differences relate to the short-term learning outcomes of undergraduate students?

4. How do the differing levels of course reform relate to the long term learning outcomes for graduated inservice K–6 teachers teaching science in their own school classrooms?

The overall research question was analyzed in four major comparisons using a research model shown in Figure 3.3. Key conclusions can be summarized as follows.

### First, conclusions based on comparison of reform vs matched comparison courses:

1. Professional development as experienced by the course faculty described in this study enabled and facilitated development of undergraduate science courses with reform characteristics.
2. Professional development and the existence of interdisciplinary collaborative faculty teams sustained course reform over time.
3. Reform efforts at more than 1/3 of the sampled universities resulted in replication by other STEM faculty at the same institutions who developed additional STEM reform courses.
4. Significant instructional differences were found between reform and matched comparison faculty teaching performance using data from RTOP ratings.
5. Differences were found between the implemented reform and matched comparison courses in all five course elements as evidenced using multiple data sources:
    a. Classroom learning environment
    b. Course structure (curriculum and sequence of learning activities)
    c. Pedagogical content knowledge
    d. Collaboration
    e. Beliefs about teaching and student learning.
6. Reform courses were characterized by significantly greater implementation of inquiry-based learning environment, instructional methods, and learning goals consistent with national documents calling for reform such as *Shaping the Future* and the *National Science Education Standards*.
7. Reform course faculty instructors as related to comparison course instructors
    a. Used a significantly higher level of science pedagogical content knowledge (PCK) in planning for and teaching their courses

b. Used significantly more student-centered instructional strategies
c. Were observed to use a significantly greater number of indicators of inquiry that related to learning environment, instructional strategies, and curriculum than comparison course faculty

8. Students in reform science courses as compared to those in comparison courses
    a. Had significantly higher positive perceptions of their classroom (course) learning environment
    b. Were not significantly different in their attitudes toward science and perceptions of the nature of science
    c. Differed among individuals but not classes
9. A subgroup of graduates of reform courses
    a. Did not differ quantitatively from teachers of comparison courses in their observed teaching performance ratings when teaching science in elementary schools
    b. Differed qualitatively from teachers of comparison courses in their use of PCK when planning and teaching science in elementary schools.

### Second, conclusions from all study courses based on level of implemented reform

1. As measured level of course undergraduate instructional reform increased, instructors demonstrated
    a. Significantly greater use of teaching practices aligned with the *National Standards for Science Education* (National Research Council, 1996)
    b. Significantly higher levels of pedagogical content knowledge before and during instruction
    c. Significantly greater use of student centered instructional strategies
2. As level of reform increased, courses had a significantly greater number of indicators of inquiry that related to learning environment and curriculum.
3. Students experiencing higher levels of course reform
    a. Described their learning course environment more positively

   b. Showed a greater, but mixed degree of ability to answer science content (achievement) questions using more complex levels of thinking

4. Higher levels of reform course experienced as an undergraduate student were associated with

   a. Higher levels of perceived science teaching efficacy as an in-service teacher, however, teaching efficacy was not found to be related to observed classroom science teaching practices
   b. Qualitatively different and more advanced PCK among inservice elementary teachers including

      i. depth of science content knowledge of the concepts taught
      ii. knowledge of how students think about science
      ii. adaption of their teaching to match students' learning needs
      ii. knowledge of science pedagogy

## IMPLICATIONS FOR THE TEACHING AND LEARNING OF UNDERGRADUATE SCIENCE

### *NSEUS Answered Important Questions*

The NSEUS study developed evidence-based responses to basic questions asked by undergraduate faculty. The questions focused on how to develop more effective entry-level science courses (often known as gateway courses) at the undergraduate level that can sustain students in their science coursework and lead them to graduation in their chosen majors. The questions investigated were:

- What are the characteristics of a reformed entry-level undergraduate science course?
- What are the outcomes of undergraduate science course reform?
- What course reform elements are most effective in changing outcomes?
- What are effective components of inquiry teaching at the undergraduate level?
- What pedagogical content knowledge is needed for faculty to be effective instructors in undergraduate courses?
- How much reform (quantity) do faculty have to implement at the undergraduate course level to show better than average gain in course outcomes?

- What level of quality of innovation and faculty expertise is needed for reform at the undergraduate course level to show better than average student gain in course outcomes?

The conclusions of this national study relate to short-term impacts on multiple undergraduate majors, long-term effects on graduated preservice teachers of science and identification of characteristics of faculty instructors and courses producing significant impacts. In summary, the model used in this study for research on reform in undergraduate science teaching can be used to predict faculty change and student outcomes.

## What Was Learned

The NSEUS research model provides evidence for an undergraduate reform process extending from initial faculty professional development to increased student learning outcomes. What we learned was that the reform process is not a short-term endeavor. It is a process that is long-term and, for most, rarely ends. Reform efforts were sustainable only within a collaborative group of faculty who had continuing administrative support. Once begun, effective undergraduate faculty professional development produced predictably favorable results when it included well-designed professional development workshops, collaborative teams working on specific course reforms, regular and continued evidence based monitoring of course strategies for teaching and learning through faculty designed action research and from third party secondary data, public reporting and sharing of data and innovative ideas supporting effective reform, adequate resources and classroom facilities to complete intended reforms, and positive administrative support and faculty rewards for research and development that involved undertaking continuous reform to successfully sustain students in graduation in their majors. Faculty professional development activities that reflect reform did profoundly affect the short- and long-term learning outcomes of undergraduate students.

Another somewhat surprising lesson learned was that a significantly high level of real reform, both in quality and quantity, in instructor knowledge and skills and in the course learning environment and curriculum was required to observe greater than expected gains in student outcomes. Adding a single innovation to a traditional course or poorly performing an innovation does not affect student outcomes. Effective reform requires significant multiple innovations by faculty over a sustained period of time to create real change. Undergraduate science course experiences did affect how students understand science on both affective and cognitive levels. Reform in the study was found not to be effective in changing student learning outcomes if one tinkered with one or two course elements. The elements of student centering, active learning, an inquiry instructional and learning approach, and continuous student feedback are interrelated in fostering real course change. These elements involve many separate components of the undergraduate science course. Use of a technique such as clickers in a traditional lecture style along with student passive attention produces little change. These course elements require an interconnected and substantial set of course components (quantity) that must be performed expertly (quality). The interconnections create a cascading effect on students that is required before significant changes in student learning outcomes become evident.

The broader impacts of this study advance the professional development, course planning, and teaching of undergraduate science faculty and the development of more effective entry-level courses. The broader impacts supported by this study can be summarized as

- A general research model supports the effectiveness of the undergraduate science course reform process studied leading to positive student learning outcomes.
- Reform efforts are sustainable with collaborative faculty knowledgeable in pedagogical innovations and administrative support.
- A significantly high level of reform both in quality and quantity is required to develop greater than expected gains in student outcomes.
- Faculty professional development activities that reflect reform profoundly affect the short- and long-term learning outcomes of undergraduate students.
- Undergraduate science course experiences (context) effect how students understand science on both affective and cognitive levels.

## AUTHOR NOTE

Throughout the duration of the project, we used multiple forms of dissemination, which proved to be important in providing faculty across the nation with research-based practices for reforming courses. Two national conferences with the theme of research-based undergraduate science teaching were held, involving over 150 researchers from 85 higher education institutions. The 2011 and 2012 national conferences were held at the University of Alabama, *Research-Based Undergraduate Science Teaching I and II,* and funded through the National Science Foundation. Our website at http://nseus.org provides resource materials from the conference presentations as well as NSEUS background research papers, presentations, and research briefs. The practices we are disseminating are especially useful for reforming undergraduate science courses involving groups of students who are not always well represented in STEM education.

Correspondence regarding this manuscript may be addressed to Dennis W. Sunal, Professor of Science Education, 205C Graves Hall, The University of Alabama, Box 870232, Tuscaloosa, AL. Email: dwsunal@bama.ua.edu

## ACKNOWLEDGMENT

This material is based upon work supported by the National Science Foundation under Grant No. 0554594. Any opinions, findings, and conclusions or recommendations expressed in this material are those of the author(s) and do not necessarily reflect the views of the National Science Foundation.

## REFERENCES

Cobern, W. W. (2000) *The thinking about science survey instrument (TSSI)—SLCSP 151*. Kalamazoo, MI: Scientific Literacy and Cultural Studies Project.

Karplus, R. (1979). Teaching for the development of reasoning. In A. E. Lawson (Ed.), *1980 AETS yearbook: The psychology of teaching for thinking and creativity* (pp. 174-215). Columbus, OH: ERIC/SMEAC.

National Research Council. (1996). *National science education standards*. Washington, DC: National Academy Press.

National Science Foundation. (1996). *Shaping the Future: Strategies for Revitalizing Undergraduate Education*. Retrieved from http://www.nsf.gov/publications/pub_summ.jsp?ods_key=nsf9873

NOVA. (n.d.). *NASA opportunities for visionary academics*. Retrieved from http://www.education.nasa.gov/divisions/higher/programs/high_ed_faculty.html

Sawada, D., & Pilburn, M. (2000). *Reformed teaching observation protocol (RTOP)* (ACEPT Technical Report No. IN00-1). Tempe, AZ: Arizona Collaborative for Excellence in the Preparation of Teachers.

Seymour, E. (1995). Revisiting the "problem iceberg": Science, mathematics, and engineering students still chilled out: Examining the causes of student attrition in science-based fields on a variety of campuses. *Journal of College Science Teaching, 24*(6), 392–400.

Siebert, E., & McIntosh, W. (2001). *College pathways to the science education standards*. Arlington, VA: NSTA Press.

Sunal, D., MacKinnon, C., Raubenheimer, C. D., & Gardner, F. (2004). A case study of a national undergraduate science reform effort. In D. Sunal, E. Wright, & J. Bland-Day (Eds.), *Research in science education: Reform in undergraduate science teaching for the 21st century* (pp. 225–240). Greenwich, CT: Information Age Publishing.

Sunal, D., Sunal, C., Ogletree, G., Steele, E., Mason, C., & Zollman, D. (2008). Research brief No. 4: What characteristics are found among institutions still offering and no longer offering reformed Undergraduate science courses? Retrieved from http://nseus.org/research_brief4.pdf

Zollman, D., Matloob-Haghanikar, M., & Murphy, S. (2012). Background Research Paper No. 34: Student's reasoning and the level of interactivity in science content courses. Retrieved from http://education.ua.edu/centers/ortd/background-research-papers/

CHAPTER 4

# INVESTIGATING THE LONG-TERM IMPACT OF UNDERGRADUATE SCIENCE REFORM COURSES ON THE PEDAGOGICAL PRACTICES OF KINDERGARTEN THROUGH SIXTH GRADE ELEMENTARY TEACHERS

**Donna Turner and Dennis W. Sunal**

## ABSTRACT

This study explored the long-term impact of faculty-created reformed undergraduate science courses on the pedagogical practices of kindergarten through sixth grade inservice elementary teachers who took these reform courses during their undergraduate programs. On-site case studies were completed with 35 faculty instructors teaching entry-level undergraduate science courses at 20 higher education institutions and 91 elementary inservice teachers. The sample was selected from a national population of diverse

*Research Based Undergraduate Science Teaching*, pp. 121–151

colleges and universities that had undergone reform in one or more of their undergraduate science courses. The data collection protocol involved classroom observations, interviews, artifact analysis, semi-structured interviews, and field notes from multiple instruments and sources. Data were collected during on-site visits from instructors and their graduated students. Quantitative and qualitative analysis identified variations in faculty instructors' as well as inservice teachers' perceptions and observations of the intended and enacted teaching goals, instruction, student difficulties, and rationale for teaching a specific science concept in observed science lessons. These perceptions and observations, identified as science pedagogical content knowledge (PCK), varied significantly among both faculty instructors and inservice elementary teachers who experienced the undergraduate reformed science courses taught by these same faculty instructors. Moreover, the following characteristics regarding the pedagogical content knowledge of faculty instructors and their graduated inservice elementary teachers emerged: (1) content knowledge regarding the science concept taught and observed during instruction, (2) orientations consistent with reform instruction advocated by the *National Science Education Standards* for effective science teaching, (3) purposeful selection of activities that best engender student understanding of specific science concepts, and (4) reflective practitioner.

Science education reform has been at the forefront of science education research endeavors for several decades. The latest reform endeavors emphasize student outcomes that emanate from inquiry understandings and abilities and provide guidance for inquiry-based science teaching and learning (American Association for the Advancement of Science [AAAS], 1990; National Research Council [NRC], 1996, 2000). These endeavors attempt to engage students in skills associated with the nature of science, which entails students predicting, investigating, testing hypothesis, finding patterns, building explanations, and communicating and justifying ideas. *A Framework for K–12 Science Education* (NRC, 2012) and *Taking Science to Schools* (NRC, 2007) suggest that these sorts of activities help to facilitate a scientific literate community. Scientific literacy is described in reform documents as "the ability to describe, explain, and predict natural phenomena as well as the capacity to pose and evaluate arguments based on evidence and to apply conclusions from such arguments appropriately" (NRC, 1996, p. 22).

The National Science Teachers Association (NSTA) standards for science teacher preparation are congruent with the vision of *A Framework for K–12 Science Education*, *Taking Science to Schools*, and *The National Science Education Standards*. According to these documents, scientific literacy for citizenship should be a primary—if not exclusive—goal of science education at the precollege level (NSTA, 2003). The *National Science Education Standards* suggests that effective science teaching requires more than knowledge of science content and possessing a host of teaching strategies; rather, it is the ability of the teacher to integrate their knowledge of science content, curriculum, learning, teaching, and students to facilitate and deepen student understanding of science content. Such knowledge allows teachers to tailor learning situations to the needs of individuals and groups. It is this specialized knowledge, called "pedagogical content knowledge," that distinguishes the science knowledge of teachers from that of scientists.

The problem is that current reform endeavors aimed at improving teacher education programs tend to emphasize either content or pedagogy without examining the influence of these same endeavors on the development of pedagogical content knowledge (PCK). The undergraduate reformed science courses examined in this study are a unique blend of content and pedagogy in which faculty instructors of the reformed science courses used their PCK expertise to represent content in a variety of ways to their students and preservice teachers. Thus, in doing so preservice teachers could develop a meaningful understanding of science content and be able to translate science content to their students in novel situations. These courses were the result of a 10-year professional project, the NASA Opportunities for Visionary Academics (NASA/NOVA), developed to facilitate change in higher education to enhance science, technology, engineering, and mathematics literacy of preservice teachers (Sunal, Wright, & Day, 2004). The research question investigated in this study is, What is the long-term impact of varying models of reformed undergraduate science courses on the science pedagogical content knowledge of inservice elementary teachers? More specifically, the following research questions were investigated:

1. Are there observed pedagogical differences between faculty teaching reformed and comparison undergraduate science courses?
2. Are there instructional differences between elementary teachers who experienced reformed instruction and those who did not based on observed pedagogical differences?
3. Is the level of pedagogical reform experienced in undergraduate science courses a predictor of the type of science instruction observed at the elementary school level?
4. How is the level of pedagogical reform experienced in undergraduate science courses related to the pedagogical content knowledge of inservice teachers?

## THEORETICAL FRAMEWORK

### Pedagogical Content Knowledge

Pedagogical content knowledge (PCK) was first described by Lee S. Shulman (1986) to explain the transformation of subject-matter knowledge into forms that are understandable and accessible to the students being taught (Abell & Lederman, 2007). Shulman suggested that PCK goes beyond knowledge of a subject to the dimension of subject matter knowledge for teaching. It includes the most useful forms of representation of science concepts and the most powerful analogies, illustrations, examples, explanations, and demonstrations that would make the subject comprehensible to others (Shulman, 1986).

Twenty years after Shulman's articulation of a conception of PCK, Loughran, Berry, and Mulhall (2006) provided additional insight by describing PCK as an academic construct that is rooted in the belief that teaching requires more than delivering subject content knowledge to students, and that student learning is considerably more than absorbing information for later accurate regurgitation. They suggested that PCK is the knowledge teachers develop over time and through experience about how to teach specific science content in ways that will lead to enhanced student understanding (Loughran et al., 2006). Successful teachers, according to Loughran et al. (2006), have a special knowledge that informs their teaching of particular content that is encapsulated in PCK. Teachers not only know content but have a variety of teaching strategies they use to teach that content for specific reasons. Effective teachers are deliberate in their selection of strategies to be used with particular content (Loughran et al., 2006). They may draw upon student alternative conceptions/misconceptions and their prior knowledge and experiences in order to shape their teaching (Loughran et al., 2006).

Pedagogical content knowledge was explicated by Gess-Newsome (1999) as a construct falling along a continuum of models of teacher knowledge that may be *integrative* or *transformative*. The integrative model implies that PCK does not exist as an exclusive domain of teacher knowledge, but rather, the extrapolation of knowledge from subject matter, pedagogy, and context domains are melded in classroom practice (Gess-Newsome, 1999). The transformative model, however, implies that these knowledge domains (i.e., subject matter, pedagogy, and context) are inextricable and result in a new form of knowledge: pedagogical content knowledge. This latter model forms the basis of our study. Within the context of this study, only content knowledge, pedagogical knowledge, and orientations toward science teaching are considered. Orientations toward science teaching are considered in order to understand the underlying goals and rationale for the specific instructional strategies used to teach specific science content.

Orientations toward science teaching are typically organized by the emphasis of instruction, so can range purely process or content to an emphasis on inquiry (Magnussen, Krajcik, & Borko, 1999). These orientations are described with regard to the goals and characteristics of science teaching that a teacher with a particular orientation would have. Nine orientations toward science teaching and the purposes and characteristics of instruction associated with that orientation were proposed by Magnussen et al. These orientations were divided into two categories by Friedrichsen and Dana (2003): (1) teacher-centered (didactic and academic rigor), and (2) reform. Reform efforts are based on either the reform efforts of the 1960s (process, activity-driven, discovery) or contemporary reform efforts and associated curriculum projects (conceptual change, project-based, inquiry, and guided inquiry). Magnussen et al. cautioned that a particular teaching strategy (e.g., use of laboratory investigations) may be characteristic of more than one science teaching orientation: "this similarity indicates that it is not the use of a particular strategy but the purpose of employing it that distinguishes a teacher's orientation to teaching science" (1999, p. 97).

Content knowledge may be defined as the concepts, principles, relationships, processes, and applications a student should know within a given academic subject, appropriate for his/her organization of the knowledge (Ozden, 2008). In earlier work, Schulman drew upon Joseph Schwab's definition of content knowledge as including both substantive and syntactic structures. The substantive structure of the content is the organization of concepts, facts, principles, and theories, whereas syntactic structures are the rules of evidence and proof used to generate and justify knowledge claims in the discipline (Schwab, 1978).

Pedagogical knowledge includes the general aspects of teacher knowledge about teaching such as learning theory, instructional principles, and classroom discipline (Abell, 2007). Pedagogical knowledge was described by Mishra and Koehler (2006) as "deep knowledge about the processes and practices of teaching and learning, encompassing educational purposes, goals, values, strategies, and more" (p. 1026). They built upon the basic description, extending it as knowledge about techniques or methods used in the classroom, the nature of the learners' needs and preferences, and strategies for assessing student understanding. Within the context of this study, pedagogical knowledge comprises knowledge of instruction and strategies, knowledge of student learning difficulties and misconceptions regarding topic specific science content, prior knowledge, and knowledge of context.

## LITERATURE REVIEW

Current research regarding the pedagogical content knowledge of K–6 elementary teachers is sparse. The dearth of research regarding PCK for K–6 elementary teachers possibly can be attributed to the idea that PCK is topic specific and is strongly influenced by content knowledge. Most elementary teachers have had fewer than four science content courses (Weiss, Banilower, McMahon, & Smith, 2001) during their undergraduate programs. Most studies regarding PCK tend to focus on secondary preservice or inservice teachers.

Studies conducted by Kaya (2009), Ozden (2008), Halim and Meerah (2002), Van Driel, De Jong, and Verloop (2002), and Van Driel, Verloop, and De Vos (1998) demonstrated that content knowledge influenced the development of pedagogical content knowledge. Moreover, the relationship among the components of pedagogical content knowledge (PCK) involving the topic ozone layer depletion for preservice science teachers' (PSTs) was examined by Kaya (2009). Within the context of the study PCK consisted of content knowledge and pedagogical knowledge. Results from this study revealed that there was a significant inter-relationship between the subject matter and pedagogical knowledge of the PSTs. Ogletree, also, found that fifth grade teachers lacked the appropriate content knowledge regarding the concept of chemical change.

Preservice teachers were found to expand their PCK by deepening their understanding of learning difficulties and student misconceptions after learning from teaching as they experienced a postgraduate teacher education program (De Jong, Van Driel, & Verloop, 2005; De Jong & Van Driel, 2004). The results of this qualitative study revealed that an extensive, one-year training course on teaching chemistry topics, which used particle models for matter, successfully contributed to the development of PCK of the preservice teachers (DeJong et al., 2005).

In a qualitative case study, Avraamidou and Zembal-Saul (2010) investigated two first-year elementary teachers to explore what aspects of teacher practices were most consistent with an inquiry-based approach, what PCK served as a mechanism for facilitating these practices, and what experiences have mediated the nature and development of these teachers' PCK.

Both teachers went through a year-long, highly mentored internship in a professional development school setting. Overall, findings revealed that an inquiry-based approach to science teaching was evident in both of the participants' instructional practices as they engaged their students in question-driven investigations, the use of observational data, making connections between evidence and claims, and communicating those claims to others (Avraamidou & Zembal-Saul, 2010).

The authors noted that Jean's PCK for scientific inquiry emphasized evidence and explanation and could be traced back to the content courses that were specifically designed for preservice teachers. No evidence in the data related to Andrea's PCK for scientific inquiry regarding her knowledge about the nature of scientific discourse (Avraamidou & Zembaul-Saul, 2009). They obtained similar results in an earlier case study with first-year elementary teachers (Avraamidou & Zembaul-Saul, 2005).

## METHOD

### Context of the Study

The setting comprised a diverse population of institutions throughout the U.S. These institutions included Historically Black Colleges and Universities (HBCUs), master's level (MA) colleges and universities, Hispanic Serving Institutions (HSI), Native American Institutions, and doctoral/research universities based on the Carnegie classification (Sunal, Sunal, Mason, & Zollman, 2008). Faculty instructors at institutions who had participated in the NASA/NOVA professional development project and who taught undergraduate science were invited to participate in the NSEUS study. Faculty were called and emailed to solicit their participation. A sample of 30 institutions was finally selected. Due to attrition and/or the dissolving of courses at some of the institutions, the final sample size resulted in 19 higher education institutions being used for data collection. The institutions that participated in this study were located throughout the nation.

### Participants

The participants comprised 35 faculty instructors and 91 K–6 grade elementary teachers. The faculty instructors of the reformed courses participated directly or indirectly in the NASA/NOVA program and were implementing reform practices in their respective undergraduate classrooms. Some faculty instructors inherited the reform course from a previous instructor who had participated in the project; thus, the course was a reformed course but the faculty had not participated directly in the NASA/NOVA project and did not necessarily implement standards-based reform practices.

## Data Collection

Data collection involved an intense week-long site visit to each of the 19 institutions. As stated earlier, this study was a part of a larger longitudinal study that was conducted over a period of five years, 2006–2011. The overarching research question investigated in this study was, "What is the long-term impact of faculty-created undergraduate science reform courses on the pedagogical content knowledge of K–6 elementary teachers?" The use of both quantitative and qualitative methods strengthened the research design and allowed for the triangulation of data. Prior to the site visits, a faculty liaison was identified to help facilitate the site visit at each campus locality. The liaison was responsible for ensuring that the investigator had contact information and schedules for the faculty instructors and inservice elementary teachers as to the times available to conduct the classroom observations and interviews. The liaison contacted the faculty instructors of reformed and comparison courses to set up a time for the classroom observations and interviews. Permission was secured from each institution to conduct the observations and interviews as required by the Institutional Review Board (IRB). Once permission was secured, faculty instructors identified former students who were now inservice elementary teachers for participation in the study. The liaison provided the dates and times as to when site visits to the University and the elementary schools could take place.

The quantitative phase of the data collection included the classroom observations of university faculty instructors and their graduated K–6 elementary teachers. Faculty instructors of reformed and comparison courses were observed at least twice during a week-long site visit using the *Reformed Observation Teaching Protocol*, a criterion referenced test designed to measure the degree of reform in a science classroom (Sawada et al., 2002). These observations lasted from 60 to 90 minutes and typically consisted of a lecture, laboratory, or a class that had a combined lecture and laboratory. At least two trained observers observed classroom instruction. The RTOPs were individually scored immediately after the classroom observation and then discussed to resolve any discrepancies between researcher ratings. The labs and lecture were combined to give an average RTOP score for each faculty. The elementary teachers were observed only once teaching a science lesson. Loughran, Mulhall, and Berry (2004) considered this a limitation because of the difficulties in trying to capture PCK in just one setting since PCK may vary across a number of lessons or one unit.

The qualitative phase of data collection used interviews and *Content Representation* (CoRe) and the *Personal and Professional experience Repertoire* (PaP-eRs) instruments. A CoRe provides an overview of how teachers conceptualize the content of a particular subject matter or topic (Loughran et al., 2006). According to Loughran et al., whether or not a particular action by a teacher is illustrative of that teacher's PCK is closely related to the thinking upon which the teacher reasons through and develops the subsequent teaching action.

Pedagogical and Professional Experience Repertoires are narrative accounts of a teacher's PCK for a particular piece of science content. Each PaP-eR then "unpacks" the teacher's thinking around an element of PCK for that content, and it is based on classroom observations and comments made by teachers during the interviews from which the CoRes were developed. Pedagogical and Professional Experience Repertoires are intended to represent the teacher's reasoning—that is, the thinking and actions of a successful science teacher in teaching a specific aspect of science content. The function of the narrative is to elaborate and give insight into the interacting elements of the teacher's PCK in ways that are meaningful and accessible to the reader, and that may serve to foster reflection in the reader about the PCK under consideration, and to open the teacher reader to possibilities for change in his/her own practice (Mulhall et al., 2003).

After the classroom observations and interviews were completed for each faculty instructor and elementary teacher, a CoRe was constructed from the interviews and a PaP-eR was constructed using the classroom observations and field notes to flesh out the goals and rationales for the observed lesson. The CoRe and PaP-eRs were evaluated using a researcher designed PCK rubric to capture rich "snapshots" of PCK for each faculty instructor regarding their orientation toward science teaching, content knowledge, and pedagogical knowledge (see Table 4.1 for PCK rubric). Since PCK cannot necessarily be gauged from one classroom setting (Loughran et al., 2006), I elicit the term *snapshots* to describe instances of specific events of PCK. Part of the initial design for the rubrics used in this study came from a careful examination of the rubrics designed and used by Ogletree (2007) in her examination of fifth grade elementary teachers' pedagogical content knowledge for states of matter. Ogletree's four rubrics included (1) content knowledge, (2) student thinking, (3) science teacher knowledge, and (4) professional development. The PCK rubric used in this study assesses PCK holistically, rather than assessing the individual components, by providing an overall PCK rating based on a combination of their content and pedagogical knowledge as evidenced in their interviews, CoRe and PaP-eR, and classroom observations.

Prior to the classroom observations and interview, I or one of my colleagues discussed the purpose of the project and the means by which privacy and confidentiality would be established. Once the faculty instructors agreed to continue with participation in the study, a consent formed was signed and stored in a secure location in which the primary investigator of the study had the only access. Each faculty instructor secured student volunteers to participate in the undergraduate student focus groups. The focus groups were interviewed after the classroom observation of the faculty instructor. I or one of my colleagues explained the nature of the study and the means by which privacy and confidentiality would be established to the undergraduate students. We also informed them that their participation in the interview was completely voluntary. Once the students agreed to continue with participation in the study, signed consent forms were received from each student.

Careful examination of the interviews, CoRe and PaP-eRs, and RTOPs using the PCK rubric yielded rich snapshots of PCK for faculty instructors. The science instruction observed, as well as the information provided in the interviews, provided insight as to the faculty instructors' orientation toward teaching science, topic specific content knowledge, and their rationale for the instructional strategy used to teach that particular science topic.

## Data Analysis

Data analysis comprised the use of various quantitative computational procedures as well as the qualitative evaluation of categories from the interviews along with the CoRes and PaP-eRs. Faculty instructors were divided into those who taught the reform course and those who taught the comparison course. Faculty of the reform course participated in the NASA/NOVA professional development project or inherited the course from someone who had previously participated in the project. Characteristics of the reform course related to the *National Science Education Standards* appropriate for science teaching and learning at the undergraduate level. The comparison course was a course of comparable subject and academic level that was not developed through NASA/NOVA professional development workshop. The comparison course was a science course that included a lecture and a lab. In order to determine if there were statistically significant differences between the reform and the comparison course, an independent t-test and a multivariate analysis of variance (MANOVA) was performed. The Statistical Package for the Social Sciences version 19 (SPSS 19) was used for all statistical procedures. In order to determine if there were statistically significant differences in the classroom instruction of those inservice elementary teachers who had taken the reform courses and those who had taken the comparison courses during their undergraduate program, a MANOVA was also performed.

Regression analysis was used to determine whether the level of reform experienced in these undergraduate science courses and was a predictor of observed pedagogical instruction in K–6 elementary classrooms. Faculty instructors were divided into high, medium, and low RTOP groups regardless of whether or not they taught the reform or comparison course based on their overall RTOP scores. The RTOP scores ranged between 0 and 100. Faculty instructors who scored between 0 and 33 were placed in the low RTOP group; faculty instructors who scored between 34 and 67 were placed in the medium RTOP group; and faculty instructors with RTOP scores of 68 or better were placed in the high RTOP group. The RTOP subscale scores were then used to predict reform instruction in elementary classrooms based on the overall RTOP scores of their graduated, inservice elementary teachers.

Interviews were coded using a set of predetermined categories to capture snapshots of the pedagogical content knowledge of faculty instructors. Salient themes emerging from the interviews were also noted. From the interviews and classroom observations, CoRes and PaP-eRs were constructed and evaluated for each faculty instructor to yield a snapshot of PCK. The interview and the CoRe and PaP-eR were analyzed using the PCK framework. The PCK framework was vetted with the literature on PCK and is based upon Shulman's (1986), Grossman's (1990), and Magnussenet al.'s (1999) work with PCK.

The PCK framework focused on *orientations toward science teaching* (OST), *content knowledge*, and *pedagogical knowledge*. The PCK framework was used to develop a PCK checklist for each faculty instructor and each K–6 elementary teacher. Based on the components of the checklist, the PCK rubric was used to provide an overall PCK rating for each faculty instructor and each inservice elementary teacher. Inter-rater reliability and validity for the PCK rubric was achieved through consultation with science educator specialists and science content specialists with teaching experience in K–16 classrooms. These individuals were experts in their respective fields and had expert knowledge in physics, physical science, biology, and chemistry. Each specialist evaluated the interviews, CoRe and PaP-eRs, and RTOPS of two teachers using the PCK framework, PCK checklist, and PCK rubric to come up with an overall PCK rating for the two teachers. The scores and comments of the specialists were the same as that of the investigator. Using the PCK checklist, interviews, RTOP field notes, and the overall PCK rating, a snapshot was written for twelve of the faculty instructors.

Currently, there is no standard means by which to capture or measure an individual's PCK (Friedrichsen, Van Driel, & Abell, 2011); however, the investigator used both quantitative and qualitative data from classroom observations, field notes, and interviews in an attempt to capture snapshots of PCK. The quantitative analysis of the RTOPs and its corresponding subscales provided insight as to the amount of reform instruction observed during the classroom observations of both faculty and inservice teachers. The RTOP alone did not provide insight as to the teacher's rationale for choosing a particular teaching strategy to teach the science concept observed during a lesson. Pedagogical content knowledge is premised on the ability of the teacher to transform subject matter into a form that is understandable and accessible to students. In part, the ability to transform specific science content is based upon the faculty instructor's knowledge of the prerequisite skills students need for the learning of a particular topic as well as their prior knowledge and experiences. The interviews, along with the CoRe and PaP-eR, allowed the investigator to examine the decision making aspect of teaching practice—that is, why faculty instructors chose certain strategies to teach specific science concepts and what they knew about their students that informed their decisions.

### *Research Question #1*

In order to answer research question one, "Are there pedagogical differences between faculty instructors of reform and comparison undergraduate science courses," an independent *t* test was used to compare the overall means of the reformed and comparison course for the total RTOP score and a multivariate analysis of variance was performed based on the subscales of the RTOP. Each faculty was observed using the RTOP instrument at least two times during the site visit. At least two research team members were present to conduct the classroom visitations. The labs and lecture were combined to give an average RTOP score for each faculty instructor. The RTOP instrument comprised five subscales: (1) Lesson Design and Implementation, (2) Propositional Knowledge, (3) Procedural Knowledge, (4) Communicative Interactions, and (5) Student Teacher Relationships. A total of 20 points can be accrued in each subscale with a maximum RTOP score of 100.

Results from the quantitative analysis of data found a statistically significant difference in the total RTOP score between faculty who were instructors of reform and comparison undergraduate science courses, $t$ (33, 2.997), $p$ =.005. Analysis of classroom observations on RTOP subscale scores between reformed and comparison courses found statistically significant group effects (Wilks = .028, $F$[5, 29] = 3.756, $p$ = .010). Univariate analyses indicated no significant group effects on the Propositional Knowledge subscale [$F$(1, 33) = 2.729, $p$ = .108]. However, univariate analyses indicated that significant group effects did occur on the Lesson Design and Implementation subscale [$F$(1, 33) = 6.242, $p$ = .018], Procedural Knowledge subscale [$F$(1, 33) = 6.978, $p$ = .029], Communicative Interactions subscale [$F$(1, 33) = 13.298, $p$ = .013], and Student Teacher Relationships subscale [$F$(1, 33) = 8.143, $p$ = .007. These results suggests that the level of reformed teaching in entry-level reformed undergraduate science courses is different from the comparison courses on all RTOP subscales other than the Propositional Knowledge subscale.

**Table 4.1 PCK RUBRIC**

| | | |
|---|---|---|
| 4 | Advanced | The teacher's responses and actions show in-depth content knowledge of the key science concept(s) of the lesson being taught. Language used is scientifically accurate, descriptive, purposeful, and appropriate for the lesson and grade level. The content discussed is relevant and current and facilitates scientific literacy. The teacher understands that science content is fluid and presents it as such. The teacher integrates information from other disciplines in teaching specific content. Emphasis is on the notion that less is more. The teacher's responses and actions show an in-depth understanding and knowledge of ways to represent content that leads to student understanding. Teacher selects specific instructional strategies that are useful for helping students comprehend specific science concepts. Teacher exhibits knowledge of ways to represent specific concepts or principles in order to facilitate student learning as well as an understanding of the strengths and weaknesses of such activities. Representations may include illustrations, examples, models, or analogies. The teacher knows whether and when a representation will be useful to support and extend the comprehension of a particular teaching situation. When planning a lesson teacher considers students learning difficulties and prior knowledge. Teacher knows about the prerequisite knowledge needed for learning specific science content, as well as, the abilities and skills that students might need. Teachers also know and understand variation in approaches to learning based on students varying development needs. In addition, the teacher knows that students have alternative misconceptions and prior knowledge about science content and use these to guide instruction. |
| 3 | Proficient | The teacher's responses and actions show solid content knowledge of the key science concept(s) of the lesson. Language used is scientifically accurate, descriptive, purposeful, and useful. The content discussed is relevant. The teacher's responses and actions show a solid understanding of the way students' think. Teachers use prior knowledge in planning or teaching only, but not both. Appropriate learning goals were set for the students. Teachers know that students have misconceptions but may or may not use these misconceptions to guide instruction. In addition they know and understand ways to represent the teaching of the lesson that leads to student understanding. Teachers have a plethora of teaching strategies but do not necessarily use chose strategies for teaching topic-specific science concepts. These strategies may subject-specific (learning cycle) rather than topic specific. They also do not necessarily select strategies based on how students learn. Use some student data but mostly experience to guide instruction. |

*(Table continues on next page)*

**Table 4.1 (Continued)**

| | | |
|---|---|---|
| 2 | Emergent | The teacher's responses and actions show partial content knowledge of the key science concept(s) of the lesson. Language used has some scientifically inaccuracies, is not always descriptive, and lacks a clear purpose. The teacher makes an attempt to make content relevant to student lives. The teacher's responses and actions show a partial understanding and knowledge of ways to represent specific science concepts. They have a number of activities that they may use to teach specific science concepts but not necessarily to facilitate student understanding. These may be activities that work and are used because they arouse student curiosity, maintain discipline, mandated by district, or they are fun. The teacher's responses and actions show a partial understanding of the way students' think. Aware that students have prior knowledge, but did not use it in planning or teaching. Some appropriate learning goals were set for the students. May or may not be aware of student misconceptions. |
| 1 | Novice | The teacher's responses and actions show very limited content knowledge of the key science concept(s) of the lesson. Language used has many scientific inaccuracies, is not descriptive, and lacks a clear purpose. Teacher relies on textbook for science content. The teacher's responses and actions show a very limited understanding of the way students' think. Prior knowledge was not considered in the planning of the lesson. Teacher demonstrates misconceptions about student thinking and their own instruction. The teacher's responses and actions show a very limited understanding and knowledge of ways to represent specific science concepts and tend to rely on textbook for teaching science concepts. |

The quantitative results of Question 1 were enhanced by qualitative findings emanating from the interviews. Faculty instructors who taught the reformed course taught in a more reformed manner. These faculty instructors were more inclined to suggest that a constructivist approach to classroom practice was the best approach to science teaching and learning. Several of the faculty instructors of the reformed course stated that students must have opportunities to engage in activities that allowed them to conduct investigations, make predictions, and formulate and test hypothesis. The following statements were taken from the interviews of faculty instructors of the reformed course.

> I teach this course differently than I teach my regular physics courses. Physics courses are lecture based.... Physics [reformed course] is inquiry based.... Elicitation addresses prior knowledge and misconceptions.... Exploration is the discovery portion.... Explanation--students give presentations. (Faculty instructor)

**Table 4.2. Means and Standard Deviations for Inservice Elementary Teachers**

| | *Course* | *N* | *Mean* | *Std. Deviation* |
|---|---|---|---|---|
| Sum | Reform | 56 | 61.1275 | 18.34317 |
| | Comparison | 35 | 60.0286 | 18.99301 |

What do you recall about the circulatory system...taking a look at blood flow ...case study on heart attack (what does heart attack have to do with blood flow—you have to know how something works—take fundamental concepts about circulatory and apply it to this disease).... Physiology—blood pressure (how does it tie it into circulatory system). Take a look at model of heart.... For each topic that I do there is some type of inquiry lesson and case studies along with viewing models. (Faculty instructor)

I teach about the process of science...to broaden their sense as to what science really is practicality of science...making connections between other disciplines and science ... connections between what they may be interested in and science.... Part of the course involves a semester long project which involves studying [science concept] ... at the very beginning of the course I take them out and let them explore to develop their own interest in the [science concept] and I am hoping that they will find ... a very, very simple question.... The reason that I do that is that I want them to find something that is theirs ... then I relate this to the process of science. (Faculty instructor)

Interviews with the undergraduate student focus groups of the faculty instructors of the reformed course corroborated the statements put forth by the faculty instructors. The following statements were taken from the undergraduate student focus groups from the reform course.

We're doing stuff ... we are not just sitting there getting lectures ... using the computer ... like what we are doing with the virtual river working in small groups ... interactive stuff rather than sitting there listening to the professor talk to us.... Even the virtual river isn't as hands-on as we get ... we go outside and do our own project, it is really independent.... [Faculty] is just overseeing it but we are really doing everything.... I feel that the experiment that we do with the [river] is much higher than what we are learning in the classroom to actually be able to do a bank profile up there is much more difficult than all sorts of other elements factor in. (Undergraduate student focus group)

I never thought of electricity and magnetism as science before this course. Yeah. I never made the link that those things were part of science. They seemed just mechanical.... I'll remember circuits from working with them.... I know it works, but now I'm starting to get a grasp on how and why it

> works.... I'm thinking more about the scientific process. In lab we always go through the process. (Undergraduate student focus group)

Faculty instructors of the comparison course taught in a more traditional manner. These faculty instructors were more inclined to spend the majority of the class time lecturing to their students. Laboratory instruction involved the use of "cookbook" labs that were separate from the lecture. In addition, faculty instructors of the comparison course did not attend professional development workshops for improving their science teaching and they did not collaborate with other colleagues about science education reform. Many of the faculty instructors stated that there was no administrative support for improving science instruction for their undergraduate students. The following statements were taken from the interviews of faculty instructors of the reformed course.

> I try to engage students as much as I can. I use Blackboard to post my notes so that they can spend class time concentrating on what I am saying rather than copying down information.... There is a lack of administrative support for faculty to collaborate with one another, and to have pedagogical professional development rather than just research support. (Faculty instructor Adam)

> 10% of the time you can make it relevant; the remainder has to be memorized ... use of the board and overheads ... online simulations and cartoons are very helpful.... I want to do group work but I don't have time.... I mainly straight lecture.... I use the online homework ... students like it now that it is in line with the lectures. (Faculty instructor Karen)

### *Research Question 2*

In order to answer Question 2—"Are there instructional differences between elementary teachers who experienced reformed instruction and those who did not based observed pedagogical differences?"—an independent $t$ test was performed to compare the overall means of the inservice teachers who experienced the reformed course during their undergraduate programs and those who did not. In addition, a multivariate analysis of variance (MANOVA) was performed to analyze the subscales of the RTOP. Results from the quantitative analysis of the data did not reveal significant differences in the total RTOP score between inservice elementary teachers who took the reform course during their undergraduate program and those who did not, $t$ (89, .274), $p$ = .785. The means and standard deviations are presented in Tables 4.2 and 4.3. Results from the quantitative analysis of classroom observations using RTOP subscale scores between reformed and comparison courses revealed no statistically significant group effect (Wilks

= .061, $F[5, 85] = 1.514, p = .194$). The level of reformed teaching in the elementary classroom, based on observations using the RTOP instrument among graduated inservice elementary teachers, is statistically similar for both groups.

**Table 4.3. Multivariate Analysis of Variance for Inservice Elementary Teachers**

| *Source* | *Dependent Variable* | *Type III Sum of Squares* | *df* | *Mean Square* | *F* | *Sig.* |
|---|---|---|---|---|---|---|
| | Lesson Design | 6.431 | 1 | 6.431 | .362 | .549 |
| | Propositional Knowledge | 8.798 | 1 | 8.798 | .712 | .401 |
| | Procedural Knowledge | 10.836 | 1 | 10.836 | .562 | .456 |
| ID | Communicative Interactions | .102 | 1 | .102 | .006 | .936 |
| | Student Teacher Relationships | 6.915 | 1 | 6.915 | .351 | .555 |
| | Lesson Design | 1 582.570 | 89 | 17.782 | | |
| | Propositional Knowledge | 1,100.231 | 89 | 12.362 | | |
| | Procedural Knowledge | 1,717.063 | 89 | 19.293 | | |
| Error | Communicative Interactions | 1,409.619 | 89 | 15.838 | | |
| | Student Teacher Relationships | 1,751.458 | 89 | 19.679 | | |
| | Lesson Design | 14,120.984 | 91 | | | |
| | Propositional Knowledge | 15,770.928 | 91 | | | |
| | Procedural Knowledge | 10,481.656 | 91 | | | |
| Total | Communicative Interactions | 14,795.056 | 91 | | | |
| | Student Teacher Relationships | 19,217.545 | 91 | | | |

*(Table continues on next page)*

**Table 4.3. (Continued)**

| *Source* | *Dependent Variable* | *Type III Sum of Squares* | *df* | *Mean Square* | *F* | *Sig.* |
|---|---|---|---|---|---|---|
| Total | Lesson Design | 15,89.001 | 90 | | | |
| | Propositional Knowledge | 11,09.028 | 90 | | | |
| Corrected Total | Procedural Knowledge | 1,727.899 | 90 | | | |
| | Communicative Interactions | 1,409.722 | 90 | | | |
| | Student Teacher Relationships | 1,758.373 | 90 | | | |

The quantitative results of Question 1 were enhanced by qualitative analyses of inservice elementary teacher's interviews and field notes from classroom observations. Based on the RTOP scores, teachers who had taken the reform course or the comparison during their undergraduate programs were statistically similar regarding science instruction. The inservice elementary teachers reported in this section had experienced the undergraduate reform course or the comparison course. The most predominant theme emerging from the interviews was the notion that hands-on activities were the best approach to teaching science with elementary students. The following statements were taken from the interviews of inservice elementary teachers who experienced the reformed course during their undergraduate programs.

> Hands-on whenever possible.... Let them see and touch the real item so that they can explore. (Elementary teacher)

> I think hands-on.... Give them that freedom of experimenting with it and then draw it into the lesson. I think that's a great way. And that's how I learned, so I'm partial to that, but I feel like that's a great way for them to actually see it and sense it and feel it. They can try it and if it fails, why. (Elementary teacher)

> Teacher asks students questions as to what will happen when you mix certain substances (oil, dish detergent) with water.... Students give their predictions ... discuss in their groups and write down their observations.... Teacher asks students is the water as clear after adding the oil ... what happened.... Students give a variety of answers...what will happen if you mix oil, soap, and water.... Students make their predictions.... Students make comparisons be-

> tween bottle with oil, bottle with soap, and bottle with soap, oil and water ... students discuss their observations. (Elementary teacher)

Inservice elementary teachers that experienced the comparison course as a part of their undergraduate program employed a hands-on approach to science teaching and learning as well. Many of them stated that the comparison course did not have an impact on their views regarding science teaching and learning. They stated in their interviews that their views on science teaching and learning had been influenced by their teaching experiences and by professional development workshops geared toward science teaching. The following statements were taken from the interviews of inservice elementary teachers who experienced the reformed course during their program.

> I saw that when my students simply read the book the information didn't stick. Students actually need to do it to make it stick so that they can form that synaptic loop and learn from that.... My university courses mostly involved memorizing stuff from books with a little bit of hands-on. It was more of my teaching experience that made me see that science was more hands-on.... I had the framework of scientific inquiry; I just didn't really experience it as an undergraduate. I gained knowledge about the concepts when I had to teach them.... I wanted to make sure to tie in today's lesson to what they did in the lab on Friday so that they had a framework to go by. I wanted to tie it to their prior knowledge.... I expanded on that activity by doing the same experiment on the smooth floor in the hallway and the rough surface of the carpet in the classroom.... I wanted to base the activity on something they had already done, but to expand on that and give them something new. (Inservice elementary teacher from the comparison course)

> The science that I learned in college was much more regimented than what I teach. In high school, science was introduced to me through a lot of different topics with a lot of exploration, but in college it was a lot more regimented and not as fun. Things might have been different if I took science education courses. Science in college was more dry. We weren't allowed to play with science in college. (Inservice elementary teacher from the comparison course)

Research Question 2 provides evidence that K–6 elementary teachers who experienced the reformed undergraduate science course during their undergraduate programs and those who did not taught in a similar manner. One of the main themes that emerged from qualitative analysis of data was that professional development geared toward science content and pedagogy enabled inservice teachers who experienced the comparison course to implement what appeared to be reformed pedagogical practices.

### *Research Question 3*

Research Question 3 asked whether the level of pedagogical reform experienced in undergraduate science courses was a predictor of the type of science instruction observed with graduated teachers at the elementary level; multiple linear regression was performed. All faculty instructors in the sample were considered, both those in the reform and comparison groups. Faculty instructors were divided into high, medium, and low groups based on their RTOP scores. The highest total composite score that could be obtained on the RTOP was 100. Faculty instructors achieving RTOP scores ranging between 0 and 33 were placed in the low RTOP group ($N$ = 5), faculty instructors achieving a RTOP score ranging between 34 and 67 were placed in the medium RTOP group ($N$ = 12), and faculty instructors who achieved a RTOP score ranging between 68 and 100 were placed in the high RTOP group ($N$ = 18). At the time of this study, an RTOP benchmark had not been established. In measuring physics PCK, MacIsaac and Falconer (2002) suggested that an RTOP score of 50 indicates considerable presence of reformed teaching in a lesson.

Initially, regression analysis was used with faculty instructors with RTOP scores ranging between 0 and 50 to predict the reform instruction observed in the elementary classrooms; however, this model was not a good predictor, [$F(5, 26) = 1.224, p = .326$]. This model produced an adjusted $R^2$ of .035 for the prediction of reformed instruction in elementary classrooms. Regression analysis was also conducted with faculty instructors who had RTOP scores ranging from 51 through 100, and this model was also found not to be a predictor of reformed instruction. Based upon the results from initial analysis, the investigator decided to divide the faculty instructors into the three groups, low, medium, and high, to determine whether the level of pedagogical reform would predict reformed instruction in elementary classrooms.

Faculty instructors with high RTOP scores were used to predict the type of instruction exhibited by inservice elementary teachers who took their science discipline courses during their undergraduate program. Overall, the model was not a significant predictor of reformed instruction in elementary classrooms [$F(5, 40) = 1.023, p = .417$]. The model produced an adjusted $R^2$ of .003 for the prediction of reformed instruction in elementary classrooms. Together, the RTOP subscales shared only .3 % of the variance in reformed instruction in elementary classrooms. The histogram and the normality plots revealed that the assumptions for this data set had been met.

### *Research Question #4*

Question 4 asked how the level of pedagogical reform experienced in undergraduate science courses related to the pedagogical content knowledge of inservice elementary teachers. Qualitative analyses conducted for research question four provides evidence regarding the extent of the pedagogical content knowledge of Kindergarten through sixth grade elementary teachers.

Qualitative analyses allowed the researcher to delineate characteristics of faculty instructors with expansive topic-specific PCK and examine whether the inservice elementary teachers who took their courses had similar characteristics regarding PCK for the science concept taught. Out of the 35 faculty instructors and the 91 elementary teachers, seven faculty instructors and 21 elementary teachers were analyzed regarding topic specific PCK. Faculty instructors were not compared to each other regarding PCK nor were the elementary teachers who took their courses. Each faculty instructor had three or more inservice teachers on which data were collected. To reduce bias, only the faculty instructors where observational data were collected on three elementary teachers were used for further analyses regarding elementary teacher PCK. Snapshots were developed for the faculty instructors and their inservice elementary teachers with the highest and lowest RTOP scores in each group of three elementary teachers per faculty instructor. Interviews, CoRes and PaP-eRs, and field notes from classroom observations from the faculty instructors and inservice elementary teachers were analyzed using a researcher-developed PCK framework based on preselected categories.

The categories comprising the PCK framework were *orientations toward science teaching* (OST), *content knowledge* (CK), and *pedagogical knowledge* (PK). Once the interviews, CoRes and PaP-eRs, and field notes of faculty instructors and inservice elementary teachers had been analyzed, a PCK checklist was completed for each faculty instructor and his or her inservice elementary teacher. Once the PCK checklist was completed, a PCK rating was given for the faculty instructors and the inservice elementary teachers using the researcher-developed PCK rubric. Snapshots were developed for the faculty instructors and inservice elementary teachers using the completed PCK checklist to determine the extent of their pedagogical content knowledge. Seven snapshots of faculty instructors and 21 snapshots of their inservice elementary teachers were developed. Each snapshot then comprised the faculty instructor and their inservice elementary teacher who had the highest and lowest RTOP scores. Again, the purpose of these snapshots is to provide an overall description as to the pedagogical content knowledge of the faculty instructors and the inservice teachers who took their courses during their undergraduate program.

Of the 21 teachers, two received a PCK rating of *advanced*, seven teachers received a PCK rating of *proficient*, eight teachers received a PCK rating of *novice*, and four of the teachers received a PCK rating of *emerging*. Inservice elementary teachers with a PCK rating of *advanced* shared the following characteristics: (1) science content knowledge regarding the specific science content taught, (2) professional development geared toward science content and pedagogy, (3) reform course or some equivalent, (4) reflective practitioners, (5) provided specific rationales for strategies used to teach content, and (6) collaboration with colleagues and science educators. Inservice elementary teachers with a PCK rating of *proficient* shared the following characteristics: (1) science content knowledge regarding the specific science content taught, (2) ongoing professional development, (3) provided specific rationales for strategies used to teach content, and (4) some collaboration. Inservice elementary teachers with PCK ratings of *emerging* or *novice* shared some of the following characteristics: (1) lacked science content knowledge regarding the science content taught, (2) constrained by school cultures that emphasized reading and math, (3) disconnect between teaching beliefs and enacted lesson, (4) lack of professional development in science content and pedagogy, and (5) constraints of state testing. An example of a faculty instructor and inservice teacher snapshot is described in the proceeding section.

## SAMPLE SNAPSHOT

### Calvin

Calvin had 30 years teaching experience and was a part of the original NASA/NOVA team. Calvin stated in his interview that he believed that an inquiry-oriented approach was the best approach to science teaching and learning. He stressed the importance of science content that is relevant and instruction that facilitates scientific literacy.

Calvin's observed lesson was about isomers and nomenclature. Calvin used models to represent isomers so that students could conceptualize this abstract concept. He stated in his interview that his students' understanding of isomers and nomenclature would dictate the direction he would take for the next lesson. In reflecting on his previous experience in teaching this topic, he knew that students had difficulties in understanding isomers in terms of molecular and structural formulas and knew that models would assist in facilitating student understanding of this topic. Calvin participated in extensive professional development workshops geared toward science content and pedagogy. He maintained collaboration with his colleagues and other science specialists and he was a reflective practitioner. Calvin achieved a PCK rating of *advanced*. His RTOP score was 89, indicating that his observed classroom instruction was reformed. Calvin's former students, Carrington, Chris, and Callie, were now inservice elementary teachers. Of the three, Callie had the highest RTOP score and Chris had the lowest RTOP score. Their snapshots are given below.

## Callie

Callie had been teaching for nine years during which time she has taught the fifth and sixth grades. She taught fifth grade for five years and sixth grade for four years. She was currently teaching the fifth grade. Callie took two science content courses during her undergraduate program as well as the reformed course.

Callie believed that a hands-on approach using experiments was the best approach to science teaching. She believed that hands-on learning activities provided students with opportunities "to see and feel [science concepts] ... that's the only way for them to remember it." The lesson observed was the water cycle. Callie stated that the big ideas of the lesson were the states of water and the notion that water is recycled throughout the environment. She had taught this lesson the previous year. She said that students do not understand that water is everywhere nor do they understand condensation. She used a simulated activity, "The Water Model," to simulate the water cycle. She said that this activity enabled her students to see that water is everywhere as well as understand that it is recycled throughout the environment. She emphasized that it is important that the water cycle be connected to the lives of her students. She began the lesson with a short review of the states of water and writes the main points on the board. Students work in groups to complete the water cycle activity. Students made predictions as to what happens to the water during each stage of the cycle. While student groups waited on their models to work, the teacher had them draw the water cycle because she knew that her students loved art. When the activities were completed, the teacher concluded with a whole-class discussion to see if student predictions were accurate. To assess student understanding of the lesson, Callie said that she will evaluate her students' lab write-ups and models and give a formal test.

Callie received a PCK rating of *emerging*. Her RTOP score was 71, indicating that the classroom instruction observed was reformed. Callie knew that her students did not understand condensation and had a difficult time understanding that water is in the air, so she chose to use the simulated water cycle activity from the textbook to model the water cycle. Callie had taught this topic for five years but does not reflect on her teaching practice. She makes no mention of how she has changed her approach to teaching the stages of the water cycle based on her experience with teaching this concept in previous years. Callie also did not discuss how her students' misconceptions or prior knowledge were taken into account when planning this lesson. The instructional strategies used were based on how she was taught in high school and during her undergraduate program. When asked what knowledge about students' thinking and/or learning influenced how she taught the main ideas or concepts, she replied, "I teach science the way I learned it."

Callie's sources of PCK were the science content courses taken during her undergraduate program, the reform course, and ongoing professional development.

## Carrington

Carrington had been teaching for nine years. She has been teaching sixth grade for the past six years. She minored in geology and had taken several geology courses along with biology, physics, and chemistry (reform course) and an elementary science methods course to satisfy her teaching requirements. Carrington stated that she had been greatly influenced by the reform course because it was a combination of science content and pedagogy. She said that her other science content courses were content that only focused on the memorization of facts. She stated:

> [The reform class] was more hands-on than my other science classes, which made science more about memorization. [The reform class] made science about doing and participating for me.... [The reform class] helped me understand a lot of the concepts in chemistry that I never felt like I understood before, like ions. The instructor broke things down and had us do things that were more hands-on. It wasn't just about reading a book. (Elementary teacher)

Her orientation toward science teaching was hands-on, as noted throughout her interview. She stated that hands-on activities helped her students to conceptualize difficult science concepts. The lesson observed was convection currents. Carrington wanted her students to understand what causes lithospheric plates to move. She said that students know that the plates move but they don't know why. Carrington began the lesson with a review of lithospheric maps (small 3-D models of the earth's tectonic plates) that they made in an earlier lesson to recall what they remembered about continents and faults. The review on plate tectonics served as a segue into the lesson on convection currents.

Carrington used a four-minute video segment on convection currents and plate tectonics boundaries. In addition, she used a simulated activity that would allow her students to generate and observe convection currents. She guided students in making predictions as to what will happen with the currents. Carrington says that she chose those specific strategies to teach the concept of convection currents to help students deepen their understanding of tectonic plates because it is difficult to conceptualize. When planning the lesson, Carrington took into account students' prior knowledge regarding convection currents. She says, "Since the students don't have a lot of science background knowledge, the pre-lab is really important to go over."

Carrington knew that convection current was an abstract concept and therefore difficult to grasp because students could not "see" it so she knew that if students had the opportunity to simulate the convection currents that they may be able to grasp it. Carrington said that the activity that she used to teach the concept was a part of the science kits provided by the district, so she does not have prior experience with this particular activity. In her transformation of the concept of convection currents, she uses videos and activities that will allow students to conceptualize their understandings of what causes the movement of the earth's plates.

Carrington's PCK rating is *proficient*. She understood that her students had a difficult time understanding abstract concepts such as convection currents and knows that her students typically have poor background knowledge regarding this concept and plans her lessons accordingly. Although Carrington knows that students have a difficult time understanding the concept of convection currents because they cannot see it, she does not mention specific learning difficulties that she knows students may have based on her previous experience in teaching this topic. She believes that hands-on activities are the best approach to science teaching and learning because they help students conceptualize difficult to understand topics, but she seems to think that hands-on activities should be used in teaching all science concepts. Carrington received an RTOP score of 71, indicating that the science instruction observed was reformed. The sources of her PCK are the number of content courses taken during her undergraduate program, the reform course, professional development, and continued collaboration with the instructor who taught the reform course during her undergraduate program.

Christopher had taken four science content courses during his undergraduate program including the reformed course. Christopher believes that science should relate to students' lives so he uses a lot of examples in during his lesson. Christopher's observed lesson was on the concept of magnetism. He had students read about the concept of magnetism using the textbook for the majority of the lesson. He provides a lot of examples that relates the concept of magnetism to the lives of his students.

Christopher's RTOP score of 49. His PCK rating was *novice*. The sources of his PCK are the number of content courses taken during her undergraduate program and the reform course. Christopher did not attend professional development workshops geared towards science teaching and learning. Christopher was not a reflective practitioner, nor did he use his knowledge or experience in teaching the concept of magnetism to assist him in selecting the appropriate teaching strategies for teaching this concept with his current students.

The faculty instructor and the inservice elementary teachers believed that a hands-on approach was the best approach to science teaching and learning. Like Calvin, Carrington and Callie thought that hands-on activities should engage students in the learning process and promote critical thinking skills. They all emphasized the importance of connecting the science content that they were teaching to their students' lives. In addition, Carrington and Callie had taken several content courses including the reform course. They each used activities that they thought would best facilitate student understanding of the specific science content that they were teaching. Christopher, however, believed that students must be given the information and then provided with an explanation in their understanding of science concepts. He stated in his interview that students may not have the capacity to explore concepts on their own depending on their level of ability. Calvin, Carrington, and Callie participated in ongoing professional development and collaborated with other colleagues and science specialists throughout the district.

## DISCUSSION AND CONCLUSIONS

The purpose of this study was to examine the long-term impact of reformed undergraduate science courses on the science pedagogical practices of Kindergarten through sixth grade inservice elementary teachers. Ninety-one inservice elementary teachers participated in this study. Of the 91 inservice elementary teachers, 56 experienced the reformed course during their undergraduate program and 35 took the comparison course during their undergraduate programs.

Quantitative and qualitative analysis of classroom observations did not reveal significant differences between the pedagogical practices of inservice elementary teachers who experienced the reform course during their undergraduate program and those who did not. Overall, these results demonstrate that the level of reformed teaching in elementary classrooms is statistically similar for both groups. The recurring theme that emerged from qualitative analysis of the interviews is the idea that professional development geared toward science content and pedagogy along with teaching experience helped to impact instructional practice.

Qualitative analysis of the 21 inservice elementary teachers revealed that two inservice elementary teachers exhibited *advanced* pedagogical content knowledge regarding the specific science concept taught during the observed lesson and five inservice elementary teachers exhibited *proficient* pedagogical content knowledge. All of the inservice elementary teachers were inclined to use a hands-on approach to science teaching and learning. Only seven of the 21 inservice elementary teachers, however, were able to provide a rationale for why they used a hands-on approach to teach the science concept during the observed lesson. These seven inservice teachers stated that students' prior knowledge, curriculum, and various learning modalities informed their decision to use specific instructional strategies to represent the science concept taught during the observed lesson. Those who exhibited *proficient* pedagogical content knowledge, however, discussed the strategies used to teach the specific science concept taught during the observe lesson from that of a more global aspect. This seemed to suggest that they believed that these strategies could be used to teach any science concept.

Fourteen teachers exhibited *novice* or *emerging* pedagogical content knowledge regarding the specific science concept taught during their observed lessons. Each of the 14 teachers stated that a hands-on approach was the best approach to science teaching and learning. These teachers' understanding of "hands-on," however, was fragmented. These teachers equated hands-on learning with activities that allowed students to "do, see, or feel" something, such as manipulating a microscope or using the Smart Board. Moreover, these teachers stated that "hands-on activities" help to keep students interested in science. These activities, however, were not necessarily used to facilitate student understanding of science content.

Also, inservice elementary teachers who exhibited *novice* or *emerging* pedagogical content knowledge attended professional development workshops that focused on reading, literacy, and science or a combination of all three. These findings are consistent with the research on the status of elementary school science teaching regarding science content courses taken during teacher education programs and the amount of emphasis given to math and reading instruction (Weiss et al., 2001). Some of these teachers attended professional development workshops that taught the teachers how to use district-provided science kits. Moreover, inservice elementary teachers with a *novice* or *emerging* PCK rating did not use their knowledge about how their students learn to inform their teaching of the specific topics observed during their classroom lesson. Overall, teachers with a *novice* or *emerging* PCK rating had taken fewer science content courses, did not use their knowledge of their students to inform teaching practice, and were not reflective practitioners.

These findings are consistent with the literature on pedagogical content knowledge: (1) content knowledge is interrelated with pedagogical knowledge and other components of PCK (Halim & Meerah, 2002; Kaya, 2009); (2) content knowledge positively influences PCK (Kapyla, Heikkinnen, &

Asunta, 2009; Ozden, 2008); (3) lack of content knowledge affects teacher trainees' awareness of likely misconceptions (Halim, Subahan, & Meerah, 2002); (4) observed differences in preservice teachers' PCK is probably due to variations in their subject matter knowledge (De Jong et al., 2005); (5) lack of science content knowledge limits pedagogical content knowledge (Appleton, 2008); and (6) lack of science content knowledge may result in elementary teachers using hands-on activities as a substitute for science PCK (Appleton & Kindt, 1999).

Several of the inservice elementary teachers in my study stated that time, lack of resources, an emphasis on math and reading, and state testing served as impediments to science teaching. These barriers were beyond the control of the teachers. The inservice elementary teachers were constrained by the context of their school environment. In addition, several of the inservice elementary teachers cited testing as a goal of science teaching and felt pressured to cover a lot of science content in a short amount of time. These barriers could serve as impediments in the development of pedagogical content knowledge.

## IMPLICATIONS

The *National Science Education Standards* and the National Science Teachers Association describe what science content elementary teachers should know and what pedagogical knowledge they should have for science teaching and learning in K–6 classrooms. Paramount to these recommendations is the notion that young children should be active learners, constructing science understanding through hands-on/minds-on inquiry experiences; engage in science investigations wherein they ask questions, collect and make sense of data, and come to conclusions that are supported by evidence; and investigate authentic questions generated from student experiences (NRC, 2000, 2007). Students should also experience science inquiry such that they develop positive attitudes, skills, and knowledge about science and the relationship between science and society (Schwarz & Gess-Newsome, 2008).

This type of instruction places a heavy demand on elementary teachers who not only must have expertise in science content, but they must have expertise in all of the subjects they teach. Teachers must go beyond knowledge of science content and traditional, generic pedagogical knowledge to ensure meaningful learning experiences for students in K–6 classrooms. More emphasis on pedagogical content knowledge is needed.

Pedagogical content knowledge regarding specific science content enables expert teachers to transform science content using instructional strategies and knowledge of students' learning difficulties, prior knowledge, and misconceptions into forms young students can understand.

Quantitative analyses revealed no significant differences in the observed pedagogical practices between inservice elementary teachers who took the reformed course during their undergraduate program and those who did not. Qualitative analyses, however, revealed that differences exist in the pedagogical content knowledge of inservice elementary teachers. Inservice elementary teachers who had PCK ratings of *proficient* or *advanced* had the higher RTOP scores. The reverse of this previous statement is not true. Teachers who had high RTOP scores did not necessarily achieve higher PCK ratings regarding the specific science concept that was taught. The following characteristics emerged from inservice elementary teachers who had high PCK ratings and high RTOP scores: (1) evidence of science content knowledge regarding the specific concepts taught during the observed lesson; (2) ongoing professional development geared toward science content and pedagogy; (3) school culture that emphasized science teaching and learning; (4) a supportive school culture; (5) instructional strategies that considered student learning difficulties, prior knowledge, and/or student misconceptions; (6) purposeful use of instructional strategies to represent science concept and facilitate student learning; and/or (7) reflective teaching practice.

Pedagogical content knowledge is a complex construct, and even more complex are the variables that mediate the development of it.

## ACKNOWLEDGMENT

This material is based upon work supported by the National Science Foundation under Grant No. 0554594. Any opinions, findings, and conclusions or recommendations expressed in this material are those of the author(s) and do not necessarily reflect the views of the National Science Foundation.

## REFERENCES

Abell, S., & Leerdman, N. (2007). *Handbook of research on science education.* Mahwah, NJ: Lawrence Erlbaum

American Association for the Advancement of Science. (1990). *Science for all Americans.* New York, NY: Oxford University Press.

Appleton, K. (2008). Developing science pedagogical content knowledge through mentoring elementary teachers. *Journal of Science Teacher Education, 19*, 523–545.

Appleton, K., & Kindt, I. (1999, March). *How do beginning primary teachers cope with science: Development of pedagogical content knowledge in science.* Paper presented at the annual meeting of the National Association for Research in Science Education, Boston, MA.

Avraamidou, L., & Zembaul-Saul, C. (2005). Giving priority to evidence in science teaching: A first-year elementary teacher's specialized practices and knowledge. *Journal of Research in Science Teaching, 42*(9), 965–986.

Avraamidou, L., & Zembaul-Saul, C. (2010). In search of well-started beginning science teachers: Insights from two first-year elementary teachers. *Journal of Research in Science Teaching, 47*(6), 661–686.

De Jong, O., & Van Driel, J. (2004). Exploring the development of student teachers' PCK of the multiple meanings of chemistry topics. *International Journal of Science and Mathematics Education, 2*(4), 477–491

DeJong, O., Van Driel, J. H., & Verloop, N. (2005). Preservice teachers' pedagogical content knowledge of using particle models in teaching chemistry. *Journal of Research in Science Teaching, 42*(8), 947–964.

Friedrichsen, P. M., & Dana, T. M. (2003). Using a card-sorting task to elicit and clarify science teaching orientations. *Journal of Science Teacher Education, 14*(4), 291–309.

Friedrichsen, P. M., Van Driel, J. H., & Abell, S. (2011). Taking a closer look at science teaching orientations. *Science Education, 95*, 358–376.

Gess-Newsome, J. (1999). Pedagogical content knowledge: An introduction and orientation. In J. Gess-Newsome & N. Lederman (Eds.), *Examining pedagogical content knowledge* (pp. 3–20). Dordrecht, The Netherlands: Kluwer.

Grossman, P. L. (1990). *The making of a teacher: Teacher knowledge and teacher education.* New York, NY: Teachers College Press.

Halim, L., & Meerah, S. (2002). Science trainee teachers' pedagogical content knowledge and its influence on physics teaching. *Research in Science & Technological Education, 20*(2), 215–225.

Kapyla, M., Heikkinen, J. P., & Asunta, T. (2009). Influence of content knowledge on pedagogical content knowledge: The case of teaching photosynthesis and plant growth. *International Journal of Science Education, 31*(10), 1395–1415.

Kaya, O. N. (2009). The nature of relationships among the components of pedagogical content knowledge of preservice science teachers: "Ozone layer depletion" as an example. *International Journal of Science Education, 31*(7), 661–688.

Loughran, J., Berry, A., & Mulhall P. (2006). *Understanding and developing science teachers' pedagogical content knowledge.* Rotterdam, The Netherlands: Sense Publishers.

Loughran, J., Mulhall, P., & Berry, A. (2004). In search of pedagogical content knowledge in science: Developing ways of articulating and documenting professional practice. *Journal of Research in Science Teaching, 41*(4), 370–391.

MacIsaac, D., & Falconer, K. (2002). Reforming physics instruction via RTOP. *The Physics Teacher, 40*, 16–22.

Magnussen, S., Krajcik, J., & Borko, H. (1999). Nature, sources, and development of pedagogical content knowledge for science teaching. In J. Gess-Newsome & N. Lederman (Eds.), *Examining pedagogical content knowledge* (pp. 95–132). Dordrecht, The Netherlands: Kluwer.

Mishra, P., & Koehler, M. (2006). Technological pedagogical content knowledge: A framework for teacher knowledge. *Teachers College Record, 108*, 1017–1054.

Mulhall, P., Berry, A., & Loughran, J. (2003). Frameworks for representing science teachers' pedagogical content knowledge. *Asia-Pacific Forum on Science Learning and Teaching, 4*(2), 1–25.

National Research Council. (1996). *National science education standards*. Washington, DC: National Academy Press.

National Research Council, 2000). Inquiry and the national science education standards, Committee on the Development of an Addendum to the National Science Education Standards on Scientific Inquiry. S. Olsen & S. Locks-Horsley (Eds.). Washington, DC: National Academy Press.

National Research Council. (2007). *Taking science to schools*. Washington, DC: National Academy Press

National Research Council. (2012). *A framework for K–12 science education*. Washington, DC: National Academy Press.

National Science Teachers Association. (2003). *Standards for science teacher preparation*. Washington, DC: National Academy Press.

Ogletree, G. L. (2007). *The effect of fifth grade science teachers' pedagogical content knowledge on their decision making and student learning outcomes on the concept of chemical change*. (Unpublished doctoral dissertation). University of Alabama-Tuscaloosa.

Ozden, M. (2008). The effect of content knowledge on pedagogical content knowledge: The case of teaching phases of matters. *Educational Sciences: Theory & Practice, 8*(2), 633–645.

Sawada, D., Piburn, M., Judson, E., Turley, J., Falconer, K., Benford, R. & Bloom, I. (2002). Measuring reform practices in science and mathematics classrooms: The Reformed Teaching Observation Protocol. *School Science and Mathematics, 102*(6), 245–253.

Schwab, J. (1978). Education and the structure of the discipline. In I. Westbury & N. J. Wilkof (Eds.), *Science, curriculum and liberal education* (pp. 229–272). Chicago, IL: University of Chicago Press

Schwartz, R. S., & Gess-Newsome, J. (2008). Elementary science specialist: A pilot study of current models and a call for participation in the research. *Science Educator, 17*(2), 19–30.

Shulman, L. S. (1986). Those who understand: Knowledge growth in teaching. *Educational Researcher, 15*, 4–14.

Sunal, D. W., Sunal, C. S., Mason, C., & Zollman, D. (2008, June). *Assessment of undergraduate courses in the sciences: A national study*. Paper presented at Enriching the Academic Experience of College Science Students National Conference, Ann Arbor, MI.

Sunal, D. W., Wright, E. L., & Day, J. B. (Eds.). (2004). *Reform in undergraduate science teaching for the 21st century*. Greenwich, CT: IAP Information Age.

Weiss, I. R., Banilower, E. R., McMahon, K. C., & Smith, P. S. (2001). *Report of the 2000 National Survey of Science and Mathematics Education*. Chapel Hill, NC: Horizon Research.

CHAPTER 5

# THE IMPACT OF REFORMED UNDERGRADUATE SCIENCE COURSES ON ELEMENTARY TEACHER SELF-EFFICACY AND SUBSEQUENT RELATIONSHIPS TO SCIENCE TEACHING PRACTICES

**Corinne H. Lardy and Cheryl L. Mason**

## ABSTRACT

This study examines the impact of reformed undergraduate science courses on elementary teachers' science teaching self-efficacy beliefs. In addition, we examine subsequent relationships among inservice teachers' self-efficacy beliefs, beliefs about their own and ideal science teaching practices, and observed science teaching practices. Data were collected using the Reformed Teacher Observation Protocol, semi-structured interviews, and the Science Teaching Efficacy Beliefs Instrument (STEBI-A). From the data set, eight case profiles of inservice elementary teachers were examined in closer detail. Participants reported that they gained more confidence in their ability to

*Research Based Undergraduate Science Teaching*, pp. 153–188

teach science effectively from courses that (1) explicitly connected the science content to the teaching of that content, (2) made students feel that they gained content knowledge, and (3) sparked students' interest in the content. Reformed courses may have influenced some individuals to have lower levels of self-efficacy by making them realize how much they do not know about science teaching. A clear relationship was not evident between science teaching self-efficacy and reformed science teaching; teachers with high STEBI scores were just as likely to be observed teaching in a reformed manner as teachers with low levels of self-efficacy. Interviews and observations revealed additional possible relationships between self-efficacy and teaching. This study demonstrates the importance of qualitative data to support quantitative data when studying self-efficacy beliefs of teachers and mechanisms for increasing efficacy.

Beliefs play a critical role in the instructional practices of teachers (Kagen, 1992; Pajares, 1992; Philipp, 2007). Therefore, if we are to improve the way that science is taught at the elementary level in the United States, we must understand which beliefs impact the ways in which elementary teachers implement instructional strategies in their science lessons, and how.

One set of beliefs consistently linked to teacher behavior in a variety of educational fields is that of teacher self-efficacy (Tschannen-Moran, Hoy, & Hoy, 1998; Wheatley, 2005). In fact, a positive correlation is so generally accepted between the level of science teaching efficacy beliefs and effective science teaching practices that increasing preservice teachers' science teaching self-efficacy has been promoted as a primary goal of undergraduate science experiences and science teacher education (Brand & Wilkins, 2007). Studying the impact of undergraduate science courses on the development of self-efficacy is therefore important if we are to effectively and positively influence these beliefs in students as they develop into teachers of science.

There is, however, some evidence suggesting that the positive correlation between self-efficacy and effective science teaching does not always hold true. A clearer understanding of the development of science teaching self-efficacy as well as the relationships between elementary teachers' science teaching self-efficacy beliefs and their actual science teaching practices is needed in order to successfully improve the ways in which science is taught at the elementary level.

## TEACHER BELIEFS AS THEY RELATE TO PRACTICE

There are many different factors that influence the behavior of teachers, such as knowledge, beliefs, attitudes, and emotions (Gess-Newsome, 1999;

Philipp, 2007; Richardson, 1996). Research from a variety of academic areas suggests that beliefs have a particularly strong influence on teacher behavior, since "the beliefs teachers hold influence their perceptions and judgments, which, in turn, affect their behavior in the classroom" (Pajares, 1992, p. 307). It is therefore important that beliefs, especially particular sets of beliefs, have a central role in research whose aim is to understand teacher behaviors (Kagen, 1992; Pajares, 1992) and to successfully help teachers to implement educational reform (Eisenhart, Shrum, Harding, & Cuthbert, 1988).

Beliefs have been found to be a strong predictor of teacher behavior in a variety of academic areas (Muijs & Reynolds, 2002; Solomon, Battistich, & Hom, 1996; Tobin, Tippins, & Gallard, 1994; Weiss, Pasley, Smith, Banilower, & Heck, 2003). Furthermore, the relationship between teacher beliefs and behavior is not a simple linear one. The relationship is much more complex (Brickhouse, 1990; Bryan, 2003; Fang, 1996; Tobin, 1993) because a teacher's various belief systems may influence his or her teaching practices in multiple ways, depending on a variety of factors (Ogan-Bekiroglu & Akkoc, 2009; Pajares, 1992); the belief systems of a teacher may even contradict one another (Bryan, 2003). In addition, not only do beliefs influence practice, but the converse is also true: teaching experiences influence teacher beliefs (Thompson, 1992).

Beliefs have also been shown to be an important part of the foundation upon which science teacher behavior is based (Bryan, 2003; Enochs & Riggs, 1990; Mansour, 2010; Tobin et al., 1994). Several studies have demonstrated how various sets of teacher beliefs may influence different facets of teacher behavior related to their science instruction. For example, Cronin-Jones (1991) identified four categories of beliefs that strongly influenced curriculum implementation of science teachers: beliefs about how students learn, the teacher's role in the classroom, the abilities of students in a particular age group, and the relative importance of content topics. In another example, Haney, Lumpe, Czerniak, and Egan (2002) provided evidence that context beliefs of elementary teachers related to science instruction (teachers' beliefs about how well environmental factors will support his or her ability to teach science) can have a large impact on these teachers' science teaching practices. Evidence has also suggested that teachers' beliefs regarding the nature of science (i.e., viewing science as content or as a process) and science as an academic subject play a large part in a teacher's science classroom practices (Brickhouse, 1990; Duschl, 1983; Levitt, 2001; Tobin et al., 1994).

Another set of beliefs that has been consistently shown to be related to (1) student achievement and motivation (Angle & Moseley, 2009; Caprara, Barbaranelli, Steca, & Malone, 2006; Guo, Piasta, Justice, & Kaderavek, 2010; Ross, 1992), (2) implementation of research-based instructional techniques (Allinder, 1994; Ghaith & Yaghi, 1997; Guskey, 1988; Scribner, 1999; Tschannen-Moran & McMaster, 2009), and (3) a commitment to effective teaching (Caprara et al., 2006; Klassen & Chiu, 2010; Skaalvik & Skaalvik, 2010; Ware & Kitsantas, 2007) is a teacher's teaching efficacy beliefs.

## SCIENCE TEACHING SELF-EFFICACY BELIEFS: AN OVERVIEW

Our study examines self-efficacy beliefs through a lens based initially in social cognitive learning theory (Bandura, 1982, 2001), further developed for the examination of teacher behavior by Gibson and Dembo (1984), and of teachers of science by Riggs and Enochs (1990).

Teacher efficacy beliefs, in general, refer to beliefs about the level of confidence individuals have in their ability to influence student learning through their teaching behaviors. This construct is composed of two specific kinds of beliefs, corresponding to the two components of Bandura's (1982) model of efficacy: personal teaching efficacy and teaching efficacy (Gibson & Dembo, 1984). According to Dembo and Gibson (1985), *personal teaching efficacy* is an individual's "belief that he or she [personally] has the skills and abilities to bring about student learning" (p. 175). *Teaching efficacy* beliefs, in contrast, are the beliefs of an individual related to teachers' abilities, as a general group, to influence student learning, or the extent to which students will learn if a teacher teaches effectively.

The construct of science teaching efficacy beliefs, introduced by Riggs and Enochs (1990), is different from general teaching efficacy beliefs in that it refers specifically to beliefs about the level of confidence individuals have in their ability to influence student learning related to science. Similar to general teaching efficacy beliefs, this construct is composed of two specific types of beliefs: personal science teaching efficacy (PSTE) and science teaching outcome expectancy (STOE). The PSTE refers to a teacher's belief in his or her own ability to effectively teach science, while STOE reflects the extent of a teacher's belief that, if teachers provide appropriate science instruction, then their students will learn.

### Self-Efficacy and Science Education Research

Since its introduction by Riggs and Enochs (1990), the construct of science teaching efficacy beliefs has developed into a popular area of research in science education, particularly in exploring the beliefs of preservice elementary teachers. In examining the related research literature, there seems to be sound reason for this popularity; a long history of evidence exists suggesting a positive link between efficacy beliefs and teacher behavior, both for general and science teaching efficacy beliefs. Research has demonstrated, for example, that teachers with low general teaching efficacy beliefs expect students to fail and place the responsibility for learning entirely on the student rather than the teacher (Ashton, 1984; Ashton & Webb, 1986). In addition, teachers with high general teaching efficacy beliefs have been shown to (1) spend less time engaged in discussion unrelated to the objectives of a lesson (Gibson & Dembo, 1984); (2) be more open to new ideas and more willing to try new instructional techniques (Allinder, 1994; Guskey, 1988; Scribner, 1999; Tschannen-Moran & McMaster, 2009); (3) employ a larger amount of planning and organization for their lessons (Allinder, 1994); (4) have greater enthusiasm for teaching (Allinder, 1994); and (5) be more committed to teaching as a profession (Caprara et al., 2006; Coldarci, 1992; Klassen & Chiu, 2010).

Since 1990, researchers have seen similar evidence connecting science teacher efficacy beliefs to science teaching behaviors. For example, Czerniak and Shriver (1994) found significant differences between preservice elementary teachers with high and low self-efficacy in their choices of instructional strategies for science lessons and the ways that they measured success of a science lesson. Specifically, high-efficacy teachers tended to choose activities in which they expected students to use higher-level thinking and problem-solving skills and were more likely than low-efficacy teachers to use teaching strategies that were based on research or theory. In addition, Czerniak and Shriver found that the teachers with high science teaching self-efficacy were oriented toward the goals of developing students' critical thinking and decision-making skills, and tended to measure success of their science lessons by whether or not they believed these goals were achieved. The teachers with low science teaching self-efficacy, in contrast, tended to measure success of a science lesson by their ability to control students and to keep the class orderly and quiet. Preservice elementary teachers with high science teaching self-efficacy also have been shown to be more likely to claim that activity-based instruction, in which students learn through cooperation and experience, is the most appropriate method of teaching science at the elementary level (Enochs, Scharmann, & Riggs, 1995).

Research regarding inservice elementary teachers also has suggested a positive correlation between science teaching efficacy beliefs and reformed teaching practices. Haney et al. (2002), for example, found that elementary teachers with higher PSTE scores

> were more likely to design lessons that: incorporated inquiry, depicted careful planning, attended to student prior knowledge and experiences, attended

> to issues of equity, utilized appropriate and available resources, encouraged a collaborative approach, and assessed students in a way that was consistent with the intended purpose. (p. 179)

The authors also found that, as compared to their colleagues with lower PSTE, these high-efficacy teachers were "more likely to convey science content appropriately by presenting content that was: significant and worthwhile, developmentally appropriate, accurate, dynamic and interdisciplinary in nature, and tied to the real world" (Haney et al., 2002, p. 179). Other evidence suggests that inservice teachers with higher levels of science teaching self-efficacy (1) claim to ask more open-ended questions, (2) do a better job of connecting science content to students' lives (Riggs, Enochs, & Posnanski, 1998), (3) teach more science per week (Desouza, Boone, & Yilmaz, 2004), (4) report using more hands-on and inquiry-based activities (Lakshmanan, Heath, Perlmutter, & Elder, 2011; Marshall, Horton, Igo, & Switzer, 2009; Ramey-Gassert, Shroyer, & Staver, 1996), and (5) exhibit more positive attitudes toward science education reform (Czerniak & Lumpe, 1996).

Evidence from research, such as that cited above, demonstrates science teaching self-efficacy to be a potentially powerful construct influencing the ways in which elementary teachers teach science. If we are to improve elementary science education, it is relevant and worthwhile to examine in greater detail how science teaching efficacy beliefs develop in teachers, and how we as undergraduate science educators can positively influence those beliefs (Brand & Wilkins, 2007).

## The NOVA and NSEUS Programs

The NASA/NOVA (NASA Opportunities for Visionary Academics) program was established in 1996 and supported by the National Aeronautics and Space Administration (NASA) to reform science courses for preservice elementary teachers (Sunal et al., 2001). The reasoning behind this project was that in order to reform elementary science education, we must reform the undergraduate science education of elementary teachers. Among the many goals of the NOVA project was to help increase preservice elementary teachers' science teaching self-efficacy through teaching undergraduate science courses in a reformed manner, thus making their science instruction more effective when they entered the elementary classroom.

Teams consisting of science, technology, engineering, and mathematics (STEM) and teacher education faculty, along with administrators from universities nationwide, participated in the NOVA project. Following professional development, team members worked with a NOVA mentor to develop an action research funding proposal designed to establish a new course or extensively reform an existing course. As a result of the NOVA project, 167 reformed undergraduate science courses were established at 103 universities across the United States.

In 2006, the National Study of Education in Undergraduate Science (NSEUS) project was funded by the National Science Foundation (NSF) to examine the short- and long-term impacts of the established NOVA courses on undergraduate students and inservice elementary teachers (Sunal, Sunal, Mason, Lardy, Zollman, & Matloob- Haghanikar, 2009). Over the course of the last six years, the NSEUS project has collected a significant amount of data nationwide regarding various ways in which the NOVA courses potentially have impacted elementary teachers and their teaching of science. Due to the apparent important influence of science teaching self-efficacy on science teaching practices, NSEUS has collected qualitative and quantitative data regarding this construct, including surveys, artifacts, interviews, and observations. The study presented here examines some of these data in an attempt to explain many of the important ways reformed courses such as those developed through the NOVA program positively influence the development of future elementary teachers' science teaching efficacy beliefs.

## Beyond the Impact of Reformed Undergraduate Science Courses on Self-Efficacy

While research results suggest that it is important and worthwhile to examine the potential impacts of undergraduate science courses, such as those developed through NOVA, on the science teaching self-efficacy of elementary teachers, further evidence indicates that this one factor alone does not tell the entire story. Although there is a great deal of evidence suggesting the important influence of science teaching self-efficacy beliefs on science teaching practices, several under-addressed issues in the related research indicate that the intricacies of the relationships between these beliefs and practices are not clear.

Several inconsistencies exist in recent research suggesting that higher self-efficacy does not always lead to more effective teaching (Wheatley, 2005). For instance, although Haney et al. (2002) found that, for five out of the six teachers in their study, greater self-efficacy scores correlated with more reformed science teaching practices, one participant did not follow this pattern. This particular participant demonstrated high self-efficacy beliefs, but observations and interviews revealed that her science teaching strategies were primarily teacher-centered and lecture-based with little to no use of inquiry. Similarly, Settlage, Southerland, Smith, and Ceglie

(2009) and Bhattacharyya, Volk, and Lumpe (2009) noted that, although the preservice elementary teachers whom they studied had relatively high levels of science teaching self-efficacy, the teaching behaviors of several subjects demonstrated a relatively low level of reform. Kind (2009) also observed that preservice secondary science teachers did a better job of choosing appropriate instructional strategies for scientific topics that they felt less confident teaching.

Much of the current research regarding science teaching self-efficacy beliefs has depended on quantitative and self-reported data. Due to the limited number of studies examining the relationship between inservice elementary teachers' self-efficacy beliefs and their observed, as opposed to self-reported, teaching practices (e.g., Haney et al., 2002; Riggs et al., 1998), many aspects of the relationship between science teaching self-efficacy and the science teaching behaviors of practicing teachers remain unclear. In order to be meaningful, future research about science teacher efficacy beliefs should include interview and observational data to support data derived from quantitative survey instruments (Pajares, 1992; Perkins, 2007; Wheatley, 2005).

It is clear that additional research is needed not only on the potential factors impacting science teaching self-efficacy beliefs, but also on the intricate relationships that potentially exist between these beliefs and science teaching practices, both perceived by teachers and observed by researchers. Thus, in our study, we take the research a step beyond simply examining the influences of the NASA Opportunities for Visionary Academics (NOVA) courses on science teaching self-efficacy by exploring how inservice elementary teachers' self-efficacy beliefs translate into their observed classroom practices and their self-described beliefs about those practices.

## PURPOSE OF THE STUDY

The purpose of this study was to examine the impact of reformed undergraduate science courses developed through the NOVA program on elementary teachers' science teaching self-efficacy beliefs. Specifically, we asked, to what extent did the NOVA courses impact elementary teachers' science teaching self-efficacy beliefs?

This study further examined subsequent relationships among participating elementary teachers' science teaching self-efficacy beliefs, beliefs about science teaching practices, and observed science teaching practices. Specifically, we investigated to what extent the science teaching self-efficacy beliefs of participating inservice teachers related to these teachers' beliefs about ideal and personal science teaching, and observed science teaching practices.

## METHOD AND ANALYSIS

### Participants

As part of the multiyear National Science Foundation-funded project, the National Study of Education in Undergraduate Science (NSEUS), a sample of 19 universities was selected by random stratification based on institutional type from the national population of 103 NOVA-participating institutions. The selected sample includes a variety of Carnegie classifications and represents a wide geographic region, encompassing Alabama, Alaska, California, Texas, Massachusetts, Michigan, Connecticut, Oklahoma, Kansas, North Carolina, Wisconsin, Virginia, Idaho, West Virginia, and Indiana.

At each of the 19 selected higher education institutions, a course developed through the NOVA program (*reform course*) and a course of comparable subject and academic level that was not developed through NOVA professional development (*comparison course*) were identified. Data were collected from the instructors and students in each of these courses. In addition, at every site, four to seven inservice elementary teachers (approximately 50% who took the *reform* course and 50% who did not) were recruited to take part in the study through coordination with a faculty member at the university acting as a NSEUS research associate ($n = 85$).

### Data Collection

During the NSEUS project, an intensive weeklong site visit was conducted at each of the 19 universities, including: (1) observations and interviews of the two identified undergraduate science course instructors and associated instructors such as graduate assistants, (2) focus group interviews with students enrolled in each of the two identified undergraduate courses, and (3) observations and interviews of each of the four to seven identified inservice elementary teachers. In addition, a variety of surveys were administered to participating undergraduate students and elementary teachers.

The overall data set collected by the NSEUS project is large and designed to answer a variety of research questions, many of which are not the focus of the research presented here. For this reason, only the data relevant to answering the immediate research questions are discussed.

## Observations

Using the *Reformed Teacher Observation Protocol* (RTOP) (Sawada & Pilburn, 2000) to measure the reform level of observed classes, at least two trained project researchers observed, at a minimum, one week's full class sessions for each undergraduate course and the science lessons taught by the elementary teachers. In cases of university courses where lecture and lab were separate, both were observed. Observed undergraduate class sessions ranged in duration from one to three hours, while observed elementary school lessons lasted 25–85 minutes. The content and strategies for all observed lessons were determined by the university instructors and elementary teachers.

Immediately following each observation, project researchers individually rated the observed lesson on each of the 25 items of the RTOP, assigning a numerical score from 0–4 for each aspect of reformed teaching, and then discussed all ratings. Overall, the inter-rater reliability was approximately 90%. In the cases where there was a discrepancy between researchers' ratings, a consensus was achieved through discussion. In cases of undergraduate courses where lecture and lab were observed separately, researchers agreed on a combined RTOP score, treating the two together as one complete class session. Detailed field notes recorded by project researchers during each observation regarding the events that occurred during the observed lessons were combined to form a descriptive report of the classroom experience.

## Interviews

Interviews were conducted with both undergraduate students and inservice elementary teacher participants in order to determine how participants perceived their undergraduate science courses, as well as other factors, to have influenced their science teaching self-efficacy. In addition, interviews with inservice teachers were used to gain information about participants' beliefs about science teaching in general as compared to their own science teaching.

Semi-structured focus group interviews were conducted with four to six undergraduate students in each class following the observed undergraduate science course sessions. During these interviews students were asked questions regarding how they felt their undergraduate science courses, and the observed course in particular, had influenced their beliefs and attitudes toward science as well as the extent of their conceptual scientific knowledge. Students also were asked questions regarding their confidence in their ability to become an effective science teacher and the factors influencing that level of confidence.

Following each elementary classroom observation, a one-on-one semi-structured interview was conducted with the elementary teacher lasting approximately 30–40 minutes. During these interviews, participants were asked a variety of questions regarding their beliefs about science teaching in general, how they believe they teach science in their classroom, and what factors they felt had influenced their beliefs. Teachers also were asked questions regarding their backgrounds such as, how many years have you been teaching and at what grade levels? Have you been involved in specialized teaching? What courses did you take as undergraduate students? and To what extent have you participated in professional development? In addition, elementary teachers were asked a series of questions regarding the observed lesson (e.g., goals of the lesson, rationale for the strategies used, mechanisms for assessment, etc.).

### STEBI-A

Following the on-site visits, each elementary teacher completed the *Science Teacher Efficacy Beliefs Instrument form A* (STEBI-A) (Riggs & Enochs, 1990). This survey was used to determine the level of Personal Science Teaching Efficacy (PSTE) and Science Teaching Outcome Expectancy (STOE) for each participant.

## DATA ANALYSES

### Impact of Reform Courses on Participants' Science Teaching Self-Efficacy

Out of the 85 inservice elementary teachers from which data were collected, 66 completed the STEBI (35 reform and 31 comparison). A preliminary test for the equality of variances indicated that the variances of the two groups were not significantly different for the overall STEBI ($F = 0.881$, $p = 0.35$), nor for the separate constructs of PSTE ($F = 1.107$, $p = 0.39$) and STOE ($F = 0.680$, $p = 0.14$). Therefore, a two-sample $t$-test was performed that assumed equal variances.

To determine the extent to which the level of reform observed in the reform courses potentially influenced the self-efficacy of the elementary teachers who took those courses, Pearson's correlation coefficient was calculated between the STEBI scores of the elementary teachers and the RTOP scores for the reform courses they took. Because it was unknown which university courses the comparison elementary teachers took, a correlation was not calculated for this group.

In order to examine some of the specific factors that participants perceived as influencing their science teaching self-efficacy, elementary teacher and undergraduate student interviews were analyzed qualitatively to uncover themes in participants' responses regarding their confidence in teaching science. Data were examined by multiple coders in order to establish inter-rater reliability (90%).

### Relationships Among Participants' Self-Efficacy, Beliefs about Science Teaching, and Practice

To determine whether a quantitative relationship existed between participating elementary teachers' science teaching self-efficacy and the level of reform in their observed teaching, STEBI (PSTE and STOE) and RTOP scores were plotted and a Pearson's correlation coefficient was calculated between the two. The correlation between STEBI and RTOP scores was examined in relationship to several different demographic factors of the teachers in order to determine whether any of these factors seemed to influence the relationship between science teaching self-efficacy and observed science teaching practices for this group of teachers. These factors included gender, years of teaching at their current grade level, years of teaching experience overall, amount of professional development, and number of science content courses.

Out of the 66 inservice elementary teachers who completed the STEBI, four at each of the extremes of high and low STEBI scores were examined more closely. Of these eight case profiles, 50% had high RTOP scores and 50% had low scores. Case profiles of all eight teachers were examined and coded according to themes to provide a more detailed picture of relationships among science teaching self-efficacy beliefs, beliefs about science teaching, and practices of the inservice elementary teachers in this study. Again, data were examined by multiple coders in order to establish inter-rater reliability (85%).

## RESULTS

The results are presented in two parts, addressing each of the two research questions for this study.

> To what extent did NOVA courses impact elementary teachers' science teaching efficacy beliefs?

Overall STEBI, PSTE subscale, and STOE subscale scores were slightly lower for teachers who took a reform course than for those who did not take a reform course (Table 5.1). These differences, however, were not statistically significant.

**Table 5.1. STEBI, PSTE, and STOE Scores for Participating Reform and Comparison Teachers**

| | *n* | *STEBI* | *PSTE* | *STOE* |
|---|---|---|---|---|
| Reform | 35 | *M* = 72.34 | *M* = 36.20 | *M* = 36.14 |
| | | *SD* = 20.13 | *SD* = 14.42 | *SD* = 7.39 |
| Comparison | 31 | *M* = 79.52 | *M* = 42.97 | *M* = 36.55 |
| | | *SD* = 21.16 | *SD* = 13.62 | *SD* = 8.82 |

Based on observations, the 38 participating undergraduate courses varied widely in their application of reform-based teaching practices; RTOP scores for the observed reform class sessions ranged from 40–90 out of a possible 100 points. In order to determine whether the elementary teachers who experienced a greater level of reform in their reform courses had higher levels of science teaching self-efficacy than those who experienced fewer reform-based teaching practices in such courses, a Pearson correlation was conducted between STEBI (PSTE and STOE) scores of the reform course elementary teachers and the RTOP scores of the reform courses they took. This comparison revealed no significant relationship between the observed reform levels of undergraduate reform courses as measured by the RTOP, and the STEBI, PSTE, or STOE scores of the elementary teachers who took those courses; teachers who took reform undergraduate courses with observed higher levels of reform did not have higher levels of self-efficacy than those teachers who experienced lower levels of reform-based teaching.

While there was no statistically significant difference between the two groups, qualitative data from interviews with undergraduate students currently enrolled in the reform and comparison courses and with inservice elementary teachers who completed these courses revealed several factors described by participants as influencing their science teaching self-efficacy. In addition, a great deal of overlap was seen between the kinds of factors most frequently cited by undergraduate students and inservice teachers as influencing their self-efficacy, as described in the following sections.

## Factors Influencing Preservice Teachers' Self-Efficacy

Three factors were found to be most often cited by undergraduate participants as increasing their science teaching self-efficacy: (1) how enjoyable they perceived the science content to be, (2) the amount of self-perceived knowledge they had about the science content, and (3) whether or not they had been shown an explicit connection of the science content to teaching and/or given the opportunity to practice teaching the content to others. All focus groups indicated that these factors were closely tied to how the science content was presented to them in their undergraduate science courses.

Results from focus group interviews with undergraduate students enrolled in reform and comparison courses indicate that students in 71.0% of the 38 focus groups stated that they were more confident teaching content that they enjoyed learning about in their courses and/or content that they were "passionate" about. For example, students stated, "[I can't become an effective science teacher] until I gain interest in it. I can't teach it if I don't like it"; and

> I was worried that if I ever had to teach science I wouldn't be interested in it. But now I'm not afraid of it. This class helps me think about the concepts in the minds of my students. Now if I had to teach science I think I could be enthusiastic about it.

Undergraduate student focus groups (55.3%) also stated that overall, they were more confident teaching science content areas in which they had taken more courses and/or felt they had more content knowledge about. In several cases, because the reform course was the only science course that they had taken, some individuals reported that they felt most confident to teach the content area(s) covered in that specific course.

Undergraduate students in 47.4% of the focus group interviews also indicated that they gained more confidence in their ability to teach science in courses that explicitly connected the content to how to teach it and/or gave students an opportunity to teach the content to children or to one another. Examples of students' quotes expressing this idea are: "We get to do it and experience it instead of just her telling us. She's modeled it for us so we have the confidence to model it for our kids"; and "Getting some experience in teaching science in this class, to the other students in the class and to elementary kids, has made me more confident that I can teach science." For students enrolled in undergraduate courses that did not offer experiences such as these, several students stated that they had more confidence since they had experiences teaching science to others outside of class, such as in a summer camp or tutoring situation.

In 23.7% of the focus groups, students noted that some of their undergraduate courses in which they had very positive experiences and learned a great deal about science education actually make them feel less confident to teach science. Students explained that the courses had made them realize how much they did not know about learning and teaching science, and, while a majority of students who expressed this opinion thought that they could become effective science teachers in the future, they did not feel confident in their current abilities.

## Factors Influencing Inservice Teachers' Self-Efficacy

During one-on-one interviews, inservice elementary teachers were asked what science content areas they felt most or least confident teaching, and why. Of the 77 teachers who described their reasons for confidence or lack thereof, a variety of reasons were cited. These reasons are listed along with example quotes from interviews in Table 5.2. As can be seen from this table, the top three reasons cited by this group of inservice teachers corresponds closely with the most prevalent factors the undergraduate participants described as having the most influence on their science teaching self-efficacy, although in a different order of frequency. For the inservice teacher participants, the most prevalent reason for high or low confidence teaching particular science content was self-perceived amount of knowledge regarding the content (54.5%), followed by personal interest in the content (41.6%), then amount of experience teaching the content (40.3%).

The contributing factors to participants' confidence were not mutually exclusive. Many teachers cited a combination of factors interacting together to influence their perceived ability to teach particular areas of science. One teacher, for example, stated:

> In college, I learned that science lessons need to be big and involved with big experiments. That really intimidated me. Now, thanks to [a colleague] I realize they don't always have to be. She's taught me a lot about how to teach science and now I like it. It makes me feel a lot more confident that I can do it. I can teach science.

The factor of "personal interest in the topic" seemed to be particularly connected to other factors. Teacher participants explained that if they are interested in a certain scientific topic, they are more likely to seek out additional knowledge on that topic from university courses, professional development, colleagues, and/or other sources. In addition, interest in a particular scientific content area also led teachers to teach it more often, thus gaining more experience teaching the content. Participants also noted that if they were enthusiastic about the content, their students would be more enthusiastic, thus further boosting their confidence for teaching that content.

**Table 5.2. Reported Reasons for Levels of Confidence in Teaching Particular Science Content (*n* = 77)**

| *Reason (in descending frequency)* | *% Teachers* | *Example(s)* |
|---|---|---|
| Amount of knowledge regarding the content | 54.5 | I have a degree in biology with a minor in physics so I feel pretty secure in my ability to teach all content areas of science.<br><br>I like to be very knowledgeable about my topic, and if I don't feel confident in it then I won't do it unless I was able to go back to school and learn more and be confident. |
| Personal interest in the content | 41.6 | I feel most confident teaching biology because it's my passion.<br><br>I feel least confident teaching physical science because I just never liked it. |
| Amount of experience teaching the content | 40.3 | I feel comfortable with what I am teaching now since I have taught it for four years.<br><br>I guess [science] is a little intimidating. I'm not really familiar with the subject, not like with reading or something else that I've taught more. |
| Judgment of whether the content was "basic" or "complicated" | 24.7 | I feel prepared to teach pretty much everything at this level because it's all pretty basic.<br><br>I don't feel comfortable teaching physics because it's just so complicated. |
| Grade-level teaching | 20.8 | At least at the third-grade level, I think I'm pretty comfortable teaching out of the science book. I think I understand everything that a third grader would know.<br><br>For what I'm teaching I'm prepared, but if I was going to teach another grade level, I would definitely take those classes so that I could learn what to do. |
| Amount of pedagogical content knowledge regarding the content | 14.3 | I know kind of where they are and what the problems are going to be as they get through the topic.<br><br>I just don't know how to bring it down to a first-grade level. |
| Perceived role of standards and/or mandated curriculum | 14.3 | I'm pretty confident about what the pacing guide tells me, to direct me to teach. I'm going to have left that to the experts to decide that. |

*(Table continues on next page)*

**Table 5.2. (Continued)**

| *Reason (in descending frequency)* | *% Teachers* | *Example(s)* |
|---|---|---|
| Support from others (colleagues, university faculty, and/or others) | 11.7 | A fellow teacher has been a good mentor. He empowers me to be confident in teaching science.<br><br>I feel most confident teaching chemistry because my wife is a bio chemist, so she serves as an extra resource for me. |
| Student reception of the content | 10.4 | I feel really confident teaching about space because the kids just love it. |

Two other factors described by teachers as impacting self-efficacy that were often linked to one another are "judgment of whether the content was 'basic' or 'complicated'" and the context of "grade-level teaching." For example, teachers stated:

> I feel pretty confident teaching the science curriculum that I'm asked to teach. It's pretty basic. I think that if I were asked to teach anything above this primary level where it gets a little more abstract and not so hands-on and not so literal I would have a problem with probably most of it.... It's really just at the very literal level at third grade.

> Chemistry is probably [the content I'm] least [confident teaching]. At the fifth grade level I'm totally comfortable with it.... I can teach what they need to know. I can teach what the standards say they need to know, but I don't feel that I have a strong grasp much beyond what the level I'm teaching is.

Approximately half of the teachers (46.8%) described their undergraduate courses as being related to one or more of the factors contributing to their confidence. Teachers, for example, described undergraduate science courses as providing or not providing content knowledge and/or pedagogical content knowledge, promoting or dissuading interest in the content matter, and/or making certain topics in science seem "basic" or "complicated." These undergraduates' science education experiences were influential factors in the development of self-efficacy beliefs related to effective science teaching and the understanding of the nature of science.

> To what extent do the science teaching self-efficacy beliefs of participating inservice teachers relate to these teachers' beliefs about general and personal science teaching and the reasons for these beliefs, and observed science teaching practices?

## Quantitative Data

Elementary teachers' Science Teaching Efficacy Belief Instrument (STEBI) scores ranged from 38 to 111 (out of a possible 125 points). Within the total STEBI, PSTE scores ranged from 13 to 63 (out of a maximum 65 points) and STOE scores ranged from 18 to 51 (out of a maximum 60 points). The RTOP scores for all elementary teachers ranged from 23 to 92 (out of a maximum 100 points). Analysis of quantitative data revealed that there was no correlation between STEBI (PSTE or STOE) scores and RTOP scores. Inservice elementary teachers with high science teaching efficacy levels as measured by the STEBI were just as likely to be observed teaching in a reformed manner as teachers with low self-efficacy.

The lack of correlation between STEBI and RTOP scores remained even when taking a variety of other factors into account, including gender, grade level, total years of experience, years of experience at current grade level, number of science content courses, and extent of professional development. The only groups of teachers for whom a statistically significant correlation could be found were (a) those teaching sixth grade and above (correlation between STOE and RTOP scores: $r(7) = 0.80$) and (b) those who had taken three or more undergraduate science courses (correlation between PSTE and RTOP scores: $r(4) = 0.86$). The number of teachers in each of these groups (nine and six respectively) is statistically small.

## Elementary Teacher Case Profiles

In order to examine potential reasons for this apparent lack of a quantitative relationship between science teaching efficacy and reformed science teaching, eight teacher case profiles were examined, which included detailed notes and transcriptions from the teachers' observations and interviews. The overall specifics of these case profiles, including teacher codes, which will be referenced subsequently, can be found in Table 5.3.

In examination of the eight teacher case profiles, several key themes related to teacher beliefs interacting with self-efficacy and science teaching behaviors emerged: (1) their meaning of "hands-on," (2) level of classroom control, (3) ongoing university faculty support, (4) beliefs regarding provided curriculum, and (5) availability of resources. These themes, discussed in the following sections, help to explain the differences in the manifestation of high or low reform levels in observed teaching with such disparate levels of self-efficacy. Because not all of these themes were relevant to all eight of the case profile teachers, only those teachers for whom each theme was pervasive are included in the following discussion.

## Table 5.3. Characteristics of Case Profile Teachers

| Teachers | T1 | T2 | T3 | T4 | T5 | T6 | T7 | T8 |
|---|---|---|---|---|---|---|---|---|
| Score Designation | High STEBI / High RTOP | | Low STEBI / High RTOP | | High STEBI / Low RTOP | | Low STEBI / Low RTOP | |
| Total STEBI-A* | 105 | 103 | 45 | 42 | 100 | 95 | 44 | 57 |
| PSTE | 58 | 63 | 21 | 15 | 52 | 54 | 17 | 24 |
| STOE | 47 | 40 | 24 | 27 | 48 | 41 | 27 | 33 |
| RTOP** | 81 | 82 | 84 | 92 | 54 | 48 | 37 | 29 |
| Reform (R)/ Comparison (C) | R | C | C | R | C | C | R | C |
| Gender | Female | Female | Female | Male | Female | Female | Female | Male |
| Type of School | Private | Public | Public | Public | Public | Private | Public | Public |
| Grade Level | 2nd Grade | 3rd Grade | 2nd Grade | 6th Grade | 4th/5th Grade | 3rd Grade | 6th Grade | 4th Grade |
| Total Years Taught | 3 | 1 | 21 | 4 | 11 | 3 | 1 | 4 |
| Years Taught at Current Grade | 3 | 1 | 16 | 2 | 1 | 3 | 1 | 1 |
| Science Background | Several undergrad science courses for preservice teachers | Extensive | Few undergrad science courses (did not remember) | Extensive | None | Few undergrad science courses | Four undergrad science courses for preservice teachers | Four required science courses as an undergrad |

*(Table continues on next page)*

**Table 5.3. (Continued)**

| | | | | | | | | |
|---|---|---|---|---|---|---|---|---|
| **Science Education Background** | One science teaching methods course | Extensive (currently earning her M.S. degree in science education) | None | Extensive (has an M.S. degree in education with a focus on science) | None | One science teaching methods course | Two elementary science education courses | One science teaching methods course |
| **Professional Development** | Some | Extensive | Some | Extensive | None | Some | Some | None |

* Maximum possible score is 125 for the total STEBI-A, 65 for PSTE, and 60 for STOE.

** Maximum possible score for the RTOP is 100.

### *The Meaning of "Hands-on"*

All eight teachers said "hands-on" was the best way to teach science at the elementary level. The teachers, however, varied greatly in their description of what the term "hands-on" means to them and, in turn, the ways that "hands-on teaching" was manifested in their observed lessons. Teachers' beliefs about the meaning of "hands-on" seemed to impact not only the ways in which they were observed teaching science, but also the beliefs about their own efficacy as teachers of science.

The two teachers who had low RTOP scores and high STEBI scores (T5 and T6) both felt they were teaching science in the way that it should be taught, in a "hands-on" manner. Observations and interviews, however, revealed that both of these teachers had images of effective activity-based science teaching that were very different from those of science education researchers. Teacher T6, for example, described working with computers and having her students answer simple questions on the SMARTBoard as "hands-on" and "interactive." Since she was confident in her ability to integrate technology into her science lessons due to her participation in professional development workshops, she felt she was teaching science effectively, both in her general teaching and during the lesson we observed. Her actual observed lesson, however, showed that her activities involved no complex student thought or interaction, nor did she incorporate any of the fundamental abilities of inquiry as described in *Inquiry and the National Science Education Standards* (National Research Council [NRC], 2000). Teacher T6 perpetuated misconceptions about the science content during her lesson, repeating multiple times that the moon is a luminous object and creates its own light. We noted that T6 seemed confused by and chose to ignore that the SMARTBoard did not register this as a correct answer. Similarly, T5 described a lesson in which her fourth and fifth graders blew bubbles as "hands-on," the kind of activity that she "should really be doing all the time for science" (interview transcript for T5). Although her students used their hands to conduct an activity in her observed lesson, little student thought or reflection in connecting the activity to the science content was apparent. With both T5 and T6, therefore, beliefs about effective science teaching, although they did not match a reformed inquiry-oriented view of science teaching, served to bolster these teachers' perceived self-efficacy while decreasing the level of reform in their actual observed teaching practices.

On the other end of the spectrum, all four teachers with high RTOP scores (T1, T2, T3, and T4) described "hands-on" science in a more inquiry-based context that much more closely aligned with current national science education guidelines (Huffman, Thomas, & Lawrenz, 2008; NRC, 1995, 2000). Their observed lessons incorporated a majority of the fundamental abilities of inquiry (NRC, 2000) including multiple lesson segments in which students were encouraged to use higher-level thinking skills and to collaborate with one another. In their observed lessons and in their own descriptions of the ways science should be taught at the elementary level, these four teachers stated that students should be given the opportunity to discuss scientific topics among themselves, make and test predictions, and investigate the answers to their own questions. Statements included:

> I think [the best way to teach science in elementary classrooms is] hands-on. When I was student teaching in the public schools, they loved it when you brought in the stuff, if they could feel it and experiment with it, whether it was a fossil or balls and ramps.... Give them that freedom of asking their questions and experimenting with it, and then draw it into the lesson.... They can try it and if it fails, why? Have them talk about why they think it wouldn't work and what they did see.

> I'm trying to do more hands-on activities like today. I introduced it and modeled it. Then they went and explored the questions that they wanted to answer, to see what happens. They wrote down their data. We came back together, we gathered the data, and we discussed it.

> [The best way to teach science] is hands-on. It's inquiry.... I'm going to go back to when the student has the question, let them go with the question and let them figure out the answer.... Like you saw today, I'm teaching the kids to argue and to discuss and to disagree with each other. And if they disagree they've got to be able to stand up and say, "This is why I disagree," and project their reasons and be able to accept when somebody else says, "Well, I don't think that works. I think it works this way." And [me] kind of stepping out of the circle and letting them come up with the answer.

Through the examination of the interviews with T1, T2, T3, and T4, the differing levels in self-efficacy among them seemed to be primarily due to personal reflection of the effectiveness of their ability to use inquiry in their classrooms. Although all four teachers with high RTOP scores had views of effective science teaching that coincided with those of a reformed inquiry-oriented approach, the two teachers with low STEBI scores were much more concerned as to whether or not they were implementing inquiry in their classrooms effectively. Teachers T3 and T4, with high RTOP scores and low STEBI scores, identified their biggest barrier in planning and teaching science as finding the best ways to effectively implement inquiry-based instruction in their classrooms while appropriately assessing student learning and understanding. On the other hand, T1 and T2, with high RTOP scores and high STEBI scores, were confident in their ability to incorporate inquiry into their science lessons, and so had high perceived self-efficacy for teaching science.

The two teachers with low RTOP scores and low STEBI scores (T7 and T8) were mixed in their description of hands-on science teaching. Teacher T7 had a view that was similar to those of T5 and T6, describing her observed teaching strategy as "hands-on note-taking" in which students silently copied notes into "foldables" she directed them to make out of construction paper. She said this strategy was demonstrative of good teaching practices since the students were "using their hands to learn about science." In the instance of T7, therefore, her lower self-efficacy seemed to be due to her concern that she was, overall, not teaching science effectively in her classroom, rather than during the specific lesson we observed. On the other hand, T8's description of ideal science teaching was much more related to the ideas of teachers with high RTOP scores. Due to his lack of classroom control and his desire to keep his students quiet and orderly, however, T8 expressed a helplessness in being able enact his beliefs about effective science teaching in his own classroom, resulting in low confidence in his ability to effectively teach science and low levels of reform in his observed science teaching.

### *Classroom Control*

The two case profile teachers with low RTOP scores and low STEBI scores (T7 and T8) seemed to be very concerned with keeping tight control of their classrooms. Teacher T7, for example, spent a great deal of time during her observed lesson correcting student behavior and keeping students quiet. Even more extremely, T8 was very uncomfortable with his inability to "control students' excitement in science." He spent a majority of his lesson unsuccessfully trying to keep his students quiet and orderly. Due to his lack of control during science activities in his classroom, T8 seemed to feel little confidence in his or his students' ability to engage in inquiry-oriented activities in the classroom. He consequently much preferred that his students be engaged in activities such as silent reading, where the students could more easily be kept quiet. He noted, "It's so nice and quiet. Reading time is so much better than science time."

All four teachers with high RTOP scores (T1, T2, T3, and T4), on the other hand, regardless of STEBI scores, seemed to be more concerned about the learning needs of their students and not as concerned about messiness or behavioral issues. These teachers were not only more comfortable with the noise level and the amount of student movement involved in inquiry-based science teaching, but also seemed more comfortable overall with handing over a greater amount of control to the students during the observed lessons and in their science classrooms in general. While describing their observed lessons, for example, these teachers stated:

> I knew that they were probably gaining so much from experimenting with the ramps themselves and from their interaction and talking. So it didn't bother me if the noise level rose or if it took longer than I expected. The students let me know what they need and I try to listen.

> We talked about it first.... I tried to just have a conversation about it to get their ideas out there so I had an idea about what they already knew. Then I introduced what we were going to be doing, and we did the experiment with the three bottles, having *them* mix, having *them* give me the predictions before we mixed. I did that so they would be comfortable with making a prediction and not worrying if it was right or wrong. "What do *you* think?" Just using those phrases.... But really letting them know that science is about observing. "What do *you* notice?" "Does it agree with what your prediction or hypothesis was?"

The two case profile teachers with low RTOP scores and high STEBI scores (T5 and T6) were mixed in their focus on classroom management. Teacher T5 seemed comfortable with student–student interaction, although this interaction was ultimately orchestrated by the teacher and focused primarily on social skills rather than science content. In the view of this teacher, it seemed that as long as the students were communicating in some way, regardless of whether or not that communication was about the scientific content of the lesson, she was confident of her effectiveness as a teacher of science. For example, she stated:

> No matter what, whether they understood that they were all circles or not really wasn't [the goal]. What I wanted with that was the interaction that went on outside and the excitement that went on out there. So I was happy with it just based on that.

In contrast, T6 kept things quiet and orderly during her observed lesson, discouraging student talking unless she called upon them. Based on T6's interview, because part of her vision of being an effective science teacher was keeping tight control over her classroom, she had high self-efficacy for her ability to effectively teach science, even though there was little to no student interaction in her science class.

### *Ongoing University Faculty Support*

During their interviews, all four teachers with high RTOP scores described the ongoing support they received from local higher education faculty through reformed-approach coursework, professional development, and/or personal contact. In three of these cases (T1, T2, and T3), this interaction was with faculty members who had participated in the NOVA professional development program. One difference between the teachers in the group with high STEBI scores and those with low STEBI scores was the extent to which higher education faculty members served as mentors to them. That is, both teachers with high RTOP scores coupled with high STEBI scores (T1 and T2) had ongoing personal relationships with local faculty members who provided them with encouragement and support. This seemed to not only boost these teachers' beliefs in their own ability to teach science, but also provided them with continuing resources and guidance to help them teach science more effectively. For the two teachers who had high RTOP scores but low STEBI scores (T3 and T4), higher education faculty had provided the resources and support to teach science but not at a high level of personal encouragement.

### *Beliefs About Provided Curriculum*

Several of the case profile teachers with varying RTOP and STEBI scores differed in their attitudes toward and beliefs about science curricular materials provided (or not provided) by their school districts. Teachers' varying attitudes seemed to impact their confidence in their ability to teach science and, to some extent, the strategies they used to teach science. In some instances teachers indicated that district-mandated kits positively influenced their ability to teach science, although this was not evident in observations of their science lessons.

In one specific example, T7, who had low RTOP and STEBI scores, felt because she did not have the district-provided activity kits for science provided to lower grades at her school., she felt that she lacked materials and guidance that would be helpful for her to teach science effectively. Teacher T5, who had low RTOP and high STEBI scores, seemed to feel empowered by the curricular materials she was given, although she had virtually no science background. She said these materials gave her direction by telling her what she should be teaching, thereby boosting her confidence, even though her observed lesson revealed low levels of reform. As she stated during her interview:

> I read what I have to do and then I do it.... I'm pretty confident about what the pacing guide tells me, to direct me to teach. I'm going to have left that to the experts to decide that.

Teachers with high RTOP scores who used district-mandated kits or other curricular materials treated these materials in a different way. Instead of seeing these materials as something that needed to be followed exactly, these teachers seemed to treat them more as suggested guidelines. Teacher T2, for example, who had high RTOP and STEBI scores, used district-mandated kits, but it was not the driving force of her science instruction. Instead, she seemed comfortable adapting the curriculum and taking the pieces that best served her instructional goals, incorporating aspects of scientific inquiry into her lesson, and encouraging a high level of student thought and reflection whenever possible.

### *Availability of Resources*

The case profile teachers also differed in their reaction to the availability of noncurricular science instructional materials in their classroom, which seemed to influence both their science teaching self-efficacy and their science teaching practices. Both T1 and T6, for example, worked at private schools with extensive resources, especially in the area of technology. Both of these teachers also described the presence of materials for science class, particularly the availability of a SMARTBoard in the classroom, as being a contributor to their perceived ability to successfully teach science. While both teachers had high confidence in their ability to teach science, they differed greatly in the reform level of their observed teaching.

Based on observations, the difference in the reform level between these two teachers seems to be partially due to differences in their background knowledge of science and science education. As an undergraduate major in elementary science and mathematics, T1 had a sound understanding of scientific concepts and science education. This understanding was apparent in the way in which she effectively utilized resources available in her classroom to support students' learning of scientific concepts. During her observed lesson, for example, she used her SMARTBoard as an effective tool to support the lesson's inquiry-based activity in which her students simulated the water pollution recently impacting a nearby river. Teacher T6, on the other hand, centered the majority of her lesson on PowerPoint slides and premade SMARTBoard activities during which she called on individual students to answer low-level questions. Throughout the lesson, T6 did not stray from the provided SMARTBoard materials. She did not expand upon them, could not answer students' questions regarding the concepts (indeed, she discouraged questions), and perpetuated student misconceptions. T6 seemed to believe that by simply using technological resources, she was teaching science in an effective "hands-on" way. Although she had little background in science or science education, T6 seemed confident that she had taught the observed lesson effectively.

## DISCUSSION

The quantitative measurements of the STEBI and the RTOP revealed no significant difference between the science teaching self-efficacy of inservice elementary teachers who took the reform courses and those who did not.

Nor did the data reveal any relationship between teachers' self-efficacy and the level of reform of the reform courses they took as undergraduates.

Several factors may have resulted in a lack of evidence for an overall quantitative relationship between undergraduate course reform and science teaching self-efficacy. One such factor is that the observed course sessions on which RTOP scores were based may not have been representative of the courses that the teachers experienced as undergraduates, some of which the participants took several years ago. Several of the science courses developed through NOVA professional development had changed instructors over time and, as a result, the content and approach of the courses may have strayed from the original concept of the designed courses. In addition, data were not available regarding the reform level of the undergraduate courses taken by those inservice elementary teachers who did not take the reform courses, or the variety of other undergraduate courses that reform and comparison participants completed.

The qualitative data available through observation notes of elementary lessons and interviews with participating preservice and inservice elementary teachers have given us additional insight into some more specific characteristics of reformed undergraduate science courses that may have influenced their self-efficacy beliefs regarding science teaching. The top three factors cited by undergraduate students as influencing their science teaching self-efficacy were essentially the same as those described by inservice teachers, although in a slightly different order: (1) perceived knowledge of the content, (2) teaching experience regarding the content, and (3) interest in the content. The first two factors have both been extensively linked to the science teaching self-efficacy of preservice and inservice teachers (Bhattacharyya et al., 2009; Brand & Wilkins, 2007; Cantrell, Young, & Moore, 2003; Carrier, 2009; Cone, 2009; Enochs et al., 1995; Gunning & Mensah, 2011; Hechter, 2011; Kind, 2009; Perkins, 2007; Ramey-Gassert et al., 1996; Swars & Dooley, 2010). Although interest or positive attitude has seldom been cited as an important factor influencing science teaching self-efficacy (Ramey-Gassert et al., 1996), this factor emerged as an important self-perceived influence on the self-efficacy of participants in our study. The level of personal interest was perceived by participants as directly increasing their confidence in their abilities to teach science and also led many of them to pursue further education and experiences related to the content. Additional education and experience could have led to an increase in both science content knowledge and science teaching experiences.

While the key factors noted by participants as being the most influential on their self-efficacy levels were the same for preservice and inservice teachers, the sources of these factors were described with slight differences for the two groups. All undergraduate focus groups cited their undergraduate courses as having at least some influence on their confidence to teach science. Among inservice teacher participants, however, less than 50% cited their undergraduate course experiences as influencing their confidence in their ability to teach science. More than half of the inservice teachers described factors they experienced after graduation as influencing their self-efficacy. This finding indicates that undergraduate course experiences, and therefore, any characteristics of reformed undergraduate science courses, were described as having had the most influence on participating elementary teachers prior to their entering the elementary classroom. Once they became inservice teachers, these course experiences were described as still having an influence on the self-efficacy of some of them, but for others, this influence faded over time or disappeared altogether as more recent endeavors, including professional development and classroom experiences, took precedence over undergraduate experiences.

Perhaps even more important than how participating elementary teachers' described undergraduate courses' impacts on their self-efficacy, however, is how self-efficacy beliefs were described as translating into behavior in the classroom. Research, therefore, is needed to investigate the question: If undergraduate science courses are successful in increasing teachers' science teaching self-efficacy beliefs, will this lead to more reformed science teaching? For our nationwide data set, we did not see an overall general correlation between self-efficacy as measured by the STEBI-A and reform level in science teaching as measured by the RTOP.

There are several credible reasons for these results. First, it is possible that single observations of each teacher were not representative of the teacher's normal science teaching practices. Researchers, however, attempted to account for this limitation by asking each teacher whether or not the observed lesson was typical of the ways in which they teach science. A second possible explanation is that STEBI-A scores are not representative of teachers' overall level of self-efficacy for teaching science. It has been argued that self-efficacy is a context-dependent construct (Bandura, 1997). The science teaching efficacy beliefs of one teacher may change depending upon the particular group of students he or she is teaching (Angle & Moseley, 2009; Ramey-Gassert et al., 1996; Raudenbush, Rowan, & Cheong, 1992) and the science content being taught (Kind, 2009; Perkins, 2007). In addition, Perkins (2007) demonstrated that preservice elementary teachers' responses to the STEBI survey changed when it was administered multiple times. The teachers who participated in this study may be more or less confident teaching science overall than is represented by the single STEBI-A survey they filled out. Indeed, results of this study indicate that participating inservice teachers had varying levels of self-efficacy for different science content areas and grade levels depending on a variety of factors (Table 5.2). Further research is needed, examining science teaching self-efficacy within these particular contexts, and across time, to gain more understanding of the development of self-efficacy in relation to teaching practices.

The interviews and observations of eight case profile teachers indicated that the relationship between science teaching self-efficacy and practice is much more complex than a quantitative correlation and is impacted by a variety of factors, such as other beliefs teachers hold. Case profile teachers, for example, differed in their beliefs about what it means for a science lesson to be "hands-on," and about the role of district-provided curriculum. Elementary teachers' beliefs regarding what effective science instruction means can have a large impact on self-efficacy beliefs, and thus have an unintended impact on science teaching behaviors, especially if these beliefs do not coincide with those of science education researchers (Bhattacharyya et al., 2009; Haney et al., 2002) or if a mismatch exists between how teachers believe they are teaching and how they are actually teaching (Roehrig, Turner, Grove, Schneider, & Liu, 2009; Wheatley, 2005). The case profile teachers in this study with high levels of confidence in their abilities to teach science, but low levels of reform in their observed teaching, believed that effective science teaching involved activities in which students "used their hands" in some way and a classroom in which the teacher had tight control of his or her students. Since these teachers' self-efficacy beliefs were based upon a different image of effective science teaching than that of proponents of reformed science teaching, these teachers' high confidence in their own abilities to teach science well was not transformed into teaching in a reformed, inquiry-oriented manner.

As observed in other studies (Czerniak & Schriver, 1994; Kind, 2009; Settlage et al., 2009), because case profile teachers with high STEBI scores and low RTOP scores believed they were using teaching practices consistent with effective science instruction, they seemed to feel little need to critically reflect upon their teaching practices. Wheatley (2002) claimed that this lack of personal reflection is one potential negative result of focusing intervention too narrowly on just the elevation of teacher efficacy beliefs. If teachers have no doubts regarding their teaching efficacy, then they will not experience a perturbation, leading them to feel no need to reform their teaching practices even if reformation would improve student learning. This lack of reflection contrasts greatly with case profile teachers who had high reform level in their observed lessons but low self-efficacy levels. These two teachers seemed to be hyper-reflective, critical concerning almost all aspects of their lessons, and particularly focused upon whether or not they were effectively implementing inquiry-based teaching strategies to promote student learning.

Teachers with varying levels of self-efficacy and reform in their science teaching also differed in their judgment of the skills required to be an effective science teacher. This difference impacted their own perceived efficacy as science teachers. Bandura (1997) claimed, "Evaluation of one's self-diagnostic skills requires not only self-knowledge of capabilities but also understanding of the types of skills needed for different activities" (p. 115). Teachers may over- or underestimate their own efficacy based on the self-perceived importance of their skill set. Teacher T5, for example, did not have any formal background in science or science education; she did not take any science content or teaching methods courses and had not participated in any professional development for improving her science teaching. In her eyes, however, these were not important criteria for her to be an effective science teacher. Instead, T5 felt high confidence in her ability to teach science due to her assurance of the effectiveness of the district-provided science curriculum and in her self-perceived ability to meet the specific needs of her students. In contrast, T3 also had very little formal science content or science education background, but saw this as a detriment to her ability to teach science well. Teacher T3 consequently lacked confidence in her ability to effectively teach science to her students, although her observed science lesson demonstrated a high level of reformed inquiry-oriented teaching. In the case of T3, her self-perceived lack of content knowledge, while decreasing her self-efficacy, also seemed to contribute to the effectiveness of her teaching practices. Because she perceived a deficiency in her own background knowledge, she felt the need to critically reflect on the effectiveness of her own science teaching and seek out aid from others regarding professional development.

Interestingly, for undergraduate students participating in this study, nearly one quarter of the focus groups indicated that reformed undergraduate science courses made them feel less confident in their abilities to teach science by making them realize how much work they still needed to do in order to learn the skills necessary to become effective teachers of science. It may be that undergraduate science courses taught in a reform-based manner help to encourage self-reflection among preservice teachers by modeling appropriate skills needed to teach science in a reformed research-based way, though more research is needed to examine this relationship in further detail.

## CONCLUSIONS

It is important and worthwhile to examine the factors of reformed undergraduate courses such as those courses developed through the NOVA professional development program and how they impact the development of elementary teachers' science teaching self-efficacy. Our study provides evidence that undergraduate students do perceive specific characteristics of reformed undergraduate science courses to influence their level of science teaching self-efficacy. In particular, both inservice and preservice teachers participating in our study reported to develop greater science teaching self-efficacy from undergraduate courses that increased their science content knowledge, explicitly connected the content to teaching, and made learning about science enjoyable. However, it is interesting that, while some of the inservice teachers in this study continued to feel a large influence from their undergraduate science courses over time, others described their undergraduate education to have little or no influence on their current self-efficacy as teachers of science. Further information is needed regarding how self-efficacy changes and develops as elementary teachers move from preservice education into inservice practice.

Perhaps even more important than the influence of undergraduate courses on self-efficacy beliefs, however, is how these beliefs translate into practice. One of the most critical outcomes of this study is that a straightforward correlation between science teaching self-efficacy and *observed* reformed inquiry-oriented teaching was not seen for this large nationwide group of teachers with widely varying backgrounds. A blanket assumption, therefore, cannot be made that increasing the efficacy beliefs of preservice and inservice elementary teachers will automatically improve their ability to effectively teach science to their students. The relationship between science teaching efficacy beliefs and science teaching behaviors is much more complex than we might assume. It is therefore important that we examine more critically not only which beliefs and how these beliefs are impacted by our courses, but also how the beliefs translate into practice.

This study highlights the importance of the use of interviews and observations in self-efficacy research and the caution to not rely solely on quantitative instruments and/or self-reported data. Further research, including longitudinal studies with multiple interviews and observations that focus on existing relationships between self-efficacy and teaching practices, is needed if we are to make accurate and informed decisions about how to develop effective elementary teachers of science.

## ACKNOWLEDGMENT

This material is based upon work supported by the National Science Foundation under Grant No. 0554594. Any opinions, findings, and conclusions or recommendations expressed in this material are those of the author(s) and do not necessarily reflect the views of the National Science Foundation

## REFERENCES

Allinder, R. M. (1994). The relationship between efficacy and the instructional practices of special education teachers and consultants. *Teacher Education and Special Education, 17,* 86–95.

Angle, J., & Moseley, C. (2009). Science teacher efficacy and outcome expectancy as predictors of students' end-of-instruction (EOI) Biology I test scores. *School Science and Mathematics, 109*(8), 473–483.

Ashton, P. (1984). Teacher efficacy: A motivational paradigm for effective teacher education. *Journal of Teacher Education, 35*(5), 28–32.

Ashton, P., & Webb, R. (1986). *Making a difference: Teachers' sense of efficacy and student achievement.* New York, NY: Longman.

Bandura, A. (1982). Self-efficacy mechanism in human agency. *American Psychologist, 37,* 122–147.

Bandura, A. (1997). *Self-efficacy: The exercise of control.* New York, NY: W. H. Freeman and Company.

Bandura, A. (2001). Social cognitive theory: An agentic perspective. *Annual Review of Psychology, 52,* 1–26.

Bhattacharyya, S., Volk, T., & Lumpe, A. (2009). The influence of an extensive inquiry-based field experience on pre-service elementary teachers' science teaching beliefs. *Journal of Science Teacher Education, 20,* 199–218.

Brand, B. R., & Wilkins, J. L. M. (2007). Using self-efficacy as a construct for evaluating science and mathematics methods courses. *Journal of Science Teacher Education, 18,* 297–317.

Brickhouse, N. (1990). Teachers' beliefs about the nature of science and their relationship to classroom practice. *Journal of Research and Development in Education, 15*(4), 13–18.

Bryan, L. A. (2003). Nestedness of beliefs: Examining a prospective elementary teacher's belief system about science teaching and learning. *Journal of Research in Science Teaching, 40,* 835–868.

Caprara, G. V., Barbaranelli, C., Steca, P., & Malone, P. S. (2006). Teachers' self-efficacy beliefs as determinants of job satisfaction and students' academic achievement: A study at the school level. *Journal of School Psychology, 44,* 473–490.

Cantrell, P., Young, S, & Moore, A. (2003). Factors affecting science teaching efficacy of preservice elementary teachers. *Journal of Science Teacher Education, 14*(3), 177–192.

Carrier, S. J. (2009). The effects of outdoor science lessons with elementary school students on preservice teachers' self-efficacy. *Journal of Elementary Science Education, 21*(2), 35–48.

Coldarci, T. (1992). Teachers' sense of efficacy and commitment to teaching. *Journal of Experimental Education, 60,* 323–337.

Cone, N. (2009). A bridge to developing effacacious science teachers of *all* students: Community-based service-learning supplemented with explicit discussions and activities about diversity. *Journal of Science Teacher Education, 20,* 365–383.

Cronin-Jones, L. L. (1991). Science teacher beliefs and their influence on curriculum implementation: Two case studies. *Journal of Research in Science Teaching, 28*(3), 235–250.

Czerniak, C. M., & Lumpe, A. T. (1996). Relationship between teacher beliefs and science education reform. *Journal of Science Teacher Education, 7*(4), 247–266.

Czerniak, C. M., & Schriver, M. L. (1994). An examination of preservice science teachers' beliefs and behaviors as related to self-efficacy. *Journal of Science Teacher Education, 5*(3), 77–86.

Dembo, M. H., & Gibson, S. (1985). Teachers' sense of efficacy: An important factor in school improvement. *The Elementary School Journal, 86*, 173–184.

Desouza, J. M. S., Boone, W. J., & Yilmaz, O. (2004). A study of science teaching self-efficacy and outcome expectancy beliefs of teachers in India. *Science Education, 88*, 837–854.

Duschl, R. A. (1983). The elementary level science methods course: Breeding ground of apprehension toward science? *Journal of Research in Science Teaching, 20*(8), 745–754.

Eisenhart, M. A., Shrum, J. L., Harding, J. R., & Cuthbert, A. M. (1988). Teacher beliefs: Definitions, findings, and directions. *Educational Policy, 2*(1), 51–70.

Enochs, L. G., & Riggs, I. M. (1990). Further development of an elementary science teaching efficacy belief instrument: A preservice elementary scale. *School Science and Mathematics*, 90(8), 694–712.

Enochs, L. G., Scharmann, L. C., & Riggs, I. M. (1995). The relationship of pupil control to preservice elementary science teacher self-efficacy and outcome expectancy. *Science Education, 79*(1), 63–75.

Fang, Z. (1996). A review of research on teacher beliefs and practices. *Educational Research, 38*(1), 47–65.

Gess-Newsome, J. (1999). Pedagogical content knowledge: An introduction and orientation. In J. Gess-Newsome & N. G. Lederman (Eds.), *Examining pedagogical content knowledge: The construct and its implications for science education* (pp. 3–17). Dordrecht, the Netherlands: Kluwer Academic.

Ghaith, G., & Yaghi, H. (1997). Relationships among experience, teacher efficacy, and attitudes toward the implementation of instructional innovation. *Teaching and Teacher Education, 13*(4), 451–458.

Gibson, S., & Dembo, M. H. (1984). Teacher efficacy: A construct validation. *Journal of Educational Psychology, 76*(4), 569–582.

Gunning, A. M., & Mensah, F. M. (2011). Preservice elementary teachers' development of self-efficacy and confidence to teach science: A case study. *Journal of Science Teacher Education, 22*, 171–185.

Guo, Y., Piasta, S. B., Justice, L. M., & Kaderavek, J. N. (2010). Relations among preschool teachers' self-efficacy, classroom quality, and children's language and literacy gains. *Teaching and Teacher Education, 26*, 1094–1103.

Guskey, T. R. (1988). Teacher efficacy, self-concept, and attitudes toward the implementation of instructional innovation. *Teaching and Teacher Education, 4*(1), 63–69.

Haney, J. J., Lumpe, A. T., Czerniak, C. M., & Egan, V. (2002). From beliefs to actions: The beliefs and actions of teachers implementing change. *Journal of Science Teacher Education, 13*(3), 171–187.

Hechter, R. P. (2011). Changes in preservice elementary teachers' personal science teaching efficacy and science teaching outcome expectancies: The influence of context. *Journal of Science Teacher Education, 22*, 187–202.

Huffman, D., Thomas, K., & Lawrenz, F. (2008). Science and mathematics instruction in a reform-based teacher preparation program. *School Science and Mathematics, 108*(4), 137–148.

Kagen, D. M. (1992). Implications of research on teacher belief. *Educational Psychologist, 27*(1), 65–90.

Kind, V. (2009). A conflict in your head: An exploration of trainee science teachers' subject matter knowledge development and its impact on teacher self-confidence. *International Journal of Science Education, 31*(11), 1529–1562.

Klassen, R. M., & Chiu, M. M. (2010). Effects on teachers' self-efficacy and job satisfaction: Teacher gender, years of experience, and job stress. *Journal of Educational Psychology, 102*(3), 741–756.

Lakshmanan, A., Heath, B. P., Perlmutter, A., & Elder, M. (2011). The impact of science content and professional learning communities on science teaching efficacy and standards-based instruction. Journal of Research in Science Teaching, 48(5), 534–551.

Levitt, K. E. (2001). An analysis of elementary teachers' beliefs regarding the teaching and learning of science. *Science Education, 86*(1), 1–22.

Mansour, N. (2010). Impact of the knowledge and beliefs of Egyptian science teachers in integrating a STS based curriculum: A sociocultural perspective. *Journal of Science Teacher Education, 21*, 513–534.

Marshall, J. C., Horton, R., Igo, B. L., & Switzer, D. M. (2009). K–12 science and mathematics teachers' beliefs about and use of inquiry in the classroom. *International Journal of Science and Mathematics Education*, 7, 575–596.

Muijs, D., & Reynolds, D. (2002). Teachers' beliefs and behaviors: what really matters? *Journal of Classroom Interaction, 37*(2), 3–15.

National Research Council. (1996). *National science education standards*. Washington, DC: National Academy Press.

National Research Council. (2000). *Inquiry and the national science education standards*. Washington, DC: National Academy Press.

Ogan-Bekiroglu, F., & Akkoc, H. (2009). Preservice teachers' instructional beliefs and examination of consistency between beliefs and practices. *International Journal of Science and Mathematics Education*, 7, 1173–1199.

Pajares, F. (1992). Teachers' beliefs and educational research: Cleaning up a messy construct. *Review of Educational Research, 62*(3), 307–332.

Perkins, C. J. (2007). *Toward making the invisible visible: Studying science teaching self-efficacy beliefs*. Unpublished doctoral dissertation, Oregon State University, Corvallis.

Philipp, R. A. (2007). Mathematics teachers' beliefs and affect. In F. K. Lester, Jr. (Ed.), *Second handbook of research on mathematics teaching and learning* (pp. 257–315). Charlotte, NC: Information Age.

Ramey-Gassert, L., Shroyer, M. G., & Staver, J. R. (1996). A qualitative study of factors influencing science teaching self-efficacy of elementary level teachers. *Science Education, 80*(3), 283–315.

Raudenbush, S. W., Rowan, B., & Cheong, Y. F. (1992). Contextual effects of the self-perceived efficacy of high school teachers. *Sociology of Education, 65*(2), 150–167.

Richardson, V. (1996). The role of attitudes and beliefs in learning to teach. In J. Sikula, T. Buttery, & E. Guyton (Eds.), *Handbook of research on teacher education* (pp. 102–119). New York, NY: Simon & Schuster Macmillan.

Riggs, I. M., & Enochs, L. G. (1990). Toward the development of an elementary teacher's science teaching efficacy beliefs instrument. *Science Education, 74*(6), 625-637.

Riggs, I. M., Enochs, L. G., & Posnanski, T. J. (1998, April). *The teaching behaviors of high versus low efficacy elementary science teachers.* Paper presented at the annual meeting of the National Association of Research in Science Teaching, San Diego.

Roehrig, A. D., Turner, J. E., Grove, C. M., Schneider, N., & Liu, Zhu. (2009). Degree of alignment between beginning teachers' practices and beliefs about effective classroom practices. *The Teacher Educator, 44,* 164–187.

Ross, J. A. (1992). Teacher efficacy and the effect of coaching on student achievement. *Canadian Journal of Education, 17*(1), 51–65.

Sawada, D., & Pilburn, M. (2000). *Reformed teaching observation protocol (RTOP).* (ACEPT Technical Report No. IN00-1). Tempe, AZ: Arizona Collaborative for Excellence in the Preparation of Teachers.

Scribner, J. (1999). Teacher efficacy and teacher professional learning: Implications for school leaders. *Journal of School Leadership, 9,* 209–234.

Settlage, J., Southerland, S. A., Smith, L. K., & Ceglie, R. (2009). Constructing a doubt-free teaching self: Self-efficacy, teacher identity, and science instruction within diverse settings. *Journal of Research in Science Teaching, 46*(1), 102–125.

Skaalvik, E. M., & Skaalvik, S. (2010). Teacher self-efficacy and teacher burnout: A study of relations. *Teaching and Teacher Education, 26,* 1059–1069.

Solomon, D., Battistich, V., & Hom, A. (1996). Teacher beliefs and practices in schools serving communities that differ in socioeconomic level. *The Journal of Experimental Education, 64,* 327–347.

Sunal, D. W., Sunal, C. S., Whitaker, K. W., Freeman, L. M., Odell, M., Hodges, J., Edwards, L., & Johnston, R. (2001). Teaching science in higher education: Faculty professional development and barriers to change. *School Science and Mathematics, 101*(5), 246–257.

Sunal, D. W., Sunal, C. S., Mason, C. L., Lardy, C., Zollman, D., & Matloob-Haghanikar, M. (2009). How are we reforming teaching in undergraduate science courses? *Journal of College Science Teaching, 39*(2), 12–14.

Swars, S. L., & Dooley, C. M. (2010). Changes in teaching efficacy during a professional development school-based science methods course. *School Science and Mathematics, 110*(4), 193–202.

Thompson, A. G. (1992). Teachers' beliefs and conceptions: A synthesis of the research. In D. A. Grouws (Ed.), *Handbook of Research on Mathematics Teaching and Learning* (pp. 127–147). New York, NY: MacMillan.

Tobin, K. (1993). Constructivist perspectives on teacher learning. In K. Tobin (Ed.), *The practice of constructivism in science education.* Hillsdale, NJ: Lawrence.

Tobin, K., Tippins, D., & Gallard, A. J. (1994). Research on instructional strategies for teaching science. In D. Gabel (Ed.), *Handbook of research on science teaching and learning*. New York, NY: MacMillan.

Tschannen-Moran, M., Hoy, A., & Hoy, K. W. (1998). Teacher efficacy: Its meaning and measure. *Review of Educational Research, 68*(2), 202–248.

Tschannen-Moran, M., & McMaster, P. (2009). Sources of self-efficacy: Four professional development formats and their relationship to self-efficacy and implementation of a new teaching strategy. *Elementary School Journal, 110*(2), 228–245.

Ware, H., & Kitsantas, A. (2007). Teacher and collective efficacy beliefs as predictors of professional commitment. *Journal of Educational Research, 100*(5), 303–310.

Weiss, I. R., Pasley, J. D., Smith, S. P., Banilower, E. R., & Heck, D. J. (2003). *Looking inside the classroom: A study of K–12 mathematics and science education in the United States*. Chapel Hill, NC: Horizon Research, Inc.

Wheatley, K. F. (2002). The potential benefits of teacher efficacy doubts for educational reform. *Teaching and Teacher Education, 18*, 5–22.

Wheatley, K. F. (2005). The case for reconceptualizing teacher efficacy research. *Teaching and Teacher Education, 21*(7), 747–766.

# PART B

## RESEARCH-BASED STRATEGIES

CHAPTER 6

# CONSTRUCTIVISM IN CONTEXT

## Factors Affecting Student Learning in Four Different Classrooms

**Emily J. Borda, Kayla Anzalone, Mathew T. Lockett, and Siri Wuotila**

**ABSTRACT**

We describe a case study in which learning and student satisfaction outcomes were compared for four courses taught by the same instructor. Three of these courses used similar versions of a constructivist chemistry curriculum, while the fourth was taught via a traditional lecture format. These comparisons have enabled us to assess the overall effectiveness of a constructivist curriculum as well as the roles of class size and structure and student population in determining the outcomes of the curriculum. Our data suggest the constructivist curriculum was more effective than the traditional curriculum in all contexts, but that its effectiveness was mitigated by larger class size and varied student motivations.

*Research Based Undergraduate Science Teaching*, pp. 191–219

## INTRODUCTION

It is well established that students have difficulty learning chemistry for deep understanding and relegate chemistry knowledge to rote processes and memorized facts (Nakhleh, 1993; Nakhleh & Mitchell, 1993; Nurrenbern & Pickering, 1987). Instructional innovations to help students learn for deeper understanding often are based on constructivism, a theoretical framework for learning that assumes students mentally build and transform ideas rather than simply absorbing them (Piaget, 1977; Vygotsky, 1978). While constructivist-based teaching approaches have been widely accepted among science educators (Bransford, Brown, & Cocking, 2000), they pose challenges when implemented in classrooms, such as lack of student understanding of constructivism, difficulty in facilitating productive discourse, and challenges with assessment (Windschitl, 2002). In higher education, these dilemmas are often heightened by larger class sizes and varying student expectations, backgrounds, and motivations. Here, we explore these challenges by comparing different courses in which a constructivist chemistry curriculum was implemented at the higher education level.

### Constructivism and Conceptual Change

Contemporary views of constructivism are rooted in the work of Jean Piaget and Lev Vygotsky. The constructivist theory developed by Piaget posits learning as a process of creating and transforming conceptions and knowledge structures rather than passively assimilating knowledge (Carey, 1987; Piaget, 1970, 1971, 1973, 1977). A student's experiences form the basis upon which to construct knowledge (Driver & Erickson, 1983), and therefore, a central role of the instructor in a constructivist environment is to provide students with experiences that trigger conceptual change (Driver & Oldham, 1986). Importantly, Piaget's theory focuses on individual cognitive processes and is essentially autonomous (Piaget, 1977).

Social constructivism is a school of thought developed by Vygotsky (Cole, 1990; Lemke, 2001; Vygotsky, 1962, 1978). Central to Vygotsky's model is the idea that knowledge is constructed within communities of practice collaborating on a learning task. Knowledge exists outside of any single individual's mind and instead, emerges as a tapestry of ideas created by a group. Though the two schools of thought differ in emphasis, Piaget's centering on the individual and Vygotsky's focusing on social interaction, they are not mutually exclusive. A unifying principle of both schools of thought is that knowledge is built through active processes rather than transmitted directly from teacher to student (Marin, Benarroch, & Gomez, 2000).

Constructivist theories describe what learning is but do not prescribe specific instructional methodologies. Constructivist curriculum and instruction can therefore take a variety of forms. Because they are rooted in the same theory of learning, however, they all have certain elements in common. Four such elements were identified by Baviskar, Hartle, and Whitney (2009) from a review of literature describing constructivist learning programs. These are: (1) eliciting prior knowledge, in which students make their incoming ideas transparent to themselves, their instructor, and/or their peers; (2) creating cognitive dissonance, in which students engage in tasks that may generate dissatisfaction with their prior knowledge; (3) application of the knowledge with feedback, in which students check the validity of their new knowledge and incorporate it into their existing conceptual framework; and (4) reflection on learning, in which students become aware of how their conceptions have changed.

The development of constructivist theories of learning initiated research into the processes by which students adopt and modify their conceptions. Theories about how these processes work are called conceptual change models (CCMs). Constructivist theories and CCMs are closely related in that constructivist theories primarily describe *what learning is* while CCMs seek to describe in detail *how learning occurs* within a constructivist framework. One of the most influential CCMs describes a rational process whereby students replace old ideas with new ones when evidence demands it (Posner, Strike, Hewson, & Gertzog, 1982). Learners essentially compare concepts and adopt the one that has higher status in terms of intelligibility (whether it is understandable to the learner), plausibility (whether it can explain a variety of phenomena), and fruitfulness (the extent to which it can suggest new questions to explore). A later CCM included four motivational factors that control the extent to which students participate in the process of conceptual change: interest and value beliefs, self-efficacy, control beliefs, and goal orientations (Pintrich, Marx, & Boyle, 1993). Of the three defined types of goal orientations—(1) social goals, in which students strive to please teachers and peers; (2) performance goals, in which students strive to attain a high score or grade; and (3) mastery goals, in which students strive to gain understanding of course content—mastery goals are most closely tied to student performance (Anderman & Maehr, 1994; Pintrich, 2000). Since the initial revision of the CCM, other researchers have added other motivational components, including experience, goals, dispositions (Demastes, Good, & Peebles, 1995; Park, 2007), and cultural factors (Aikenhead & Jegede, 1999).

## Personal Epistemology

Classrooms are laden with implicit understandings of what teaching and learning should look like, the types of discourse that are valued, and what assessment should look like, among other factors (Bruner, 1996; Joseph, Bravmann, Windschitl, Mikel, & Green, 2000). Many classroom cultures in higher education cue students to adopt nonconstructivist views of learning, by featuring the instructor as a transmitter and students as receptors of knowledge, and assess students according to the correctness of their factual knowledge (Heckman, 1987; Sirotnik, 1983; Tyack & Cuban, 1995). When an instructor introduces a constructivist approach into such an environment, he or she must negotiate what Windschitl (2002) terms a "cultural dilemma," in which students' beliefs about learning differ from the constructivist model. Part of negotiating a cultural dilemma involves addressing students' tacit understandings of what knowledge is and what learning should look like. These understandings are defined as a learner's personal epistemology (Hofer & Pintrich, 1997; Elby, 2009). Commonly cited dimensions of one's personal epistemology include how he or she thinks about (1) the certainty of knowledge, ranging from unchanging to tentative; (2) the source of knowledge, ranging from authority to empirical evidence and reasoning; and (3) the structure of knowledge, ranging from isolated facts to integrated knowledge (Bendixen & Rule, 2004; Elby, 1999; Hammer, 1994; Hofer & Pintrich, 1997; Perry, 1970; Schommer-Aikins, 2004). Studies suggest that a sophisticated epistemology is positively correlated to conceptual change (May & Etkina, 2002).

Many undergraduate students, especially those in introductory courses, have been shown to hold naïve epistemological beliefs that work against their learning in a constructivist curriculum (Elby, 1999, 2001; Hammer, 1994; Perry, 1970). The term "epistemological framing" was defined by Hutchison and Hammer (2010) as the way students view a classroom activity, ranging from the unproductive view of "classroom game," in which the goal is to come to a correct answer predetermined by the instructor, to a productive view of "sense-making," in which the goal is to construct meaning from a physical evidence, as scientists do. Evidence from their study suggests students can revert to unproductive framing if the instructor uses even subtle cues asserting him or herself as an authority figure looking for a specific answer. Other studies suggest personal epistemologies to be context-dependent (Elby, 2009; Hammer & Elby, 2002; Leach, Millar, Ryder, & Sere, 1999; Louca, Elby, Hammer, & Kagey, 2004), suggesting it is possible to encourage epistemological sophistication under the right conditions.

## Chemistry for the Informed Citizen Curriculum

The goal of this study is to assess the implementation of a constructivist chemistry curriculum, *Chemistry for the Informed Citizen* (CIC), in different classroom settings. Initially designed to be implemented in a chemistry course for preservice elementary teachers, CIC has learning targets identified with the aid of national standards (American Association for the Advancement of Science, 1990, 1993; National Research Council, 1996). Constructivist learning activities specifically designed to meet the chemistry learning targets were adapted from existing materials (Berkheimer, Anderson, & Blakeslee, 1988; Goldberg, Bendall, Heller, & Poel, 2006; Paradis, 2007) that target known misconceptions about the particulate nature of matter and chemical change (Kind, 2004; Gilbert, De Jong, Justi, Treagust, & van Driel, 2002; Nakhleh, 1992; Yezierski & Birk, 2006; Zoller, 1990). Table 6.1 illustrates how each of the CIC activities meet the four constructivist criteria identified by Baviskar et al. (2009).

**Table 6.1. Constructivist Curriculum Elements**

| *Constructivist Element* | *Description* | *Example for Learning Target: Matter is Made of Small Particles* |
|---|---|---|
| Prior knowledge | Initial ideas sections contain questions carefully crafted to probe students' existing small particle models | Students: (a) draw representations of what various substances would look like on a scale smaller than can be seen using the most powerful microscope and (b) make a prediction about the total volume of a mixture made from 20 ml each of ethanol and water. |
| Cognitive dissonance | In investigation sections, students collect evidence, instructor performs demonstration, or students are provided evidence in curriculum | Students weigh one graduated cylinder containing 20 ml ethanol and another containing 20 ml water, then combine the two liquids into one cylinder and weigh the mixture and both cylinders. They see that the volume of the mixture is less than 40 ml, but that its weight is additive. |
| Application | Constructing explanations sections contain questions to aid students in interpreting evidence and prompt them to construct small particle models to explain the evidence | Students mix 20 ml of sand and 20 ml of pebbles, observe a volume reduction, and then are asked how the sand and pebbles are similar to water and ethanol. They are then prompted to draw a representation of what the water and ethanol look like alone then mixed, on a scale smaller than can be seen using the most powerful microscope. |
| Reflection | In revisiting initial ideas sections, students are prompted to comment on how their ideas have changed | Students look back at their initial drawings of air, water, and salt, and comment on whether or not they drew small particles. |

After several iterations with preservice teachers, the CIC curriculum was adapted for Chemistry 101, which serves a broader population of non-science majors and has a maximum enrollment of 96 students. Chemistry 101 students met five hours per week, three in a large lecture hall and two in smaller laboratory sections. The activities in the science education version of CIC were transformed into laboratory activities, to be completed in the lab portions of Chemistry 101, and inquiry-based classroom activities (ICAs), to be completed in the lecture hall. The investigations and sense-making discussions were structured differently for each context (Table 6.2), but each type of activity met the four constructivist criteria (Baviskar et al., 2009) by incorporating questions that guide students to construct ideas based on evidence.

**Table 6.2. Pedagogies Employed in Each Type of Activity**

| | *SCED* | *CHEM* | |
|---|---|---|---|
| | *Activities (classroom/laboratory)* | *Labs (laboratory)* | *ICAs (lecture hall)* |
| Investigations | Carried out by students | Carried out by students | Carried out by instructor as demonstrations, or data are supplied to students |
| Sense-making discussions | Instructor led whole-class discussions about selected CIC prompts after small group discussions of same prompts | Check-ins in with instructor or TAs at prescribed points. Instructor and TAs use prompts designed to probe for common misconceptions and difficulties | Instructor led whole-class discussions or interactive lectures about selected CIC prompts after small group discussions of same prompts |

Alongside these pedagogical changes, the chemistry content was adapted to meet the broader needs of the general education audience in Chemistry 101. More emphasis was placed on explaining and classifying chemical bonds and using ideas about chemical bonding to explain physical and chemical properties. Because chemical bonding does not appear in elementary or middle school science standards (AAAS, 1993; National Research Council, 1996), it was deemphasized in the science education course. Students instead spent the last two weeks of the course conducting independent investigations in order to gain familiarity with open inquiry methods they might use with future students. A "fossil fuels" strand was introduced in Chemistry 101 to help that audience make connections to current societal issues. This strand consisted of additional ICAs in which students applied chemistry and nature of science concepts to issues like oil spills, oil refining, and interpreting the Intergovernmental Panel on Climate Change (IPCC) report on climate change (Solomon, Qin, Manning, Chen, Marquis, & Averyt, 2007). It is certainly true that all students would benefit from engaging in these types of discussions; however, the science–technology–society content of these activities is not aligned with elementary standards. The inclusion of an open inquiry project was therefore deemed more important for the preservice teachers given the limited time for the course. Finally, both curricula included explicit instruction on the nature of science (NOS), in which students discussed the difference between observation and inference, the role of creativity in science, and the lack of a universal scientific method among other NOS tenets (Lederman, Abd-El-Khalick, Bell, & Schwartz, 2002).

## Research Questions

This study describes a comparison of four different courses: one traditionally taught Chemistry 101 course and three offerings of CIC. One CIC course took place in Chemistry 101 and two in courses for preservice elementary teachers. The three CIC courses differed by size, structure (separate vs. integrated lab and discussion activities), and types of students enrolled. We compared content knowledge and student perception data to address two research questions: (1) How does a constructivist curriculum impact undergraduates differently from traditional lecture-based instruction? and (2) How do class size, structure, and student population influence the learning of undergraduates in a constructivist environment? Although these questions are not new, the implementation of constructivist curricula in undergraduate science courses lags far behind K–12 classrooms and often appears in only a portion of the course (laboratory, tutorials, and/or recitation sections). We believe the holistic integration of constructivism described here adds to the current literature on constructivist curriculum and methods. Furthermore, the availability of data from different types of courses with the same instructor is rare at the higher education level and gives us a unique ability to observe how different groups of students interact with such curricula in different circumstances. It is important to note the purpose of this study is not to undertake a detailed evaluation of the effectiveness of the CIC curriculum. Evaluation and revision of the curriculum are ongoing components of the CIC project. Instead, we are interested in studying the influence of contextual factors on how a reasonably mature version of a constructivist curriculum is received.

## METHOD

### Context of the Study

The courses chosen for this study are a subset of courses taught using the CIC curriculum at a mid-sized university on the west coast. They were selected in order to control as many variables as possible. All four courses were taught by the same instructor (the first author). The CHEM Trad course was the most recent offering of the traditional lecture-style Chemistry 101 curriculum before CIC was implemented. The CHEM CIC course was the most recent offering of the CIC version of Chemistry 101 that was taught by the first author at the time this study was undertaken. The SCED CIC A and B courses were the most recent offerings in the chemistry course for preservice teachers, with B occurring after A. The SCED CIC A course had a typical enrollment in which most students were preservice elementary teachers, while B had a much lower preservice teacher enrollment, with a significant subset of students taking the course to fulfill a general education requirement. Table 6.3 summarizes the enrollment of each course, as well as the types of comparisons that can be made with this selection of four courses.

**Table 6.3. Participants in Each Course and Possible Comparisons**

| | | | | *Comparisons (effects of the following on student learning)* | | |
|---|---|---|---|---|---|---|
| | *n* | *% pre-Service Teachers* | *% Upper-Classmen* | *Curriculum* | *Class Size/ Structure* | *Student Population* |
| CHEM | | | | | | |
| Trad | 86/75[a] | 6/7[b] | 21/21[b] | X | | |
| CIC | 91/44 | 15/16 | 16/23 | X | X | |
| SCED | | | | | | |
| CIC A | 25/22 | 96/95 | 96/95 | | | X |
| CIC B | 13/12 | 62/58 | 69/67 | | X | X |

a. Number of students enrolled in the course/number of students who gave permission to participate in the study.
b. Percent of students in each category enrolled in the course/percent of students in each category who gave permission to participate in the study.

Because the two chemistry courses had approximately the same student populations but different curricula, their comparison helped us answer the first research question: How does a constructivist curriculum impact students differently from traditional lecture-based instruction? The other courses were compared to answer the second question: How do class size, structure, and student population influence the learning of undergraduate students in a constructivist environment? Because SCED CIC B was more similar in student population than A to the constructivist chemistry course, those two courses were compared to assess the effects of class size and structure on student learning. The populations of these two courses were still, to be sure, significantly different: SCED CIC B still had a higher percentage of preservice teachers, most of whom were upperclassmen, than CHEM CIC. We justify this comparison by noting that SCED CIC B represents the lowest preservice teacher to general education student ratio in the history of SCED CIC offerings to the date of this manuscript's publication (SCED CIC, after all, is marketed almost exclusively to those with an education interest). As we shall see, student satisfaction data for SCED CIC B compare more closely to CHEM CIC than SCED CIC A, suggesting the inclusion of only a few non-preservice teachers can change the dynamics of a small class quite dramatically. Although this comparison is not perfect, we believe important qualitative insights can be extracted from it. Finally, the two SCED courses consisted of significantly different populations in terms of the number of preservice teachers taking the course versus those taking the course to fulfill a general education requirement, as well as how far along the students were in their academic programs (e.g., underclassmen vs. upperclassmen). Their comparison therefore yielded information about the effects of motivational factors and general academic maturity on students' learning and perceptions of CIC.

A comparison of the content between the four courses allowed us to further contextualize our results. Table 6.4 describes the approximate number of hours spent on each major topic in each course, as well as the number of questions on the assessment instrument (Chemical Concept Questionnaire, CCQ), that target each topic.

## Data Collection and Analysis

Two factors were measured in all courses. The first, conceptual understanding of chemistry content, is directly related to the goals of the CIC curriculum. The second, students' perceptions, gives us information pertaining to how students felt about what, to them, may have been a significant departure from traditional undergraduate-level curricula.

### *Conceptual Understanding*

Selected items from two instruments, the *Chemical Concept Inventory* (CCI) (Mulford & Robinson, 2002) and the *Particulate Nature of Matter* (ParNoMa) questionnaire (Yezierski & Birk, 2006), were combined to create the *Chemical Concept Questionnaire* (CCQ), which was used to assess students' conceptual understanding of chemistry content. Cronbach's alpha was determined to be 0.76 for the instrument.

**Table 6.4. Content of Each Course and Alignment to Assessment Questions on the CCQ**

| | *# Hours of Instruction* | | | | *# Questions* |
|---|---|---|---|---|---|
| | *CHEM*[a] | | *SCED* | | *CCQ* |
| | *Trad* | *CIC* | *CIC A* | *CIC B* | |
| Particulate nature of matter | | 2 (L) | 5 | 7 | 3 |
| Size of small particles | | 1 (C) | 2 | 2 | 1 |
| Atomic structure | 1 (C) | 2 (C) | 2 | 3 | |
| Kinetic theory of gases | | 2 (L) | 4 | 4 | 1 |
| Periodic table | 2 (C) | 1 (C) | 3 | | |
| Chemical bonding | 3 (C) | 4 (L, C) | | | |
| Intermolecular forces | 2 (C) | 3 (L, C) | 5 | 6 | |
| Phase transitions | | 2 (C) | 4 | 4 | 5 |
| Solutions and dissolving | 4 (C) | 3 (L) | 4 | 5 | |
| Conservation of matter | 4 (C, L) | 4 (L, C) | 6 | 6 | 8 |
| Chemical equations | 1 (C) | 1 (C) | 1 | | |
| Heat transfer in chemical reactions | | 5 (C, L) | | | 1 |
| Stoichiometry | 4 (C) | 2 (C) | | | 1 |
| Acid/base chemistry | 5 (L, C) | | | | |
| Organic chemistry | 11 (L, C) | | | | |
| Nature of science | | 6 (C, L) | 10 | 9 | |
| Fossil fuels | 7 (C) | 6 (C) | | | |
| Total | 44 | 44 | 46 | 46 | 20 |

a. For Chemistry 101 courses, hours are reported as classroom (C) or laboratory (L). Topics that spanned both were reported in chronological order of the context in which they were addressed.

Normalized gain scores were calculated as a fraction of the total possible gain (Hake, 1998) for the following sets of CCQ items: (1) all items; (2) items related to course content only (these items differ from course to course, see Table 6.4), and (3) items targeting students' understanding of the conservation of matter. This concept enabled meaningful comparison because it was given a similar amount of instructional time in all courses, and the CCQ contained a significant number of questions targeting it. Gain scores were calculated using average pre- and post-assessment scores rather than by averaging individual gain scores, as the latter would necessitate eliminating students scoring 100% on the pretest from the calculation (Hake, 2001). Due to nonparametric distribution of data, the Wilcoxon signed ranks test was used to test the significance of gains. No distribution information is available for the average class gain scores, so no tests of statistical significance could be performed between classes.

### *Student Perceptions*

Two sets of data were collected to gauge students' perceptions of the CIC curriculum. The first data set, from a student evaluation questionnaire developed by the institution at which this study took place and administered at the end of each course, provided information on overall student satisfaction. Each questionnaire consisted of several Likert items and three open-response items. Here, we report scores on four common Likert items for each course. Results from the Likert portion were calculated on a 5-point scale and reported in aggregate form. Open-ended responses were reported anonymously to the instructor.

On the qualitative portion of the student evaluation questionnaire, students in all classes were asked to respond to three prompts: (1) "What aspects of the teaching, or content, of this course do you feel were especially good?" (2) "What changes could be made to improve the teaching, or the content, of this course?" and (3) "Other comments." Responses were coded according to common themes. Most codes from the CHEM courses were derived only from the lecture evaluations, as many of the same students completed both the lecture and the lab forms and a larger sample of students completed the lecture forms (the primary instructor taught only one of the two lab sections in each CHEM course). With one exception, we observed similar codes on both forms, lending validity to the lecture forms as a close approximation of overall CHEM student perceptions. The exception was that more comments about ICAs appeared on the lecture forms, and more comments about labs appeared on the lab forms. The lab forms, therefore, were used only to tabulate comments pertaining specifically to labs.

The second set of student perception data provided an understanding of what the students *thought* they learned in each class. A "metacognition question" at the end of the CCQ asked, "Did any of the class activities change your understandings of chemistry or the nature of matter? If so, which one or two activities were the most effective in changing your understandings?

Why? Please explain below." Due to a switch in data collection procedures midway through the study, this question was only asked in CHEM CIC and SCED CIC B. Responses to the metacognition question were coded according to: (1) the extent to which students claimed their ideas changed, (2) the type(s) of ideas they claimed had changed as a result of instruction, and (3) the instructional elements students reported to have affected their learning. Responses indicating the extent of a student's learning were compared to their scores on the CCQ to ascertain to what extent students' perceptions matched the extent of their actual learning.

## RESULTS

### Conceptual Understanding

Gains in conceptual understanding, as measured by the CCQ, are summarized in Table 6.5. All gains were statistically significant except the conservation of matter gain for SCED CIC B. Lack of statistical significance on this score, as well as lower significance levels on the other scores for SCED CIC B, is most likely due to the small sample size for this class (Table 6.2). We used the Wilcoxan Signed Ranks Test due to having a small sample. Gain scores for CHEM Trad on the course content and conservation of matter questions match typical gain scores of about 20% on conceptual assessments given in other traditional science lecture courses (Coletta, Phillips, & Steinert, 2007). Gain scores are higher for all of the CIC courses.

### Student Perceptions

#### *Student Evaluations*

Table 6.6 summarizes the scores for the four common items on the quantitative portion of the student evaluations for the lecture and lab component of each CHEM class and for each SCED class. The first two items evaluate students' perceptions of the course with respect to the specific criteria of clarity of expectations and fairness of assessments, whereas the latter two are more general measures of students' satisfaction with the course and the instructor's contribution as a whole. General trends in the scores suggest SCED CIC A students were the most satisfied, while SCED CIC B and CHEM CIC students were the least satisfied. The latter group was consistently less satisfied with the classroom portion than the lab portion of the course. This pattern was true for the CHEM Trad group as well, although the differences between the classroom and laboratory scores were not as large for these students.

**Table 6.5 Mean Scores and Normalized Class Gain Scores on the CCQ**

| | | Pre (%, SD) | Post (%, SD) | Gain Score (%)[a] |
|---|---|---|---|---|
| | | | *All 28 questions* | |
| CHEM | Trad | 47.48 (15.10) | 51.33 (16.15)** | 7.34 |
| | CIC | 42.69 (13.86) | 50.97 (14.11)** | 14.45 |
| SCED | CIC A | 42.37 (16.51) | 55.03 (18.03)** | 21.97 |
| | CIC B | 46.73 (17.46) | 59.52 (14.79)* | 24.02 |
| | | | *Course Content Only* | |
| CHEM | Trad | 52.48 (21.27) | 62.18 (20.93)** | 20.41 |
| | CIC | 49.04 (21.11) | 67.22 (19.95)** | 35.68 |
| SCED | CIC A | 50.27 (21.07) | 73.53 (13.18)** | 46.77 |
| | CIC B | 58.33 (21.64) | 75.98 (16.91)* | 42.35 |
| | | | *Conservation of Matter (8 questions)* | |
| CHEM | Trad | 61.00 (26.71) | 72.50 (23.97)** | 29.49 |
| | CIC | 53.41 (27.14) | 71.02 (26.24)** | 37.80 |
| SCED | CIC A | 49.43 (30.73) | 76.70 (18.21)** | 53.93 |
| | CIC B | 51.46 (35.14) | 78.13 (21.40) | 43.24 |

a. Normalized gain scores were calculated as a percentage of the possible gain: [(post-pre)/100-pre] X 100. *$p < 0.05$; **$p$ , 0.01 in Wilcoxon signed-ranks test.

In order to aid our interpretation of the student evaluation results, responses to the open-ended prompts were coded for common themes. Three major categories were uncovered, about which students wrote both positive and negative comments: (1) the constructivist activities (labs and ICAs for CHEM, activities for SCED), (2) course content, and (3) clarity of explanations. Four types of specific comments were frequent enough to merit their own categories, all of which expressed dissatisfaction with a component of the course. These were: (1) unanswered questions, in which students expressed frustration with the instructor's perceived lack of willingness to give direct answers; (2) problems with pacing, in which students complained about activities moving too quickly or slowly; (3) busy work, in which students expressed dissatisfaction with the amount of time writing or repetitiveness of questions in the course pack (most of these responses included the words "busy work" verbatim); and (4) too easy, in which students thought the tasks were not challenging enough. The frequency of these themes in student comments for each course is summarized in Table 6.7.

**Table 6.6. Mean Scores for the Common Items on Quantitative Portion of Student Evaluations**

| | *CHEM* | | | | *SCED* | |
|---|---|---|---|---|---|---|
| | *Class* | | *Lab* | | *CIC A* | *CIC B* |
| *Item* | *Trad* | *CIC* | *Trad* | *CIC* | | |
| Clarity of student responsibilities and requirements | 3.75 (0.87)[a] | 3.61 (0.81) | 3.76 (0.97) | 3.69 (1.11) | 4.33 (0.64) | 3.08 (1.04) |
| Fairness of evaluation procedures | 3.37 (1.07) | 3.19 (0.99) | 3.55 (1.28) | 3.43 (1.04) | 4.40 (0.65) | 3.00 (1.41) |
| Effectiveness of the course / lab overall | 3.19 (1.16) | 2.72 (1.12) | 3.67 (0.96) | 3.64 (1.26) | 4.12 (0.73) | 2.69 (1.18) |
| Instructor's overall contribution | 3.73 (1.07) | 3.19 (1.06) | 3.85 (0.87) | 3.84 (1.01) | 4.48 (0.51) | 3.69 (0.85) |

a. Data entered as score (standard deviation).

### *Metacognition*

Analysis of students' responses to the questions, "Did any of the class activities change your understandings of chemistry or the nature of matter? If so, which one or two activities were the most effective in changing your understandings?" enabled us to gain more specific insight into how the content and pedagogy in the CIC curriculum was viewed by the students. As mentioned previously, these questions were only asked in two of the four courses in this study: CHEM CIC and SCED CIC B. Therefore, the data presented in this section can be viewed as an exploration of how the least satisfied students viewed their learning in the constructivist curriculum. Because of the diversity in responses to this question, we only report quantitative data for the broader themes we observed (Table 6.8). A finer-grained quantitative analysis did not prove useful in identifying general trends in the data.

We first examined students' responses to the metacognition question in terms of how much content, if any, they felt they learned. About 20% of the responses from each course include claims the student learned nothing as a result of the CIC curriculum. More SCED students (30%) than CHEM students (15%) claimed that some, most, or all their ideas changed. The higher CCQ gain for SCED students is consistent with this pattern. When we matched the individual student responses to their CCQ gains, only two of the participants in CHEM CIC and one in SCED CIC B who claimed

**Table 6.7. Percentage of Responses Including Common Themes from Qualitative Portion of Student Evaluations**

| | | | *Percentage of Responses Including Theme* | | | |
|---|---|---|---|---|---|---|
| | | | *CHEM* | | *SCED* | |
| *Theme* | | | *Trad* | *CIC* | *CIC A* | *CIC B* |
| Constructivist Activities: | ICAs | + | N/A | 19.2 | 34.8[a] | 18.2[a] |
| | Labs | + | 26.3[b] | 21.6[b] | | |
| | ICAs | - | N/A | 25.0 | 4.3[a] | 9.1[a] |
| | Labs | - | 10.5[b] | 18.9[b] | | |
| Course content | | + | 15.1 | 1.9 | 4.3 | 27.3 |
| | | 5.7 | 0.0 | 0.0 | 0.0 | |
| Clarity of explanations | | + | 13.2 | 13.5 | 0.0 | 0.0 |
| | | 0.0 | 1.9 | 0.0 | 18.2 | |
| Unanswered questions | | | 7.5 | 9.6 | 17.4 | 54.5 |
| Problems with pacing | | | 9.4 | 3.8 | 0.0 | 45.5 |
| Busy work | | | 9.4 | 26.9 | 4.3 | 9.1 |
| Too easy | | | 11.3 | 11.5 | 0.0 | 0.0 |

a. The SCED curriculum was not split into lab and classroom portions.
b. Lab comments were from the laboratory evaluation sheets.

**Table 6.8. Summary of Students' Responses to Metacognition Question at the End of the CCQ**

| | *Percentage of Responses Including Theme* | |
|---|---|---|
| | *CHEM CIC* | *SCED CIC B* |
| No/not many ideas changed | 20.0 | 20.0 |
| Some/most ideas changed | 15.0 | 30.0 |
| Learned specific content | 37.5 | 50.0 |
| Learned through constructivist modes of instruction | 57.5 | 50.0 |

they learned nothing actually had a negative or 0 gain score. The rest of the students who made such statements displayed normalized gains of 33–80%.

Most responses did not contain references to specific activities. The most frequent reference among the small number of those that did was to the first lab/activity in both courses establishing the particulate nature of matter through the volume reduction of an ethanol-water mixture (Table 6.1) and tying temperature to the speed of motion of small particles through observation of Brownian motion in milk. Although few specific activities were cited, a fair number of responses from each course listed specific content. The most frequent content areas cited were the particulate nature of matter, intermolecular bonds, phase transitions, solutions and dissolving, and conservation of matter, all of which are closely aligned with CIC learning goals (Table 6.4). This result suggests the big ideas at the core of the CIC curriculum were more memorable to students than specific examples or types of problems. We observed a fair number of references to different modes of instruction in each course, alongside references to content. The constructivist activities (labs and ICAs for CHEM CIC; activities for SCED CIC B) were often cited as being generally helpful in these responses (again, with few references to specific activities), in that they enabled students to better visualize and discuss phenomena and concepts.

## DISCUSSION

### Conceptual Understanding

Comparison of CCQ gains for the two CHEM courses (Table 6.5) suggests the constructivist curriculum was more effective at supporting development of conceptual understanding than the traditional lecture-style curriculum. Especially important is the difference between the two courses on the conservation of matter questions, which represent similar instructional time in both courses.

These data also suggest the smaller class size and more fluid structure of the SCED courses were more effective at promoting student learning than the larger CHEM courses with separate laboratory sections. Comparison of CHEM CIC with SCED CIC B reveals larger gain scores in every category for the SCED course. Though the conservation of matter gain is not statistically significant for SCED CIC B, the overall content of the two courses was similar enough (Table 6.4), especially as it aligned with the items on the CCQ, to warrant a comparison of the course content scores which showed statistically significant gains in both courses.

Finally, comparison of the two SCED courses suggests that students' motivational factors and/or level of academic experience are important in determining the learning outcomes of a constructivist curriculum. The SCED CIC B course enrolled a lower percentage of preservice teachers than A and showed lower gains in course content and conservation of matter. Although the latter was not statistically significant for section B, the overall content of the two offerings was nearly identical (Table 6.4), warranting a comparison of the course content scores. It is important to note here that the preservice teachers and general education students had taken about the same number of college science courses prior to SCED B, so the difference in conceptual understanding cannot be explained by previous science training.

## Student Perceptions

### *Student Evaluations*

Comparison of the two CHEM groups sheds light on student perceptions of a constructivist versus traditional curriculum. The CHEM CIC students were less satisfied on all quantitative measures than the Trad students. This difference is more marked for the classroom portions than the lab portions. The similarity between scores on the lab evaluations for both CHEM courses suggests students were just as satisfied with a constructivist approach versus a more traditional verification approach in the lab setting.

Responses to the open-ended questions did not reveal many obvious differences between the two CHEM courses. While the traditional lab received slightly better reviews than the constructivist lab, this difference was not large enough to call into question the similar scores on the quantitative portion of the evaluation. Most positive comments about the traditional lab referenced the interest level of the experiments, while most of those in the constructivist lab discussed the role the labs played in the students' learning. Comments pertaining to course content were more frequent among Trad versus CIC students, which may be a reflection of the more numerous topics covered in the Trad course. Specific comments about busy work appeared at a greater frequency in the CIC course, most citing the amount of writing required in the course pack (ICAs and labs). Busy work comments in the Trad class pertained most frequently to the number of homework assignments given.

To explore the role of class size and structure in a constructivist curriculum, SCED CIC B can be compared to CHEM CIC. Comparison of the quantitative evaluation scores reveals a higher degree of satisfaction on all measures for the laboratory portion of CHEM CIC. Furthermore, the classroom portion of CHEM CIC received higher scores than SCED CIC B on three out of the four items. The exception is the item, "Instructor's contribution overall to the course/lab," which received the lowest scores on the CHEM CIC classroom form and similar scores for CHEM CIC lab and SCED CIC B. Mirroring the differences in student satisfaction between the classroom and lab portions of CHEM CIC, this result suggests that implementation of the constructivist activities was the most problematic in the lecture hall, whereas they were received with a higher degree of satisfaction in laboratory settings.

Comparison of open-ended response codes reveals higher ratings for the constructivist activities by the SCED CIC B students compared to the CHEM CIC students. The SCED students expressed a greater degree of dissatisfaction with the clarity of explanations compared to the CHEM students, some of whom positively rated the clarity of the instructor's lectures, even though lecture was rarely employed in any of the CIC courses.

Comparison of the two SCED groups, which can lead to hypotheses about the roles of motivational factors and level of academic experience, reveals some interesting differences. Students in SCED CIC A reported at least a full point of greater satisfaction than SCED CIC B students on the first three items of the quantitative portion of the student evaluations, and almost a full point for the last item. On most of the quantitative items, SCED CIC B students were the least satisfied of the four groups, and SCED CIC A students were the most satisfied. While the smaller SCED class sizes increase the likelihood of large variations in the data, the magnitude of this difference is striking considering the two courses were taught by the same instructor using the same curriculum. By comparison, the largest difference between the two CHEM courses, which had entirely different curricula, was 0.76. These data suggest the curriculum was received very differently by the SCED CIC B students compared to the SCED CIC A students.

Examination of responses to the open-ended questions adds more specificity to the origins of SCED CIC B students' dissatisfaction. The category that received the highest proportion of negative comments in B versus A was clarity. Many of these comments cited the constructivist nature of the curriculum as inadequate in providing students with confidence in their knowledge. In addition, SCED CIC B students complained about unanswered questions and pacing at a significantly higher frequency than SCED A students. Comments about unanswered questions mirror complaints about clarity, in that students expressed a lack of confidence in their knowledge and wanted more feedback from the instructor. Students who complained about the pacing often commented that discussions were too drawn out. Consistent with all of these results is the observation that the constructivist activities in general received more positive reviews in SCED A than in B.

### *Metacognition*

Comparison of responses to the metacognition question for CHEM CIC and SCED CIC B revealed more similarities than differences, suggesting that students had similar perceptions of the curriculum's effect on their learning despite differences in the delivery of the curriculum in each course. This may be due in part to the general nature of the prompt, though the mismatch in some cases between "I learned nothing" responses and actual learning gains on the CCQ suggest some students may have underdeveloped metacognition skills.

## CONCLUSIONS

In this study we examined how contextual factors influenced the effectiveness of a constructivist curriculum, both with respect to content knowledge and student perceptions. We selected four courses taught by the same instructor, three of which used a constructivist curriculum and one that used a traditional lecture approach, for a case study. Pairwise comparisons of these courses allowed us to make inferences about factors that influenced student learning and perceptions. Although small sample sizes and lack of random sampling make generalizations to larger student populations problematic, the hypotheses generated here suggest a range of possible outcomes of constructivist curricula and lead to suggestions that can positively impact all students.

### Impacts of Constructivist Versus Traditional Curriculum

To answer the first research question—How does a constructivist curriculum impact undergraduates differently from traditional lecture-based instruction?—we compared two large chemistry courses, one of which was taught as a survey course using lecture as the main pedagogical tool (CHEM Trad) while the other used a constructivist curriculum in both the classroom and lab (CHEM CIC). Comparison of CCQ gains suggests the constructivist curriculum was more effective at advancing conceptual learning than the traditional curriculum. Student evaluations, however, revealed the CIC students were less satisfied with classroom instruction than the Trad students. Many of the students in the constructivist course saw the classroom activities as "busy work," instead of instruments to help them construct ideas.

The incongruity between student satisfaction and learning data may stem from a cultural dilemma (Windschitl, 2002), a mismatch between students' personal epistemologies and the kind of learning the curriculum asks of the students. This hypothesis is consistent with research suggesting that college undergraduates often enter introductory science courses with a view of knowledge and learning that centers on rote memorization of disconnected facts and mathematical problem solving, rather than conceptual understanding (Elby, 1999, 2001; Hammer, 1994). The minimization of rote or algorithmic activities in the CIC curriculum may have led some students to view the constructivist activities as a waste of time, even though the CCQ results showed they led to conceptual learning for most students. Although the basic tenets of constructivism were introduced on the first day of each CIC course through discussion and interactive lecture, more continuous justification of constructivism was probably necessary.

Higher quantitative student evaluation scores for the laboratory portion of CHEM CIC than for the classroom portion suggests either that students perceived the constructivist activities as more effective in lab than in the classroom, or that these activities matched their expectations better for the laboratory portion of the class than for the classroom portion. This may be because didactic teaching methods were less of an expectation in lab and the constructivist activities were therefore more in line with students' expectations, or that there was simply more support in lab for engaging with the material, as the presence of TAs lowered the instructor-to-student ratio. This result is consistent with previous research that establishes personal epistemologies as being context-dependent (Elby, 2009; Hammer & Elby, 2002; Leach et al., 1999; Louca et al., 2004).

## Impacts of Class Size and Structure

The second research question was, How do class size, structure, and student population influence the learning of undergraduates in a constructivist environment? To answer the first part of this question about instructional environment, we compared a large course that was divided into classroom and laboratory sessions (CHEM CIC) with a small course that moved fluidly between laboratory investigations and discussions (SCED CIC B). Data from the CCQ suggest the smaller, more fluid class was a more effective environment in which to implement a constructivist curriculum in terms of learning outcomes. However, student evaluation results suggest that CHEM CIC students were more satisfied than SCED CIC B students were. A likely explanation for these data is that the smaller size of the SCED course made it more possible to hold students accountable for meaningful engagement with the curriculum, which allowed for more learning to take place but may have also provoked more frustration.

The metacognition data are consistent with the inference above, in that they revealed a higher percentage of SCED students claiming their ideas changed. However, the specific content cited as having changed, as well as the modes of instruction that brought about those changes, were similar for the two groups of students. Therefore, the content and pedagogy seem to have been perceived relatively similarly in the two implementations, regardless of how the students felt it impacted their learning.

## Impacts of Student Population

Comparison of the two SCED courses allowed us to make inferences about the role of motivational factors in mitigating or enhancing the outcomes of the constructivist curriculum (second part of research question 2). The largely preservice teacher population of SCED CIC A made a higher average gain than the mixed population of SCED CIC B on CCQ questions related to course content and conservation of matter. This result supports the inclusion of extrarational factors such as beliefs and motivations into conceptual change models (Demastes et al., 1995; Park, 2007; Pintrich et al., 1993), lending credence to the argument that: "Students can and do adopt different goals and purposes for their school work, and becoming cognitively engaged in the myriad of classroom academic tasks is really a choice they can make for themselves" (Pintrich et al., 1993, p. 168).

How might motivational factors interact with the cognitive factors, dissatisfaction, intelligibility, plausibility, and fruitfulness posited in the original conceptual change model (Posner et al., 1982)? One hypothesis is that factors such as goal orientation or level of interest impact students' "depth of processing" (Pintrich et al., 1993, p. 192). Depth of processing influences (1) the extent to which students identify and attend to discrepancies between their incoming ideas and evidence that might be contradictory to those ideas (cognitive dissonance) and (2) to what extent students are able to make sense of and evaluate a new concept (intelligibility, plausibility). Preservice teachers may have seen more value in the curriculum for their future careers and/or had a better understanding of constructivism from previous education courses, therefore engaging in deeper processing than students who did not.

An even larger gap in student satisfaction was observed between SCED CIC A and B than in learning outcomes. Examination of comments on the qualitative portion of the student evaluations reveals sense-making discussions as a source of dissatisfaction in SCED CIC B. Evidence from these comments suggests some of these students failed to see these discussions as constructing the "answers" they were looking for, perhaps because of a need for knowledge to be validated by an authority (e.g., expert, instructor, textbook) (Elby, 2001; Hammer, 1994; May & Etkina, 2002). Thus, the students may not have engaged deeply in or valued the results of these discussions because they represented knowledge constructed by non-authority figures. The lack of effective engagement in these discussions by some students may have lowered the quality of the consensus ideas as a result.

It cannot be ignored that a higher percentage of SCED A than B students were upperclassmen. Upperclassmen have more experience managing their time, mastering study techniques, and working in groups than underclassmen. This factor may also explain, in part, the greater learning and satisfaction of SCED A students. In reality, motivation, personal epistemologies, and academic experience are probably all at work in explaining differences in learning and satisfaction and likely influence each other; for example, upperclassmen have had more opportunities to develop mastery goal orientations (Pintrich, 2000) than underclassmen.

## SUMMARY

Our data strongly suggest that the constructivist curriculum was effective in advancing students' conceptual knowledge of chemistry, regardless of the varying perceptions of the students in this study. The most successful implementation on all measures was in SCED CIC A. This course represented a combination of favorable variables: small class size, fluid structure, and a high percentage of preservice teachers/upperclassmen. Students in the other two constructivist courses, SCED CIC B and CHEM CIC, appear to have gained more conceptual understanding than the CHEM Trad students but lagged behind SCED CIC A students. Students in SCED CIC B and CHEM CIC were also the least satisfied. Our data suggest that for the three constructivist courses, large class size was the least favorable factor in students' conceptual growth, while differences in student population appeared to be the strongest determining factor in student satisfaction with the CIC curriculum. Reduced learning in the larger class may have been due to weaker accountability and feedback mechanisms but also due to difficulty in leading consensus-building discussions in a large lecture-style classroom.

Student motivations and experience appear to have had a greater impact on students' satisfaction with the curriculum than with their overall learning. Our hypothesis for this relationship is that constructivism may interface well with the possibly more sophisticated epistemologies of preservice teachers, many of whom had been introduced to constructivism in previous courses, minimizing the cultural dilemma described by Windschitl (2002). Although we have some qualitative evidence of these more sophisticated epistemologies from student evaluation data, a future study would employ an instrument to measure these epistemologies more directly (Magolda, 2001; White, Elby, Frederiksen, & Schwarz, 1999)

## IMPLICATIONS

Our data suggest that a constructivist curriculum is worthwhile even if it is received differently in different contexts. Below we make a number of recommendations about how to enhance the effectiveness of a constructivist curriculum, both with respect to learning and satisfaction, in the least favorable contexts, potentially closing the gap in outcomes between less and more favorable constructivist settings.

## Strategies for Implementing Constructivism in Large Courses

1. Strict accountability mechanisms must be built into the evaluation scheme for large courses. Since constructivist methods rely on discussions in which claims are built upon evidence by students, participation is crucial. Although attendance and participation were part of CHEM CIC students' grades, it probably needed to be attached to a higher percentage of the points. We hypothesize that learning gains would be positively affected but that student satisfaction might be negatively affected by this change if accountability did not come with strategies to create student buy-in for constructivism. Such strategies will be discussed in the next section.
2. Large courses need an effective structure for facilitating sense-making discussions. The instructor must use strategies to ensure not only that students understand the content of the ideas being discussed, but also that consensus ideas represent "answers" in a constructivist curriculum. Simple techniques such as keeping a written record of the discussion at the front of the room or constructing a written summary of the discussion at the end of the class (McKeachie & Wilbert, 2010) can be powerful ways of organizing the outcomes of a discussion so that students don't feel overwhelmed and can see a clear thread. Classroom response systems have also been shown to be effective for structuring discussions when used skillfully (Gauci, Dantas, Williams, & Kemm, 2009). In addition, closely monitored online discussion forums can serve as an additional venue for sense-making (Choi, Land, & Turgeon, 2005; Hara, Bonk, & Angeli, 2000; Palmer, Holt, & Bray, 2008). Finally, discussions of video cases showing how students interact in effective sense-making discussions can lead towards improved engagement.
3. Students need more contact with instructors in classroom portions of large lecture/lab courses. Learning in a constructivist curriculum relies on students' ability to interpret evidence. Although the CIC curriculum includes questions based on known misconceptions (Kind, 2004; Gilbert et al., 2002; Nakhleh, 1992; Othman, Treagust, & Chandrasegaran, 2008; Yezierski & Birk, 2006) to help students interpret their data, it cannot realign all possible lines

of student thinking that are off track. Small group or one-on-one interaction with an instructor is therefore essential in helping students make sense of their data and construct sensible claims. In one example of a holistic adoption of a constructivist curriculum in a large science class, Massachusetts Institute of Technology created a studio classroom to enable students to work together around tables, with multiple TAs acting as support, in introductory physics courses (Dori, Belcher, Bessette, Danziger, McKinney, & Hult, 2003; Rimer, 2009). In many institutions, such holistic reform is unfeasible. However, moderate success with peer-assisted team learning models, in which peers who have successfully completed earlier iterations of a course led teams of students in learning activities, has been reported (Hockings, DeAngelis, & Frey, 2008; Quitadamo, Brahler, & Crouch, 2009).

## Strategies for Working With Varied Student Populations

4. Students unfamiliar with constructivist-based instruction methods (arguably most college undergraduates with the possible exception of some preservice teachers) require explicit instruction, including descriptions of the roles of students and instructor as well as data showing the effectiveness of constructivism. Such framing may shift students' epistemologies toward knowledge construction and away from knowledge received by authority. Strategies shown to be effective include written prompts within a curriculum to encourage reflection on the process of learning, a grading system that heavily weights effort, assessment questions emphasizing explanation over "correct" answers, and reduced use of a textbook (Elby, 2001).
5. Metacognition instruction is recommended for all students, especially those less familiar with constructivism. Responses to the metacognition question suggest many of the students in CHEM CIC and SCED CIC B lacked sophistication in their ability to reflect on the extent to which their ideas changed. Students should not only have multiple opportunities to reflect on their learning, but should be guided to do this in a structured way in which they learn about and practice key components of metacognition, such as checking their ideas for internal consistency and coordinating evidence with their claims.

## ACKNOWLEDGMENTS

This study was funded through a grant from the National Science Foundation (DUE- 0737551). The claims in this chapter do not necessarily reflect the views of the National Science Foundation. The authors gratefully acknowledge assistance with review of this manuscript from Dr. Andrew Boudreaux and assistance with statistical analysis from Dr. Daniel Hanley.

Correspondence concerning this article should be addressed to Dr. Emily Borda, Western Washington University Department of Chemistry, 516 High St, Bellingham, WA, 98226.

## REFERENCES

American Association for the Advancement of Science (AAAS). (1990). *Science for all Americans*. New York, NY: Oxford University Press.

American Association for the Advancement of Science (AAAS). (1993). *Benchmarks for science literacy*. New York, NY: Oxford University Press.

Aikenhead, G. S., & Jegede, O. J. (1999). Cross-cultural science education: A cognitive explanation of a cultural phenomenon. *Journal of Research in Science Teaching, 36*(3), 269–287.

Anderman, E. M., & Maehr, M. L. (1994). Motivation and schooling in the middle grades. *Review of Educational Research, 64*(2), 287–309.

Baviskar, S. N., Hartle, R. T., & Whitney, T. (2009). Essential criteria to characterize constructivist teaching: Derived from a review of the literature and applied to five constructivist-teaching method articles. *International Journal of Science Education, 31*(4), 541–550.

Bendixen, L. D., & Rule, D. C. (2004). An integrative approach to personal epistemology: A guiding model. *Educational Psychologist, 39*(1), 69–80.

Berkheimer, G. D., Anderson, C. W., & Blakeslee, T. D. (1988). *Matter and molecules teacher's guide: Activity book*. East Lansing, MI: The Institute for Research on Teaching, College of Education, Michigan State University.

Bransford, J. D., Brown, A. L., & Cocking, R. R. (Eds.). (2000). *How people learn: Brain, mind, experience, and school*. Washington, DC: National Academy Press.

Bruner, J. S. (1996). *The culture of education*. Cambridge, MA: Harvard University Press.

Carey, S. (1987). *Conceptual change in childhood*. Cambridge, MA: MIT Press.

Choi, I., Land, S. M., & Turgeon, A. J. (2005). Scaffolding peer-questioning strategies to facilitate meta-cognition during online small group discussion. *Instructional Science: An International Journal of Learning and Cognition, 33*(5), 483–511.

Cole, M. (1990). Cognitive development and formal schooling: The evidence from cross-cultural research. In L. Moll (Ed.), *Vygotsky and education: Instructional implications and applications of sociohistorical psychology* (pp. 89–110). Cambridge, UK: Cambridge University Press.

Coletta, V. P., Phillips, J. A., & Steinert, J. J. (2007). Interpreting force concept inventory scores: Normalized gain and SAT scores. *Physics Review Special Topics–Physics Education Research, 3*(010106), 1–5.

Demastes, S. S., Good, R. G., & Peebles, P. (1995). Students conceptual ecologies and the process of conceptual change in evolution. *Science Education, 79*(6), 637–666. doi:10.1002/sce.3730790605

Dori, Y. J., Belcher, J., Bessette, M., Danziger, M., McKinney, A., & Hult, E. (2003). Technology for active learning [Editorial]. *Materials Today, 6*(12), 44–49

Driver, R., & Erickson, G. (1983). Theories-in-action: Some theoretical and empirical issues in the study of students' conceptual frameworks in science. *Studies in Science Education, 10*(1), 37–60.

Driver, R., & Oldham, V. (1986). A constructivist approach to curriculum development in science. *Studies in Science Education, 13*, 105–122.

Elby, A. (1999). Another reason that physics students learn by rote. *Physics Education Research, 67*(7), S52–S57.

Elby, A. (2001). Helping physics students learn how to learn. *Physics Education Research, 69*(7), S54–S64.

Elby, A. (2009). Defining personal epistemology: A response to Hofer & Pintrich (1997) and Sandoval (2005). *Journal of the Learning Sciences, 18*(1), 138–149.

Gauci, S. A., Dantas, A. M., Williams, D. A., & Kemm, R. E. (2009). Promoting student-centered active learning in lectures with a personal response system. *Advances in Physiology Education, 33*(1), 60–71.

Gilbert, J. K., De Jong, O., Justi, R., Treagust, D. F., & van Driel, J. H. (Eds.). (2002). *Chemical education: Towards research-based practice*. Dordrecht, The Netherlands: Kluwer Academic Publishers.

Goldberg, F., Bendall, S., Heller, P., & Poel, R. (2006). *Interactions in physical science*. Armonk, NY: It's About Time Herff Jones Education Division.

Hake, R. (1998). Interactive-engagement versus traditional methods: A six-thousand-student survey of mechanics test data for introductory physics courses. *American Journal of Physics, 66*(1), 64–74.

Hake, R. (2001, 5/18/01). Suggestions for administering and reporting pre/post diagnostic tests. Retrieved from http://physics.indiana.edu/~hake/

Hammer, D. (1994). Epistemological beliefs in introductory physics. *Cognition and Instruction, 12*(2), 151–183.

Hammer, D., & Elby, A. (2002). On the form of a personal epistemology. In B. Hofer & P. R. Pintrich (Eds.), *Personal epistemology: The psychology of beliefs about knowledge and knowing* (pp. 169–190). Mahwah, NJ: Erlbaum.

Hara, N., Bonk, C. J., & Angeli, C. (2000). Content analysis of online discussion in an applied educational psychology course. *Instructional Science, 28*(2), 115–152.

Heckman, P. (1987). Understanding school culture. In J. Goodlad (Ed.), *The ecology of school renewal* (pp. 63–78). Chicago, IL: University of Chicago Press.

Hockings, S. C., DeAngelis, K. J., & Frey, R. F. (2008). Peer-led team learning in general chemistry: Implementation and evaluation. *Journal of Chemical Education, 85*(7), 990–996.

Hofer, B. K., & Pintrich, P. R. (1997). The development of epistemological theories: Beliefs about knowledge and knowing and their relation to learning. *Review of Educational Research, 67*(1), 88–140.

Hutchison, P., & Hammer, D. (2010). Attending to student epistemological framing in a science classroom. *Science Education, 94*(3), 506–524.

Joseph, P. B., Bravmann, S. L., Windschitl, M. A., Mikel, E. R., & Green, N. S. (2000). *Cultures of curriculum*. Mahwah, NJ: Lawrence Erlbaum.

Kind, V. (2004). *Beyond appearances: Students' misconceptions about basic chemical ideas*. London, UK: Royal Society of Chemistry.

Leach, J., Millar, R., Ryder, J., & Sere, M.-G. (1999, April). *An investigation of high school and university science majors' epistemological reasoning in the context of empirical investigations*. Paper presented at the Annual Meeting of the American Educational Research Association, Montreal, PQ.

Lederman, N. G., Abd-El-Khalick, F., Bell, R. L., & Schwartz, R. S. (2002). Views of nature of science questionnaire: Toward valid and meaningful assessment of learners' conceptions of nature of science. *Journal of Research in Science Teaching, 39*, 497–521.

Lemke, J. L. (2001). Articulating communities: Sociocultural perspectives on science education. *Journal of Research in Science Teaching, 38*(3), 296–316.

Louca, L., Elby, A., Hammer, D., & Kagey, T. (2004). Epistemological resources: Applying a new epistemological framework to science instruction. *Educational Psychologist, 39*(1), 57–68.

Magolda, M. B. (2001). A constructivist revision of the measure of epistemological reflection. *Journal of College Student Development, 42*(6), 520–534.

Marin, N., Benarroch, A., & Gomez, E. J. (2000). What is the relationship between social constructivism and piagetian constructivism? An analysis of the characteristics of the ideas within both theories. *International Journal of Science Education, 22*(3), 225–238.

May, D. B., & Etkina, E. (2002). College physics students' epistemological self-reflection and its relationship to conceptual learning. *American Journal of Physics, 70*(12), 1249–1258.

McKeachie, W. J. (2010). *Teaching tips* (13th ed.). Belmont, CA: Wadsworth.

Mulford, D. R., & Robinson, W. R. (2002). An inventory for alternate conceptions among first-semester general chemistry students. *Journal of Chemical Education, 79*(6), 739–744.

Nakhleh, M. B. (1992). Why some students don't learn chemistry. *Journal of Chemical Education, 69*(3), 191–196.

Nakhleh, M. B. (1993). Are our students conceptual thinkers or algorithmic problem solvers? *Journal of Chemical Education, 70*(1), 52–55.

Nakhleh, M. B., & Mitchell, R. C. (1993). Concept learning versus problem solving: There is a difference. *Journal of Chemical Education, 70*(3), 190–192.

National Research Council (NRC). (1996). *National science education standards*. Washington, DC: National Academy Press.

Nurrenbern, S. C., & Pickering, M. (1987). Concept learning versus problem solving: Is there a difference? *Journal of Chemical Education, 64*(6), 508–510.

Othman, J., Treagust, D. F., & Chandrasegaran, A. L. (2008). An investigation into the relationship between students' conceptions of the particulate nature of matter and their understanding of chemical bonding. *International Journal of Science Education, 30*(11), 1531–1550.

Palmer, S., Holt, D., & Bray, S. (2008). Does the discussion help? The impact of a formally assessed online discussion on final student results. *British Journal of Educational Technology, 39*(5), 847–858.

Paradis, J. (2007). *Explorations in conceptual chemistry: A student activity manual*. San Francisco, CA: Pearson/Benjamin Cummings.

Park, H. (2007). Components of conceptual ecologies. *Research in Science Education, 37*(2), 217–237. doi:10.1007/s11165-006-9023-8

Perry, W. (1970). *Forms of intellectual and ethical development in the college years: A scheme*. New York, NY: Holt, Rinehart & Winston.

Piaget, J. (1970). Piaget's theory. In P. H. Mussen (Ed.), *Carmichael's manual of psychology* (pp. 703–732). New York, NY: Wiley.

Piaget, J. (1971). *Biology and knowledge*. Edinburgh, UK: Edinburgh Press.

Piaget, J. (1973). *To understand is to invent: The future of education*. New York, NY: Grossman.

Piaget, J. (1977). *The development of thought: Equilibration of cognitive structures* (A. Rosin, Trans.). New York, NY: Viking Press.

Pintrich, P. R. (2000). Multiple goals, multiple pathways: The role of goal orientation in learning and achievement. *Journal of Educational Psychology, 92*(3), 544–555.

Pintrich, P. R., Marx, R. W., & Boyle, R. A. (1993). Beyond cold conceptual change: The role of motivational beliefs and classroom contextual factors in the process of conceptual change. *Review of Educational Research, 63*(2), 167–199.

Posner, G. J., Strike, K. A., Hewson, P. W., & Gertzog, W. A. (1982). Accommodation of a scientific conception: Toward a theory of conceptual change. *Science Education, 66*(2), 211–227.

Quitadamo, I. J., Brahler, C. J., & Crouch, G. J. (2009). Peer-led team learning: A prospective method for increasing critical thinking in undergraduate science courses. *Science Educator, 18*(1), 29–39.

Rimer, S. (2009). At M.I.T., large lectures are going the way of the blackboard. *New York Times*, 12.

Schommer-Aikins, M. (2004). Explaining the epistemological belief system: Introducing the embedded systemic model and coordinated research approach. *Educational Psychologist, 39*(1), 19–29.

Sirotnik, K. (1983). What you see is what you get: Consistency, persistency, and mediocrity in the classroom. *Harvard Educational Review, 53*, 16–31.

Solomon, S., Qin, M., Manning, Z., Chen, M., Marquis, M., & Averyt, M. (2007). *Contribution of Working Group I to the Fourth Assessment Report of the Intergovernmental Panel on Climate Change*. Cambridge, UK and New York, NY: Cambridge University Press.

Tyack, D., & Cuban, L. (1995). *Tinkering toward Utopia*. Cambridge, MA: Harvard University Press.

Vygotsky, L. S. (1962). *Thought and language*. Cambridge, MA: MIT Press.

Vygotsky, L. S. (1978). *Mind in society: The development of higher psychological processes*. Cambridge, MA: Harvard University Press.

White, B., Elby, A., Frederiksen, J., & Schwarz, C. (1999). Epistemological Beliefs Assessment for Physical Science (EBAPS). Retrieved from http://www2.physics.umd.edu/~elby/EBAPS/home.htm

Windschitl, M. (2002). Framing constructivism in practice as the negotiation of dilemmas: An analysis of the conceptual, pedagogical, cultural, and political challenges facing teachers. *Review of Educational Research, 72*(2), 131–175.

Yezierski, E. J., & Birk, J. P. (2006). Misconceptions about the particulate nature of matter. *Journal of Chemical Education, 83*(6), 954–960.

Zoller, U. (1990). Comments and criticism: Students' misunderstandings and misconceptions in college freshman chemistry (general and organic). *Journal of Research in Science Teaching, 27*(10), 1053–1065.

CHAPTER 7

# A CONTRAST OF THE SCIENCE TEACHING PRACTICES OF TWO PRESERVICE EARLY CHILDHOOD EDUCATORS

**Deirdre Englehart**


## ABSTRACT

Preservice teachers need time and experiences to support their teaching practices as they move from internships into teaching in their own classroom. Many early childhood teachers do not have a strong background in science to support inquiry-teaching practices. This case study research investigated the teaching practices of two preservice early childhood educators during their junior and senior internship experiences while teaching science. The Views of Science Inquiry instrument was used at the beginning and end of the research along with observations and interviews to explore their science teaching practices. Results of this research investigation confirm that although teachers may have access to the same curriculum, their own background, experiences, and beliefs influence their classroom practices, especially regarding the implementation of scientific inquiry.


*Research Based Undergraduate Science Teaching*, pp. 221–245

Preservice teachers are provided internship experiences to support their growing knowledge of teaching practice. Such practical experience provides them with a platform to learn in action. It provides opportunities to experience what they have learned about in the college classroom setting and encourages putting the theories into practice (Black, 2004). A solid foundation of theory is intended to be taken to the classroom and practiced during this time to complete the final portion of the formal learning process of the undergraduate degree program.

The overall goals for the teacher of science include developing students' understanding of science content and using instructional strategies to support student learning. Previous research suggests that beginning teachers lack the skills necessary to manage the classroom and the pedagogical content knowledge (PCK) that supports them in the specific teaching of science. Shulman (1986) defined pedagogical content knowledge as the subject matter knowledge supporting appropriate teaching practices. Teachers must have a solid foundation in content and teaching practices in order to support their students in learning science. Since early childhood teachers are considered "generalists," they may not have a strong science background with which to support teaching for conceptual understanding, including science inquiry practices (Weiss, Banilower, McMahon, & Smith, 2001). Lastly, many teachers lack experiences in scientific inquiry and therefore may not practice this in their teaching of science (Davis, Petish, & Smithey, 2006). All of these factors influence a teacher's ability to integrate inquiry practices in the classroom. This study explores the journey of learning to teach science experienced by two early childhood students during their junior and senior internship experiences.

## REVIEW OF LITERATURE

Teachers have years of classroom experiences as a student that influence their thinking and practice. They tend to teach in ways that are comfortable to them, which generally involves a replay of past experiences. Everett, Luera, and Otto (2006) indicate that teaching is the only profession where people have been involved in observations for years prior to their professional studies in the field. Plevyak (2007) explains that most experiences the preservice PreK–3 teachers have had in science involved direct instruction. These previous experiences may not have involved the learner in science or promoted higher-order thinking skills or inquiry practices.

## Teaching Skills

In applying their knowledge, teachers develop many skills through their own experiences of teaching. These skills are related to pedagogical knowledge and pedagogical content knowledge. Pedagogical knowledge refers to general strategies for teaching. Pedagogy is the art of being a teacher, so pedagogical knowledge includes the style or methods of teaching. The use of social interaction in learning, building instruction on prior knowledge, and problem solving and higher level learning as important components of pedagogical knowledge has been emphasized by Gee, Boberg, and Gabel (1996). Different aspects of pedagogical knowledge that support new teachers in teaching science were also identified by Hudson (2004). Some skills included in pedagogical knowledge are planning, timetabling, preparation, implementation, classroom management strategies, and questioning skills.

One challenge for teachers is to create a classroom learning environment that is productive and manageable. Research by Davis et al. (2006) states that although new teachers want their classrooms to be student-centered, the concerns they have about classroom management may work against that goal and lead them into engaging in less reform-oriented practices. The active nature of science teaching requires skillful teaching practices to maintain control in the classroom. These pedagogical strategies are not always present for preservice teachers and develop through time and experience in the classroom (Mulholland & Wallace, 2003, 2005).

Research on teaching abilities has expanded to include more than just general teaching practices, or pedagogical knowledge, to support classroom learning. Shulman (1986) identified three categories of content knowledge including subject matter content knowledge, pedagogical content knowledge, and curricular knowledge. Content knowledge refers to the amount and organization of knowledge, which goes beyond knowledge of facts and concepts into an understanding of the structures of the subject matter. Pedagogical content knowledge (PCK) deals with the subject matter knowledge for teaching. The ways of representing and formulating the subject so it is understandable are included in PCK. Curricular knowledge has to do with the curriculum and the range of programs designed for teaching different subjects or topics at various grade levels. In this category, teachers must consider the curricular alternatives appropriate for instruction when teaching specific subjects of the curriculum. The elements of PCK affect teacher practice and teachers' ability to implement scientific inquiry in classrooms.

## Scientific Inquiry

Inquiry is a teaching method supported by research on how children learn and is a central component of the content standards for science (National Research Council, 2000). The National Research Council explains that curiosity is the basis for inquiry and that "from birth, children employ trial-and-error techniques to learn about the world around them" (2000, p. 5). Research from Yoon and Onchwari (2007) sees the connection between developmentally appropriate practices and constructivism. They promote constructivism, emphasizing that this is how each child interacts with his or her environment in a distinctive way. These unique approaches help the child to construct his or her unique understandings of the world. According to the National Research Council (2000), inquiry "encompasses not only an ability to engage in inquiry but an understanding of inquiry and of how inquiry results in scientific knowledge" (p. 13). Inquiry is advocated as an important practice in teaching science. The National Research Council (NRC, 1996, 2000) has identified five essential features of inquiry: engaging in scientifically oriented questions, gathering evidence, developing explanations based on evidence, evaluating explanations in light of alternative explanations, and communicating and justifying proposed explanations. Some inquiry activities will not be considered full inquiry because they will be missing some of these essential features of inquiry and will, therefore, be considered partial inquiries. "The degree to which teachers structure what students do is sometimes referred to as 'guided' verses 'open' inquiry" (NRC, 2000, p. 29). These distinctions are made according to the degree of control or responsibility the students have in regard to asking questions, developing investigations, and communicating their findings. Guided inquiry indicates that the teacher has more control in the structure of the lessons while in open inquiry the students gain more control in the overall inquiry process.

## Teaching Inquiry

Many early childhood educators are not equipped to teach through inquiry, as suggested by the National Science Education standards. Previous research from Davis et al. (2006) discussed that to teach inquiry-oriented science, teachers must have strong understandings of inquiry and abilities to teach inquiry. Her analysis of findings emphasized that many preservice teachers have unsophisticated understandings of inquiry that do not facilitate teaching with this approach. The National Research Council emphasizes the need to prepare teachers for inquiry-based teaching. "For students to understand inquiry and use it to learn science, their teachers need to be well-versed in inquiry and inquiry based methods" (NRC, 2000, p. 87). The NRC further notes that most teachers have not experienced science through inquiry or conducted scientific inquiry themselves. Research by Sadaghani and Costley (2009) suggests that preservice elementary teachers who participated in field experiences involving them in teaching science inquiry were less confident in their conceptual understandings of content and of their ability to teach through science inquiry. The researchers concluded that the field experiences had created a venue for students to understand the realities of teaching through this method, so they needed to overcome perceived difficulties with this approach prior to implementing it. This research also aligns with research from Schwartz and Crawford (2005), which showed that many teachers had a limited view of science inquiry because they did not have a background or strong experiences with it.

Survey results from Weiss et al. (2001) indicated that out of all elementary teachers in the study, 65% of teachers reported a moderate or substantial need to learn more about how to use inquiry-oriented teaching strategies, and 63% indicated they needed support in understanding students' thinking in science. These pedagogical gaps relate to areas specific to the teaching of science. Methods courses can provide opportunities for preservice teachers to deepen their understanding of inquiry practices, but they need more time and interactions with young children to fully support the implementation of these practices (Plevyak, 2007). These teachers indicated that time, classroom management, and the ability of young children were obstacles to implementing inquiry practices. In a research study conducted by Gilbert (2009) with early childhood teachers, he found that philosophy statements allowed participants to formulate positive ideas about teaching science and science inquiry. When these preservice teachers began to teach science in the classroom, however, there was a disconnect between their philosophy and their teaching practices, which often ended with more traditional practices. The teachers again emphasized that classroom management, student behavioral issues, and time were limiting factors when teaching science inquiry. Of those preservice teachers participating, students who came to Gilbert's class with a strong aversion to science inquiry were not impacted by the philosophy statements. Preservice teachers with research experiences and science content background were more likely to use inquiry in the classroom, according to a study by Windschitl (2003). He further emphasizes the need for preservice teachers to conceptualize science as a way of knowing.

My study attempted to explore preservice teachers' ability to understand and teach through science inquiry. The goal was to investigate how specific curriculum materials can impact and support preservice teachers' ability to implement inquiry teaching in their internship.

## METHOD

My study included two early childhood preservice teachers and utilized a case study approach. Both preservice teachers were nontraditional students enrolled in a Bachelor of Science degree in early childhood education at a regional campus in the south. Students in this program had science coursework as a part of their general education requirements and one science methods course during their junior internship experience. The teacher candidates were involved in two semesters of internship experiences in public schools. The first internship consisted of two days a week in a primary grade classroom while the second was a full-time classroom teaching experience in which they took over full teaching responsibility for up to six weeks.

An overview of my study's method is as follows. The preservice teachers completed the *Views of Science Inquiry* instrument (Lederman & Ko, 2003), followed by an interview that encouraged them to clarify their own ideas as reported on the instrument (see Appendix A). Second, each teacher was observed teaching science in her junior internship site. They were initially observed teaching lessons of their choice with the curriculum guided by their internship teachers. Observations were conducted using the *Field Supervisor Observation Instrument* developed by Windschitl (2004). During the initial science teaching, the preservice teachers were asked to reflect upon their teaching. Third, the preservice teachers were observed with the Windschitl instrument teaching a science unit based upon the Curriculum Access System for Elementary Science (CASES, n.d.) educative curriculum materials. The CASES curriculum was developed by the CASES research group at the University of Michigan. While teachers were using the CASES curriculum, they were interviewed using a different protocol to ascertain their use of the educative curriculum materials and how they were supportive in their teaching (see Appendix B). Preservice teachers also reflected upon their teaching. At the conclusion of the research, the *Views of Science Inquiry* was once again administered along with an interview to obtain changes in teachers' knowledge, beliefs, and understandings of science and teaching inquiry.

### Views of Scientific Inquiry Instrument

The *Views of Scientific Inquiry—Elementary School Version* (VOSI-E) (Lederman & Ko, 2003) seeks to understand views about what work scientists do and what doing science entails (see Appendix A). The purpose is to seek an understanding of perceptions related to science inquiry. This instrument specifically assesses the development of the following five ideas related to scientific inquiry. One, the teacher understands investigations have multiple methods and purposes. Two, there should be consistency between evidence and conclusions in investigations. Three, scientific data can be interpreted in multiple ways. Four, there is a difference between data and evidence. Five, data analysis involves the development of patterns and explanations.

**Table 7.1**

| *Timeline* | *Activities* |
|---|---|
| Beginning of the Research | 1. Preservice teachers complete the VOSI-E and are interviewed based on their answers |
| Preservice Teachers teach science in their junior internship site | 2. Observed using the Field Supervisor Observation Instrument, Reflections on Teaching |
| Preservice Teachers teach science in their senior internship site using the CASES curriculum | 3. Observations using the Field Supervisor Observation Instrument, Reflections on Teaching<br>CASES interview |
| End of the Research | 4. Preservice teachers complete the VOSI-E and are interviewed based on their answers |

## The CASES Curriculum

The CASES website has two K–2 early childhood science units that have been developed with educative features and eight unit/lessons that have been developed for grades 3–5. The educative features of these curriculum units include an introduction, driving questions, standards, science background, students' alternative ideas, unit lessons, assessment, and ideas and resources. The CASES curriculum units have outlined educative supports including Use of Inquiry Practices, PCK Instructional Strategies, PCK Curricular Rationales, PCK Alternative Ideas, and Subject Matter Knowledge.

Data analysis for my study involved an evaluation of the VOSI-E. Analysis also included examining, categorizing, and tabulating the data from the *Field Supervisor Observation* instrument (Windschitl, 2004) along with the preservice teacher reflections and interviews. Specific techniques for analysis included pattern matching and explanation building. Data were organized topically and sorted to look for patterns. Merriam indicates that "in addition to the participants' own categories, classification schemes can be borrowed from sources outside the study at hand" (1998, p. 137). Categories were established that connected to scientific inquiry to ascertain how the preservice teacher was integrating them into teaching practices. The five essential features of inquiry specifically were used to categorize data in lessons to indicate if teachers were able to implement inquiry with students: engaging in scientifically oriented questions, gathering evidence and developing explanations based on evidence, evaluating explanations in light of alternative explanations, and communicating and justifying proposed explanations (NRC, 2000).

## RESULTS

### Case One: Pippa

Pippa (pseudonym) is a Caucasian female in her early forties. She was a nontraditional student who earned her degree over an extended period of time. At the beginning of the study, Pippa was in her junior internship experience in a third grade, self-contained public school classroom. In the second part of the study, Pippa was participating in her senior internship experience in a kindergarten classroom. Pippa had prior experiences with teaching in the school system. In the science methods course, she participated in inquiry practices and conducted her own science inquiry. The inquiry that she conducted in the methods course occurred at the same time that she was in her junior internship placement. She had positive attitudes about science but was not comfortable with the content.

A review of her initial responses on the VOSI-E indicates she had naive views of scientific inquiry and was not confident in her ability to teach science. On the VOSI-E, Pippa responded in a general way to the question, "What types of activities do scientists do to learn about the natural world?" She explained that scientists attempt to explain and discover events that occur in nature. She further explained that scientists are guided by the science process skills in doing their work but, in the pre-interview, added that this could "follow a scientific method format but could be broader." She was not aware of how this method might change, only that it may not be a strict lock-step method. The VOSI-E instrument provided an example of a scientist who observed birds and made decisions based upon her observations, asking if this work was scientific. Here, Pippa indicated the work was not truly scientific because she did not investigate the birds' beaks. Her response shows a more limited view of scientific inquiry because she believes that, in an experiment, you would need identification and manipulation of variables and controls, which aligns with previous research by Schwartz and Crawford (2005). Pippa believed that it was not an experiment, and she explained that it was observation. She didn't have an understanding that, even though it was not an experiment, it can still be scientific in terms of a broader view of scientific inquiry.

### Science Teaching

***Pre-CASES Lessons:*** Pippa was placed in a third grade classroom for her junior internship. She taught three science lessons with each lasting approximately 30 minutes. The first lesson was a review for a science test on the topic of matter. The second lesson dealt with turkeys and was taught close to Thanksgiving with students creating a science fact book on them. For the last lesson, Pippa worked with a small group of students to read a science text and create a What I know, What I want to know, and What I learned (KWL) chart on the topic of energy, and specifically light.

In the lessons Pippa implemented prior to using the CASES curriculum, she indicated that science content was a difficulty encountered during her preparation and implementation of the lessons she taught. She also indicated that some of the lessons would not really qualify as science lessons but were only a reading lesson with a science text. The observations of Pippa's lessons did not show strong components of scientific inquiry. The five areas identified by the National Research Council (2000) and Windschitl (2004) related to inquiry were used to identify patterns. In the first two lessons, Pippa engaged in scientifically oriented questions but did not delve further into inquiry. Questions were at the lowest level of questioning according to the *Field Supervisor Observation Instrument*: "Learner engages in question provided by teacher, materials, or other source" (Windschitl, 2004). Also, the questions were not driving questions but were only used to interact with students and to ascertain their knowledge of a topic. The focus of the lessons was not inquiry but matter and animals. The last lesson prior to the CASES curriculum was based on reading a science text selection on light, and it did not include a science investigation. In this lesson, Pippa used a KWL chart, which allowed her to move from the lower end of the spectrum where the learner engages in questions provided by the teacher to a higher level in which the learner poses the questions. This was the first lesson to include the higher level of inquiry, where children were allowed to consider their own questions. Students developed a number of questions on the topic of light such as: Would a light bulb melt ice? Would a bulb burn out if you kept it on for an hour? Why are stars so bright at night? And Why does light bounce off mirrors?

In her reflections about teaching, Pippa indicated that science content was a concern for her. She explained that she had to really work hard, and review information, in order to be able to present it to the students. On her first lesson, she reflected,

> I obviously don't feel that the review I did was mine alone. I quickly realized that I should have taken more time to look over the questions. If I had done so, I would have made sure I knew the answers to the ones that I was unsure of.

When Pippa reflected on the turkey lesson, she said that she was not really teaching science or inquiry: "This was not a good science lesson! It was a lesson in following directions and doing steps in order; it was not a lesson based on inquiry by any means."

***CASES Lessons***: Pippa taught five lessons from the CASES plant unit during her senior internship in a kindergarten classroom with 16 students. In the CASES interview, Pippa indicated she was excited about teaching the CASES lessons. Her senior internship site was a kindergarten classroom, which did not implement much science. She taught the CASES science lessons on the topic of plants one day a week over a five-week period. Pippa followed the CASES lessons in a very systematic manner. She enacted them very closely to the way they were developed. Pippa's teacher was supportive of her teaching science using the CASES curriculum.

Pippa seemed very comfortable and even excited about the lessons and the student involvement occurring during them. Her own excitement was visible as she introduced each lesson and engaged children in the topic of seeds. The children in the classroom showed their own excitement about the lessons.

> Whether I did a good job teaching or not, I really enjoyed doing [science] with those poor science starved children. They seemed to have a good time too, which is even more rewarding. (It really does concern me that this area of the curriculum is so neglected.) (CASES interview)

Several aspects of scientific inquiry were visible in the CASES lessons. "I feel that they were involved with inquiry, but at its most basic level of guided inquiry due to the questions begin provided for them. This, I think is an appropriate level for kindergarten students due to their limited exposure to science and consequently limited background knowledge" (CASES interview). When referring to the essential features of inquiry, many aspects were apparent in the CASES lessons taught by Pippa. Each lesson was developed around a central question. These questions were evident in each lesson. Questioning was a key feature of scientific inquiry. The CASES lessons used generally provided the question students pursued. In the lessons taught by Pippa, students did not explore any questions they developed themselves. Another aspect of inquiry investigated was evidence gathering. Students collected evidence in each lesson that was taught in the plants unit. Communication was another aspect of the CASES lessons that was integral in Pippa's teaching. All of the lessons Pippa taught included at least four or five of the five essential features of inquiry. Overall, Pippa implemented the science lessons in ways that facilitated children's thinking and inquiry practices. She proved to be thoughtful regarding the management of lessons and prepared carefully and thoroughly.

Crawford (1999) felt that most teachers would have difficulty with inquiry because they lack the understandings of pedagogical content knowledge and the ability to make the learning of specific science content accessible for students. Pippa used resources available to her in the process of teaching through inquiry practices and was able to use the curriculum to meet the needs of her students. She carefully planned and prepared for the science topics to develop her own understanding. For example, she collected and analyzed seeds to fully experience and understand the lessons prior to her implementation. She also spoke to other educators she considered to be teaching experts to consider the implications of instruction and how different aspects of the instruction would play out in the enactment of lessons.

Pippa commented that the CASES curriculum was a thinking curriculum: "You don't need to bribe kids to get involved because it is engaging" (CASES interview). Pippa felt that using the CASES curriculum was her first real experience in teaching science. She taught the pre-CASES lessons but did not really consider them science lessons. "They were mostly reading or review lessons with some science topics but not actual science" (CASES interview). Two features she felt were very valuable to her in planning and teaching with the CASES curriculum were the background information and children's alternative ideas related to the content. "If you don't have an understanding of what most children think on the topic, you can't really address those ideas in your lessons" (CASES interview). The CASES curriculum helped her to understand what students might be thinking regarding the topic of plants. It also helped her to know how to respond if the alternative ideas came up in the lesson.

At the end of the classroom lesson experiences, Pippa was given the VOSI-E instrument again. Initially, her answers to this instrument seemed quite similar to the initial instrument, but her interview gave insight into Pippa's developing beliefs regarding scientific inquiry. She was asked if there was just one scientific method that should be followed in investigations. Her initial response indicated that the scientific method should be followed in using inquiry. Later in the interview, she clarified her thinking: "Inquiry is broader. I was thinking about the two and confused them. I don't think it [science inquiry] followed step-by-step method. The scientific method is portrayed that way in science projects." Then, referring to her science investigations with students, she said:

> We were not going in a linear fashion but rather coming around revisiting things. You see connections and you go back to something again and revisit

not step-by-step— more learning takes place. The terminology [scientific method] remains from science projects in the past.

This response indicated that Pippa, through her teaching experiences, expanded and broadened her view of scientific inquiry. Pippa also had more time to teach the CASES lessons as she was able to teach five lessons from the plants unit.

## Case Two: Alexis

Alexis is a Caucasian female in her late twenties. She was not a traditional student but earned her degree over an extended period of time, as did Pippa. Alexis was in her junior internship experience in a first grade team teaching classroom in a Central Florida public school. Two teachers co-taught in this room with a total of 28 children. Her first experience teaching took place in her junior internship classroom when she first taught science documented in this study. Alexis' senior internship placement was in a kindergarten classroom.

Alexis's overall responses on the VOSI-E showed that she had a variety of ideas related to scientific inquiry, some accurate and some not. When asked about the kind of work scientists do, Alexis provided a general answer: "They do experiments to find a better way for us to live longer and healthier." When provided with the example of a scientist observing animals, she believed the work was scientific but not an experiment. Here, she indicated the scientist only observed. When further asked how the observation could be made into an experiment, she said she didn't know how. She added that experiments were more fun but observation was fine also. Here, her belief seems to be that experiments were not necessary to gather scientific evidence. This response shows some understandings of a broader view of the multiple methods used to conduct scientific inquiry but she is not able to clearly articulate those additional methods. The comment she made regarding experiments being more fun does not adequately express an understanding that observations can be the basis of scientific information. Observations are a part of experiments, are conducted over time, can be analyzed to develop patterns, and can be considered experiments.

Regarding the question of why scientists have different ideas about the extinction of dinosaurs, she emphasized that this was due to religious beliefs. She believed some scientists are moving beyond scientific facts and are influenced by their religion. In the interview, she said that evidence was needed for drawing conclusions. When considering the development of a new medicine, she said, "Scientists need hard evidence and to be able to back it up" (pre-interview). On the one hand, she saw the need for hard evidence yet indicated that some scientists may not use hard evidence, or facts, due to their religious background. These views appear to be contradictory.

Alexis also provided an example of her daughter, who performed experiments without even knowing it, as opposed to what might be done in a classroom or by a scientist. It seems Alexis realizes to some extent that there are different methods for doing science but she appears to still be clarifying her own understandings of scientific inquiry.

### Science Teaching

**Pre-CASES lessons**: Overall, Alexis taught three lessons during her junior internship. Alexis taught two lessons related to energy to a group of 28 first graders: the first one about sunlight and the second about food. Her last lesson was on insects. Her lessons lasted between 20–25 minutes. Alexis taught her first two science lessons from a scripted curriculum used by the school. The lessons included a flip chart with pictures and facts for students to review. In her lesson reflection, she indicated that, in the first lesson, she was not prepared to teach the lesson and needed more information regarding the science background of the concept of energy. Alexis also was overwhelmed trying to provide hands-on activities to 28 children in the lessons she taught. The science lessons were Alexis' first teaching experiences, as she did not have prior experience in the classroom environment. In all of the three pre-CASES lessons Alexis taught, scientific inquiry was not a strong component. Questioning was used in most lessons but at the lowest level where the learner engages in questions provided by the teacher, materials, or other sources (NRC, 2000: Windschitl, 2004). Most questions were used in a question/answer format to facilitate discussions. In the last lesson taught by Alexis, she also integrated communication at the lowest level: "Learner given steps and procedures for communication" (Windschitl, 2004). In this lesson on insects, children shared their ideas about their favorite insect. Alexis' implementation of the pre-CASES lessons did not demonstrate scientific inquiry, strong pedagogical knowledge, or pedagogical content knowledge. Some factors that influenced the struggle with these lessons were a general lack of support from her teachers and classroom management issues. Additionally, Alexis did not have any prior teaching experience when she began her junior internship.

*CASES Lessons*: For her senior internship experience, Alexis interned in a kindergarten classroom. Science was not a strong focus in this classroom but Alexis had a certain amount of autonomy and flexibility to include science in her whole-group activity time and circle time. The inclusion of science in circle time resulted in a direct instruction approach to science along with center-type activities. The direct instruction occurred when Alexis reviewed important concepts with the whole group and then during center time, when students were involved in activities related to the concept. Alexis taught four lessons from the CASES unit on plants in her senior internship classroom, with the lessons ranging from 22–28 minutes. Schneider and Krajcik (2002) found evidence from classroom trials that indicated teachers used the educative features in the lessons. Examples from Alexis's lessons also indicate she was using the inquiry features provided in the CASES curriculum to support her teaching practice. She used a number of strategies designed to support teachers in the implementation of the CASES lessons. Scientifically oriented questions, data collection, and communication of findings were the three essential features of scientific inquiry most apparent in Alexis' teaching. She also implemented explanations to some level in her teaching of the CASES lessons. The extent to which the CASES curriculum materials were supportive in this setting was dependent upon "how the opportunity is used by the individual" (Davis & Krajcik, 2005, p. 4). This setting appears to indicate Alexis was limited in her acceptance and implementation of the curriculum materials because of other classroom factors, such as placement of the science time during the circle time in the classroom and the management of students during the science lessons.

Alexis indicated that she really liked the CASES curriculum and felt it was supportive in her efforts to teach science inquiry. "I actually like science, I do like it better, and I feel more comfortable with it now that I have some experience. The program [CASES] does give you all the facts [so] you don't have to do much digging. To me it was a hard subject to teach because you need to know all the facts, you don't have to do too much digging" (CASES interview). Alexis made adaptations to the curriculum in the CASES lessons during her teaching. Beyond the materials that were provided for her, Alexis found that she used the Internet to support her own background knowledge when teaching the lessons. She mentioned that she searched the topic of plants to support her own learning prior to teaching students. Alexis decided her students needed additional support or background knowledge in her lessons. In most CASES lessons, she added a review of pertinent concepts to support her children's ideas. She also included a literature selection at the beginning of one lesson, indicating that the book helped children to understand how seeds might move. These adaptations display Alexis' belief that children need support in their learning, and that the scientific activities will not provide sufficient experiences to build their understandings in science. Forbes and Davis (2007) discussed the tension that many teachers have between textbook-guided lessons and inquiry-based activities. They indicated that teachers used curriculum materials in various ways that reflected their own orientations to teaching as well as the influence of the school setting. Although Alexis indicated that she believes inquiry is important, she also believes the textbook or other sources have vital information the students should know up front. She mentioned that she had read the book and changed other aspects of some of the lessons. When asked if this took away from the lessons as being inquiry-oriented, she said, "At this age level, no, because I think they need some background knowledge. They wouldn't have predicted the way seeds moved" (CASES interview).

Alexis' ideas in the pre- and post-VOSI-E responses were similar. Her main belief is that experiments are the central avenue for scientists to do their work but indicated they could use observations and research. In the pre- and post-VOSI-E, she shared that observations can be made into an experiment but also indicated that she thought the scientific method was a "more like a procedure and that inquiry is just looking and digging for stuff without an explanation at the end, without knowing the outcome" (interview). Although she does seem to understand that science could involve more than just experiments, she did not clarify how observations would be considered research. Also, her initial definition of inquiry indicated it was like it is just messing around or playing around (pre-interview). She clarified her definition in the post-interview to say that in inquiry you don't know the outcome. She did not illuminate her stance on data and evidence and why they are significant in inquiry. On the whole, her beliefs demonstrated some understanding of inquiry but not in-depth knowledge.

## CONTRASTING THE TWO TEACHERS

These two preservice teachers used the CASES curriculum materials in a variety of ways. The ways in which materials were used contributed to the level of inquiry and the inquiry features included in the lessons. Davis and Forbes (2007) stated, "Curriculum materials possibly exert the most direct influence on day-to-day classroom activity in which teachers and students engage" (p. 2). The CASES curriculum materials showed influence on the specific classroom practices of these preservice teachers. It is not clear if they actually changed the teachers' science teaching practices. Pippa implemented the CASES curriculum in ways that were aligned with the intention of the materials. She did not appear to experience conflict regarding the implementation. Alexis was constrained in her use of the materials by time and classroom management. Science was not a priority in her class but integrated into the broader learning goals of circle time. She took away from other subject areas to accomplish the lessons that she taught. She also looked to find lessons that would fit into the time frame, which further influenced her enactment of the curriculum. She disregarded the overall building of ideas through the curriculum, instead, picking and choosing lessons that fit the criteria.

The extent to which the CASES curriculum materials were supportive of the two preservice teachers in teaching science inquiry was also dependent upon "how the opportunity is used by the individual" (Davis & Krajcik, 2005, p. 4). Pippa clearly strived to reach understanding of the materials and of how they would be supportive of children in their inquiry about plants. She further used the materials along with her own Internet research to pursue a deeper understanding of the science content and the lessons. Alexis was willing to teach the science lessons but adapted them based on her own ideas about science, how children learn, and their abilities. For each lesson she taught, she did not allow the children an opportunity to develop their ideas about science concepts on their own, but began each lesson with the information up front.

One interesting observation that arose in this study was teachers' ability to implement scientific inquiry. In order for preservice teachers to implement scientific inquiry, they needed stronger knowledge and experience in how to manage the classroom during the active learning that took place. Pippa did have more classroom experience before beginning her internship experiences in this study which may have influenced her practice. Preservice teachers need support in teaching science and, in particular, teaching through scientific inquiry. Previous research has shown that "preservice teachers, given certain caveats and adequate support can feasibly create inquiry-based environments similar to those advocated in the standards" (Crawford, 1999, p. 37). In this study, Pippa did have a teacher and other educators who supported her in learning and exploring the CASES materials.

Other research by Gado (2005) advocated for support for early childhood teachers to develop their competence and comprehension of teaching inquiry. The CASES curriculum provides the built-in supports for teaching inquiry. The educative features of the CASES unit includes an introduction, driving questions, standards, science background, students' alternative ideas, unit lessons, assessment, and ideas and resources. Additionally, preservice teachers who use the CASES materials have access to educative supports that explain the background for science inquiry. These include: Use of Inquiry Practices, PCK Instructional Strategies, PCK Curricular Rationales, PCK Alternative Ideas, and Subject Matter Knowledge.

During this research study, both preservice teachers included more components of science inquiry when using the CASES educative curriculum compared to when they taught their own science lessons (see Table 7.2). Additionally, the preservice teachers had more experience in teaching science when they used the CASES curriculum; the lessons also provided specific supports that are built in so teachers can better implement scientific inquiry. The findings align with previous research by Petish (2004) regarding the use of educative materials that served to expand the teachers' repertoire of instructional strategies related to inquiry. The five essential areas in science inquiry teaching include questions, evidence, explanations, connecting explanations to scientific knowledge, and communicating/justifying explanations (Windschitl, 2004). Questions were a big part of the curriculum and inquiry. Pippa included the use of driving questions in her lessons. Driving questions are ones that provide a focal point for lessons. Teachers and students elaborate, explore, and answer the questions during lessons. The CASES curriculum also emphasizes referring back to the questions throughout the lessons to maintain the focus. Both Pippa and Alexis used scientifically oriented questions as a focus for lessons. Scientifically oriented questions are ones that focus on science aspects of learning and can be the same or similar to the driving questions.

**Table 7.2**

| | *Pre-CASES Lessons* | | *CASES Lessons* | |
|---|---|---|---|---|
| *Features of Inquiry* | *Pippa—3 Lessons* | *Alexis—3 Lessons* | *Pippa—5 Lessons* | *Alexis—3 Lessons* |
| Scientifically oriented questions | √√√ | √√√ | √√√√√ | √√√√ |
| Evidence/Data analysis | | | √√√√√ | √√√ |
| Scientific Explanations | | | √√√ | |
| Connects explanations to scientific knowledge | | | √√ | √√ |
| Communication | | √ | √√√√√ | √√ |

Other features of inquiry present in the teaching included evidence gathering. Both teachers implemented some evidence gathering practices in the CASES lessons, mainly in the form of observations and classification. Pippa worked with students to emphasize the need for accuracy when making observations and encouraged children to make detailed drawings and use the correct colors, "so that if someone else looked at your drawing they would know what you had drawn" (*Field Supervisor Observation Instrument*). Alexis provided information to students at the beginning of the lesson instead of allowing them to gather their own evidence. If students made observations, she did not always have them record their findings.

Neither of the teachers went further in the area of evidence gathering to support students in analyzing information. This analysis might have included double checking information, repeating experiments, or verifying accuracy as suggested by the NRC (2000). The development of scientific explanations is an important aspect of inquiry because it helps children make sense of science (NRC, 2000).

The third essential feature of inquiry is formulating explanations. The CASES curriculum emphasizes the process of having children make a claim using their evidence to explain why. Pippa regularly asked children to explain their conclusions using evidence. When the students indicated they thought a seed would travel by water, for example, she asked, "What was it about the seed that made you think that?" (*Field Supervisor Observation Instrument*). Alexis asked students to explain their ideas during science lessons, but she also began the lessons with explanations in the form of a direct instruction mini-lesson of the content.

The fourth feature of scientific inquiry that was a part of the CASES curriculum was "learners connect explanations to scientific knowledge" (NRC, 2000). In the CASES curriculum, this aspect of inquiry was incorporated into the formation of scientific knowledge. The NRC (2000) indicates that investigations provide the context for developing definitions and the associated science understandings. Pippa was comfortable with allowing her students to construct their knowledge on their own. She brought in scientific knowledge when it was a part of the curriculum. She, for example, shared a diagram of the seed and its parts with students at the end of the lesson. In one lesson Alexis began by reading a book about how seeds travel, students then examined the seeds. Next, she asked the students to explain their ideas into a microphone. She would respond to the students saying, "Why do you think so?" (*Field Supervisor Observation Instrument*). She emphasized the need for children to consider their ideas in this lesson, and she also provided information before the students participated in the lesson. This contradicted the plan for students to develop the science ideas during the lesson.

Aligned with the development of explanations as a feature of inquiry is communication. As mentioned above, Alexis encouraged this when students shared their ideas by talking into a microphone. Communication also can support clarification of ideas for students as they share their ideas with others. Pippa included a variety of ways in which children could communicate their understandings including drawings, verbal explanations in class, class charts, and seed charts as ways for children to share their ideas and understandings of the content.

Classroom management skills and pedagogical skills were a factor when comparing the two preservice teachers and their ability to teach science. The observations indicate that Pippa was better able to manage the classroom in order to teach science. Science may be a more active subject area when children are engaged in their learning, and therefore, harder for new teachers to manage (Mulholland & Wallace, 2003, 2005). Pippa had previous experience with children in schools, which appeared to support her ability to manage the science lessons based upon the Field Supervisor Observation Instrument. Alexis struggled to maintain control of the students while teaching science. She worked through each lesson she taught in order to find ways to manage and facilitate the active nature of teaching science.

Research reported by Crawford indicates that many beginning teachers will have difficulty with inquiry because of their lack of pedagogical content knowledge: "The ability to adapt and mold instruction in response to student-centered inquiry appears a likely stumbling block for novice teachers who have difficulty with improvisation during interactive teaching" (Crawford, 1999, p. 7). In this study, Alexis was focused on general management strategies and, therefore, not able to really consider the most effective strategies to support children's learning of science. This finding may indicate that preservice teachers may need to reach a certain level of competency in management prior to moving forward in inquiry practices and the strong consideration of how they support children's learning. Pippa was one exception to Crawford's ideas in this case study. She was able to overcome obstacles in order to fully accept and understand how to teach using this curriculum including the use of many inquiry practices.

## CONCLUSIONS

Although previous research has indicated the difficulty preservice and inservice teachers face in implementing science inquiry practices, the teachers in this study did incorporate a number of inquiry features into their teaching practice. Preservice teachers need support in teaching science and, in particular, teaching through scientific inquiry (Weiss et al., 2001). The CASES curriculum provides support for many aspects of inquiry, as demonstrated by Pippa and Alexis. Both Pippa and Alexis implemented more features of science inquiry when using the CASES curriculum (see Table 7.1). Although the CASES curriculum provided support for these two preservice teachers in implementing scientific inquiry, the extent to which they enacted the inquiry appeared to be based on additional factors. The type of inquiry that was demonstrated by the early childhood students was guided inquiry.

Some factors that seemed to interfere with the teachers' ability to implement the science inquiry practices included their beliefs about inquiry, classroom management, and time. The research by Plevyak (2007) was supported in this study, as Alexis appeared to need more time and interactions with young children to fully support her implementation of the inquiry practices. Additionally, Alexis and Pippa were exposed to scientific inquiry in their science methods course, but neither student came with a background or strong experiences related to inquiry which made the transition to these teaching practices more difficult (Gilbert, 2009; Windschitl, 2003). The National Research Council (2000) also emphasizes that "it can take a significant amount of time to make transformational changes in teaching" (p. 91), reinforcing the idea that Alexis and Pippa did not have a strong background in teaching science and therefore need more time to fully implement inquiry.

The findings of this research study bring new questions to consider. When two preservice teachers used the CASES curriculum, they did include more features of scientific inquiry. Further research on the use of the CASES materials with preservice and inservice teachers would be recommended. Future research could investigate the effects of the CASES curriculum over an extended period of time on preservice and classroom teachers' science teaching practices. For instance, if the teachers are not guided by the specific CASES curriculum, will they still implement the five aspects of inquiry into their science teaching after they have used the curriculum? Also, when looking at the features of inquiry that were included in the CASES lessons, it would be beneficial to see if preservice teachers could move toward higher levels of the essential features and if they would regularly include all five features of inquiry into science teaching practices. It would be beneficial to include the CASES curriculum in the science methods course as an additional way to introduce curriculum that supports science inquiry. Lastly, it would be important to explicitly teach more about science inquiry in the science methods course and to link the course activities with the first internship experience to see if support can be provided to students as they bridge the gap between theory and practice.

## SUMMARY

This case study investigated how two preservice early childhood teachers implemented science inquiry practices in their internship sites during their junior and senior internship experiences. In the field of early childhood education, we realize that each child is unique and brings a wide range of experiences that impact their understandings of content. Teachers also bring to their teaching a wide variety of experiences, which have formed their views of teaching (Everett et al., 2006). We must look to these views and beliefs to help us understand and support teachers in their movement towards inquiry-based teaching practices. Teacher beliefs and understandings of science inquiry teaching can be addressed to some extent in science methods courses (Gilbert, 2009; Plevyak, 2007). Preservice teachers may need additional time to develop their pedagogical content knowledge through ongoing teaching experiences. This study shows that although the CASES curriculum does provide support for the two preservice teachers when implementing scientific inquiry, other factors must also be considered. A teacher's willingness and openness to learn and teach inquiry can also be influenced by the teachers' classroom management and beliefs about children's abilities and how they learn. Additionally, classroom management in more active science lessons also affected the preservice teachers' implementation of science inquiry. Teachers' ideas and understandings of scientific inquiry also influence how they use curriculum materials in practice.

## APPENDIX A

*Views of Scientific Inquiry—Elementary School Version* (VOSI-E) (Lederman & Ko, 2003).

### Instructions

- We want to know what you think about science and how it is done.
- Some questions may have more than one part. Please make sure you put answers for each part.
- There are no "right" or "wrong" answers to the following questions.
- If you want, you can draw pictures to help explain your ideas.

1. What kind of work do scientists do?
2. Explain how scientists do their work.
3. A scientist studied many different kinds of birds. She noticed that birds that eat the same types of food usually have the same shaped beaks. For example, many birds that eat hard nuts have short, strong beaks, and many birds that pick worms out of the ground have long, thin beaks. So, the scientist decided that there is a connection between beak shape and the type of food birds eat.

a. Do you think her work was scientific? Why or why not?
b. Do you think her work was an experiment? Please explain why or why not.

4. A long time ago all the dinosaurs died. Many scientists are trying to find out why this happened.

   a. Do you think they will all come up with the same reasons for why this happened?
   b. Explain why or why not.
   c. What information do scientists use to explain their reasons for why this happened?

## APPENDIX B

Protocol for Science Teaching Interview (Pre-Interview).

The VOSI-E should be reviewed prior to the interview period. Questions and topics will be guided by the responses provided but the general topics for the interview are outlined below.

*General Topics will include:*

Terminology used

Explanation of responses

Verbal clarification of ideas

Participants general comments about instruments, teaching science or feelings about science

***CASES Implementation Interview Questions:***

- Do you feel that you successfully implemented science lessons in your internship site?
- Do you feel you were successfully able to implement the CASES curriculum?
- What support was helpful in teaching science in your internship site?
- How did you use the CASES curriculum?
- Can you explain how you used the CASES resources to plan your science unit? When planning specific lessons?
- Do you think the students learned what you intended them to learn in the CASES unit? Why or Why not?
- Do you think the CASES website has influenced your thinking about teaching science? Explain.

- How did you use the information on the CASES website? What aspects did you use? What parts did you choose not to use? Why?
- When you were preparing to teach using the CASES lessons, what did you find helpful?
- How did the CASES curriculum support inquiry?

Protocol for Science Teaching Interview (Post-Interview)
The protocol for the post interview will be the same as the pre-interview with the following general questions added:

- Overall, how have your ideas and feelings regarding the teaching of science changed since the beginning of this project?
- How has your understanding of inquiry changed since the beginning of this project?
- What is your overall philosophy of education and science education?
- What are your beliefs related to how children learn?
- What do you believe is the best way for children to learn science?

## REFERENCES

Black, K. (2004). Science in the trenches: An exploration of four pre-service teachers' first attempts at teaching science in the classroom, *International Journal of Science and Mathematics Education, 2,* 25–44.

Crawford, B. A. (1999). Is it realistic to expect a pre-service teacher to create an inquiry-based classroom? *Journal of Science Teacher Education, 10*(3), 175–194.

Curriculum Access System for Elementary Science. (n.d.) University of Michigan. Retrieved from http://cases.soe.umich.edu/

Davis, E. A., & Krajcik, J. S., (2005). Designing educative curriculum materials to promote teacher learning. *Educational Researcher, 34*(3), 3–14.

Davis, E. A., Petish, D., & Smithey, J. (2006). Challenges new science teachers face. *Review of Educational Research, 76*(4), 607–651.

Everett, S. A., Luera, G. R., & Otto, C. A., (2006). Pre-service elementary teachers bridge the gap between research and practice. *International Journal of Science and Mathematics Education, 6*(1), 1–17.

Forbes, C. T., & Davis, E. A. (2007, April). *Beginning elementary teachers' learning through the use of science curriculum materials: A longitudinal study.* Paper presented at the annual meeting of the National Association for Research in Science Teaching, New Orleans, LA.

Gado, I. (2005). Determinants of K–2 school teachers' orientation toward inquiry-based science activities: A mixed method study. *International Journal of Science and mathematics Education, 3*, 511–539.

Gee, C. J., Boberg, W. S., & Gabel, D. L. (1996, April). *Pre-service elementary teachers; Their science content knowledge, pedagogical knowledge, and pedagogical content knowledge.* Paper presented at the annual meeting of the National Association for Research in Science Teaching, St. Louis, MO.

Gilbert, A. (2009). Utilizing science philosophy statements to facilitate K–3 teacher candidates' development of inquiry-based science practice. *Early Childhood Education Journal, 36*(4), 431–438.

Hudson, P. (2004). Toward identifying pedagogical knowledge for mentoring in primary science teaching. *Journal of Science Education and Technology, 13*(2), 215-225.

Lederman, J. S., & Ko, E. K. (2003). *Views of scientific inquiry—Elementary school version.* Unpublished paper. Illinois Institute of Technology, Chicago, IL.

Merriam, S. B. (1988). *Case study research in education: A qualitative approach.* San Francisco, CA: Jossey-Bass.

Mulholland, J., & Wallace, J. (2003). Facilitating primary science teaching: A narrative account of research as learning. *Teachers and Teaching: Theory and Practice, 9*(2), 133–155.

Mulholland, J., & Wallace, J. (2005). Growing the tree of teacher knowledge: Ten years of learning to teach elementary science, *Journal of Research in Science Teaching, 42*(7), 767–790.

National Research Council. (1996). *National science education standards.* Washington, DC: National Academy Press.

National Research Council. (2000). *Inquiry and the national science education standards: A guide for teaching and learning.* Washington, DC: National Academy Press.

Petish, D. A. (2004). *Using educative curriculum materials to support new elementary teachers' practice and learning.* Unpublished doctoral dissertation University of Michigan, Ann Arbor.

Plevyak, L. H. (2007, July). What do pre-service teachers learn in an inquiry-based science methods course? *Journal of Elementary Science Education, 19*(1), 1–13.

Sadaghani, H. R., & Costley, S. N. (2009). *The effect of an inquiry based early field experience on pre-service teachers' content knowledge and attitudes toward teaching.* Paper presented at the Physics Education Research Conference.

Schneider, R. M., & Krajcik, J. (2002). Supporting science teacher learning: the role of educative curriculum materials, *Journal of Science Teacher Education, 13*(3), 221–245.

Schwartz, R. S., & Crawford, B. A. (2005). Authentic scientific inquiry as a context for teaching nature of science: Identifying critical elements for success. In L. B. Flick & N. G. Lederman (Eds.), *Scientific inquiry and nature of science* (pp. 331–355). Dordrecht, The Netherlands: Springer.

Shulman, L. S. (1986). Those who understand: Knowledge growth in teaching. *Educational Researcher, 15*(2), 4–14.

Weiss, I., Banilower, E., McMahon, K., & Smith, P. (2001). *Report of the 2000 National Survey of Science and Mathematics Education.* [online]. Retrieved from www.horizon-research.com

Windschitl, M. (2003). Inquiry projects in science teacher education: What can investigative experiences reveal about teacher thinking and eventual classroom practice? *Science Teacher Education, 87,* 112–143.

Windschitl, M. (2004). Folk theories of "inquiry": How pre-service teachers reproduce the discourse and practice of an atheoretical scientific method. *Journal of Research in Science Teaching, 41*(5), 481–512.

Yoon, J., & Onchwari, J. A. (2007). Teaching young children science: Three key points. *Early Childhood Education Journal. 33*(6), 419–423.

CHAPTER 8

# EARLY SCIENCE TEACHING EXPERIENCES FOR UNDERGRADUATES IN AN ON-CAMPUS, HANDS-ON LABORATORY

**Sarah Tanya Heaston and Bev Marcum**

## ABSTRACT

Research has shown that undergraduates planning to teach elementary grades do not feel confident about the prospect of teaching science. Reasons given included: insufficient content knowledge, lack of enjoyment of science classes in school, lack of positive experiences in science during their schooling, and an expressed fear of science. This study focused on the effect of an early science teaching experience for undergraduates in an on-campus hands-on laboratory where undergraduates had the opportunity to interact with elementary school children in a controlled, supervised classroom setting at California State University–Chico. Practicing using real materials with small groups of children for short periods of time and having the opportunity to reflect on the teaching experience and to repeat it after further discussion and revision has proven to be an effective way of increasing positive attitudes toward teaching science and increasing content knowledge in science for

*Research Based Undergraduate Science Teaching*, pp. 247–274

the participating undergraduates. Pre-post attitude surveys indicated that undergraduates experienced increases in their perception of being prepared to teach science, knowing what to do to help their students like science, and feeling confident that they have the skills to help their students learn science. All undergraduates in the early science teaching experience also showed gains in science content knowledge. In addition, the Hands-on Laboratory classroom setting provided opportunities for inservice teachers to participate in professional development opportunities as an extension of bringing their classroom students to campus.

There is general agreement among educators, as expressed in the *Taking Science to School* report from the Committee on Science Learning: Kindergarten Through Eighth Grade, that improvements in science education in the United States in the last 15 years have lagged in comparison to students in other countries (National Research Council [NRC], 2007). This committee states that teachers need "opportunities to learn how to teach science as an integrated body of knowledge and practice—to teach for scientific proficiency" (NRC, 2007, p. 7). The committee also recommends that teachers learn how to teach science to diverse populations, including English learners and students in low socioeconomic groups, "to provide adequate opportunities for all students to learn science" (NRC, 2007, p. 7). We report, here, on the impact on attitude change and content learned via an early teaching opportunity for undergraduates planning to become teachers in grades K–8.

With the implementation of standards in all disciplines and particularly with an emphasis on standardized testing in language arts and mathematics, teachers at the elementary level spend less time teaching science. The Center for Education Policy reported that science instruction time in elementary schools surveyed has decreased by an average of 33% of total instructional time in science since the No Child Left Behind (NCLB) legislation was instituted in 2002 (McMurrer, 2008). Local teachers have reported the same effect: that increased instructional time requirements for English language arts and mathematics have led to decreased time spent on other subjects, especially science. This finding is particularly true for the primary grades, K–3, in California, since science testing in California occurs only in the fifth grade and only covers standards from the fourth and fifth grades.

Other factors compounding the problem for science education in the elementary grades include inadequate resources for laboratory and field based science investigations, teachers' lack of science content knowledge, and the negative self-efficacy of many elementary teachers about their ability to teach science (Czerniak & Chiarelott, 1990). It also has been reported that the negative attitudes of teachers can be correlated to students' negative attitudes about science (Czerniak & Chiarelott, 1990). More locally, and with unpublished results, a fourth grade teacher who brought her classes each year to California State University–Chico's science lab surveyed her incoming fourth graders and found that approximately 40% of the students report that they either (1) think science is more difficult than other subjects, (2) say they do not like science as much as other subjects, or (3) think that they are not as good in science as they are in other subjects. This is a remarkable finding for students coming from the third grade since the school where this survey was conducted focused almost exclusively on English language arts and mathematics in grades K–3 and science was not part of the instructional program for the fourth grade students being surveyed.

Even when elementary school teachers are given time for science, they may not be enthusiastic about teaching it. One of the most overlooked problems with the implementation of science education standards was that many teachers were not prepared for them. Downing, Filer, and Chamberlain (1997) found teachers who were already in the classroom were less confident about teaching science compared to teachers who are just completing their training programs and that some teachers avoided science instruction as much as possible in order to avoid having to deal with a subject they feel uneasy teaching. These teachers did not possess the appropriate science background to be able to teach the content, nor did they have the necessary experience in using the materials that are required to teach science effectively (Ritter, Boone, & Rubba, 2001).

Teachers' under-preparation for teaching science at the elementary level reflects negatively on teacher preparation programs. Undergraduate science classes for future teachers typically are taught by science faculty in the disciplines of biology, chemistry, physics, and/or earth science with an emphasis placed on content specific to each discipline. Science courses for undergraduates planning to teach often are part of the program of general education science for non-science majors and consist of an overview of the discipline. In many cases, the classes consist of a collection of factual information with little relevance or mechanism of transfer for the future teacher.

That the majority of undergraduates planning to become future elementary teachers express a dislike of science and a lack of confidence in their ability to learn science and, consequently, in their ability to teach science to elementary children is well documented (Jarrett, 1999; Jones & Carter, 2008; Lederman, 2008; Tosun, 2000; Watters & Ginns, 2000). Jarrett (1999) emphasizes that lack of interest in science leads to lack of coursework in science leading to further lack of interest and confidence

Science content is being emphasized in some teacher preparation programs, as are instruction in pedagogy and teaching strategies (Johnston, 2003; Kelly, 2000). Because teachers teach the way they were taught, much research is focusing on the need for college faculty in teacher preparation programs to model effective teaching styles in undergraduate classrooms (Gibson, Bernhard, Kropf, & Van Strat, 2001). In one study, the confidence of undergraduate preservice teachers was shown to improve after teaching in elementary classrooms (Murphy, Beggs, & Carlisle, 2003). Tosun (2000) focused on prior science experiences and achievement in science classes and the effect on preservice elementary teachers' attitudes toward science. One interesting finding of Tosun's research was that no matter how high students scored on content exams, they still were likely to have negative feelings toward science and the teaching of science. Students with low scores and minimal science experience showed a lack of confidence toward sharing science content. Tosun recommended frequent successful field experiences for preservice elementary teachers to help improve their self-efficacy.

Another project documented the development of a constructivist-based elementary science methods course for preservice teachers over a four-year period. The course included science methods content, journal writing, curriculum development, creating science inquiry activities for a science museum, evaluating lessons, and participating in a teaching experience (Kelly, 2000). Results from this project showed a significant change in the preservice teachers' attitudes toward teaching science. Prior to the course, the students did not like science and felt that it was hard to master, or that it was boring. After the course, the students' views of science improved. Results also showed a significant improvement in their confidence to teach science.

A strong case is made by Marzano, Pickering, and Pollock (2001) for a direct connection between the teacher and the success of elementary students. Variables affecting the ability of elementary school children to learn were identified and studied; the most important factor among all variables was found to be the teacher. An effective teacher can transcend any number of barriers to learning to create effective learning environments. Undergraduate students with positive attitudes and an appropriate level of content knowledge seem likely to become more effective teachers. The attitude of the preservice teacher toward science is strongly correlated with his or her ability and attitude toward teaching. Having successful science experiences helps create positive attitudes. Positively affecting students' belief about their abilities to learn and teach science may be increased by "doing" the science activities (Cochran, 1997).

To counter some of the negative impacts of typical science courses for future teachers and provide relevance to teaching science in the elementary grades, an early science teaching experience for undergraduates at California State University–Chico was instituted through a (National Aeronautics and Space Administration /NASA Opportunities for Visionary Academics, or NASA/NOVA) integrated science teaching grant in 2002. This course, the *Internship in Science Teaching*, is now a required class for liberal studies students planning to enter the multiple subject credential program in the university's College of Communication and Education. A multiple subject credential is the credential of choice for most elementary teachers and, in the state of California, qualifies them to teach grades K–8. Additionally, science majors who are considering teaching as a career also may take the internship class to gain teaching experience.

In the Internship in Science Teaching class, undergraduates are the facilitators for elementary and middle school students who visit campus and participate in the Hands-on Lab science education experience. In the safe and supervised environment of the Hands-on Lab, undergraduates have the opportunity to experiment with teaching strategies, obtain practical experience managing science materials, learn to ask strategic questions, and assess for understanding. Children are organized into small groups and rotate to different, but related, mini-lessons on specific topics. Topics originally were selected from the California Science Content Standards for fourth and fifth grades to provide additional support to inservice teachers of grades tested in science.

The same science content lessons are repeated in a second week of the course with a different visiting class of inservice teachers' students. In addition to teaching experience, the lab includes a debriefing and training component during which the science faculty facilitates discussions on pedagogy and content related to the teaching experience. The discussions and journal entries that are part of the course provide undergraduates with the opportunity to practice reflective teaching and learn from the experiences of their peers as well as from their own experiences. Since initiating the program in 2002, nearly 1,800 undergraduates, 350 inservice teachers, and 27,000 school children have experienced science teaching and learning in the Hands-on Lab. Teachers within an 80-mile radius have brought their classes to campus for this experience. We have a waiting list of classes that would like to participate. We also have been able to utilize separate professional development and partnership funding for teachers and specific school districts to ensure a diverse student population with whom our undergraduates interact.

The research investigation on this class was undertaken to provide data on the effectiveness of the class in reaching two of the goals: improving attitudes toward science and increasing content knowledge in science. The research project constituted partial fulfillment of a master's degree in science teaching for the first author, Tanya Heaston. The study group consisted mostly of undergraduate liberal studies majors who ranged from sophomore to senior level based on the number of units of college coursework completed. The intent of this study was to determine if a guided teaching experience in science positively influenced undergraduate student attitudes toward science and toward science teaching. Previous research has shown that the attitude toward science of future teachers improved when they participated in presenting science curriculum to an elementary classroom (Gibson et al., 2001; Kelly, 2000). The instrument used in the current study was developed by McGinnis (1998) and used for attitudinal and belief studies of future teachers in college science courses. This study focused on the teaching experience in a controlled setting in a laboratory on campus with no more than six students per undergraduate student. In addition, standards-based content knowledge for each undergraduate was measured at the beginning and at the end of the class.

The current study is unique in that college undergraduates have the opportunity to learn and to teach science in a small group setting to visiting elementary students multiple times. Because the short lab activities are repeated as groups of elementary students move through the lab, undergraduates facilitating the mini-lessons can quickly reflect on and change their approach to presenting materials and asking questions to assess understanding. The same theme is taught during two consecutive weeks, which allows the undergraduate students an opportunity to discuss the activity with their instructors and peers.

## DEVELOPMENT OF THE INTERNSHIP IN SCIENCE TEACHING COURSE

California State University–Chico had offered local elementary school teachers the option of bringing their classes to campus for science experiences for many years before it was developed into an early teaching experience course for undergraduates. Historically, two different laboratory experiences had been developed by faculty in the chemistry department to expose children to some basic techniques and concepts in science. Two to four undergraduate chemistry majors each semester had volunteered to conduct the science experiments, Purple Cabbage Chemistry and Mystery Powders, for the visiting children. Teachers brought their students to campus either by school bus, by city bus, by parent drivers, or by walking. Some laboratory supplies for the program were obtained through small grants.

Funding from the NASA/NOVA program made it possible to establish a special science class for a selected group of undergraduates interested in teaching, and this special class utilized a similar science teaching experience as part of the laboratory. This project was originally entitled *Lowering the Learning Barriers in Science for Pre-service Elementary Teachers*. The goals of this project were to:

1. Positively impact the motivation of future elementary school teachers to understand biological concepts pertaining to their health, and the health of the environment, by providing a stimulating venue that is viewed as relevant to the preservice teachers.
2. Increase knowledge of the nature of scientific investigation by incorporating investigative procedures into the course.
3. Enhance confidence in communicating science content accurately by providing a forum for information exchange through written and oral language.
4. Provide experiences that foster lifelong learning of science.
5. Provide early experiences for teaching science to children.

Undergraduate students from an accelerated program called *Teaching Links* constituted the first group of undergraduates for the science teaching experience. The *Teaching Links* program was funded by a Title VII grant awarded to the College of Communication and Education to recruit, advise, and financially support bilingual undergraduates to become credentialed teachers to serve English learners, children for whom English is a second language. The liberal studies Advising Office and the Center for Bilingual and Multicultural Studies placed the *Teaching Links* freshmen in the integrated NASA/NOVA science class the first semester it was taught in the fall, 2002. In subsequent semesters, additional liberal studies undergraduate students were enrolled into the course, and the class became a required course for liberal studies majors in the fall, 2004. This course now has 200 undergraduate students enrolled each year.

An additional, and important, aspect of this program was the existence of a partnership between the university and the inservice teaching community via the presence of a professional development network, the California Science Project, in the College of Natural Sciences. In 1991 the California legislature authorized a group of Subject Matter Projects to provide professional development in various content areas for K–12 teachers. Science was one of the subjects, and Bev Marcum, a co-principal investigator for the NASA/NOVA grant, was the director of the regional site of the California Science Project. The existence of the California Science Project's network of teachers provided a source of experienced inservice teachers to advise and participate in the new course in science teaching. Information presented in the lab was completely aligned to the Science Content Standards for California Public Schools (California Department of Education, 2000) as well as the National Standards (NRC, 1996).

## CONDUCTING THE COURSE

Undergraduates involved in the study were taking part in the early teaching experience class, the Internship in Science Teaching, also known as the Hands-on Lab (HOL). The semester-long class met twice a week; two hours each week were devoted to a lecture/activities seminar, and one hour each week was spent teaching elementary students in the teaching lab. During the seminar portion of the class, the undergraduates were provided with background on science content, briefed on teaching strategies, encouraged to make changes in their own strategies, and provided with writing prompts to help them reflect upon their experiences in their journals. During these two-hour periods, the undergraduates also were introduced to a variety of standards-based laboratory activities. Each activity was designed to last 15 minutes. Undergraduates were expected to learn the science behind the activities and were given time to practice teaching the labs to one another.

From week three through the rest of the semester, elementary school students visited the lab. The HOL lab consisted of six lab tables/stations, each with a different science activity for a particular science unit. The elementary school class was split into six groups of 4–6 students each. After 15 minutes at the initial starting table/station, the whole class rotated to the next station until each student had experienced all six activities. The rotation to all six stations took one hour and 45 minutes. Each group of undergraduates stayed in the lab for one hour. The lab units/themes changed every two weeks, and in the remaining twelve weeks of the semester, the undergraduates had taught six separate science units two times each.

Each undergraduate was able to teach his or her particular lab station three times during each teaching hour. Undergraduates were able to work with small groups of students and experience different learning styles. This method of teaching the same lab multiple times allowed undergraduates to test different techniques for teaching a particular science concept. If there were problems helping elementary students understand a concept, the undergraduates would either self-correct or ask for help from the lab instructor. Undergraduates also had to adjust their lessons for time, difficulty of content, age of the elementary students, and language abilities.

## CONDUCTING THE STUDY

Assessment of the undergraduates from semesters previous to the term in which this study occurred showed that they did not feel comfortable with the idea of teaching science as future elementary school teachers, their stated career choice. Analysis of the undergraduates' journals showed that interns:

- don't have much experience teaching elementary school students
- are uncomfortable teaching science
- lack science content knowledge
- prefer teaching reading over science
- have had few positive science experiences

After a semester learning and teaching in the Hands-On Lab (HOL) the interns reported that they:

- have built confidence after teaching the same HOL lab 2–3 times
- understand science content after teaching labs 2–3 times
- plan to teach science in their elementary classrooms
- believe that science can be fun, interesting, and worthwhile

Based on these findings, it was hypothesized that if the students are allowed to teach science labs as undergraduates, as early in their coursework as possible, and if it is a good experience for them, then they will have a better attitude toward science, see the relevance in learning the science content in their college classes, and possibly teach more science when they are credentialed teachers. The first of these, an improved attitude toward science, was the focus of this study.

## UNDERGRADUATES IN THE STUDY

Seventy undergraduates participated in the HOL during the fall 2005 semester, when the study was conducted. Fifty-six of these were liberal studies majors; seven were child development majors, and the remaining seven were majoring in biology or another science. Of the 70 participants, 20 were sophomores, 8 were freshmen, 10 were juniors, and 32 were seniors. Twenty-seven of the undergraduates were associated with a cohort of students who were on an accelerated, fast track to complete their bachelor's degree and teaching credential within four years, instead of the usual five required for a preliminary credential in California. This group is known as the ITC- LINKs (Integrated Teaching Corps Linked classes) group. Twenty-six of the ITC-LINKs students were female, one was male. Seventeen of these students were sophomores and 10 were juniors.

Altogether, 8 males and 62 females participated in the study. The majority of the group, ($n$ = 56) were education majors. The rest of the group consisted of four mathematics education, four science education, two mathematics/science education, and four other majors. The survey showed that, of the 70 students, 65 intended to pursue an elementary teaching credential, two intended to enter the secondary math or science teaching credential program, and three students planned to go on to higher degree programs. The pre-surveys showed 40 students intended to teach grades K–3, 20 were interested in teaching grades 4–8, one student intended to teach secondary education, and nine students were undecided as to which grade level they would like to teach. The posttest showed a slight change, with 23 students wanting to teach 4th–8th grades and 37 wanting to teach kindergarten–third grade. Undergraduates' age and socioeconomic background were not considered in this study, although grade ranking (freshman, sophomore, junior, senior) was a consideration.

Most of the students involved in the study had taken at least two science classes in high school. The fewest number of science classes taken by any single student was one and the most was five. High school science classes taken by the undergraduates ranged from biology, chemistry, and physics, to astronomy, human anatomy, health, and geology (see Figure 8.1). Only three of the undergraduates had taken advance placement (AP) classes through their high school education. The range of college science classes students had taken was not quite as varied as were their high school courses. Students entered the Hands-on Lab (HOL) having taken an average of 1.2 science courses; the majority had taken two college science classes. The liberal studies program requires four specific science courses for completion of the degree; two in the life sciences (biology and field biology), and two in the physical sciences (physical science and Earth science). The science backgrounds of the non-liberal studies majors in the class ranged from one student who had taken eight science intensive courses, including two upper division courses, to a student with only a field biology class.

## SURVEYS

During the first week of the semester, prior to the presentation of any content material, the interns enrolled in the HOL course were given an attitudinal survey. The instrument for the attitudinal survey was taken from the Maryland Collaborative for Teaching Preparation (MCTP). The MCTP Survey Instrument, the *Attitudes and Belief about the Nature of and the Teaching of Mathematics and Science* (McGinnis, 1997), has been tested and found to be robust. The survey was modified to include only those questions regarding science teaching and attitudes. Student responses were measured on a 5-point Likert scale.

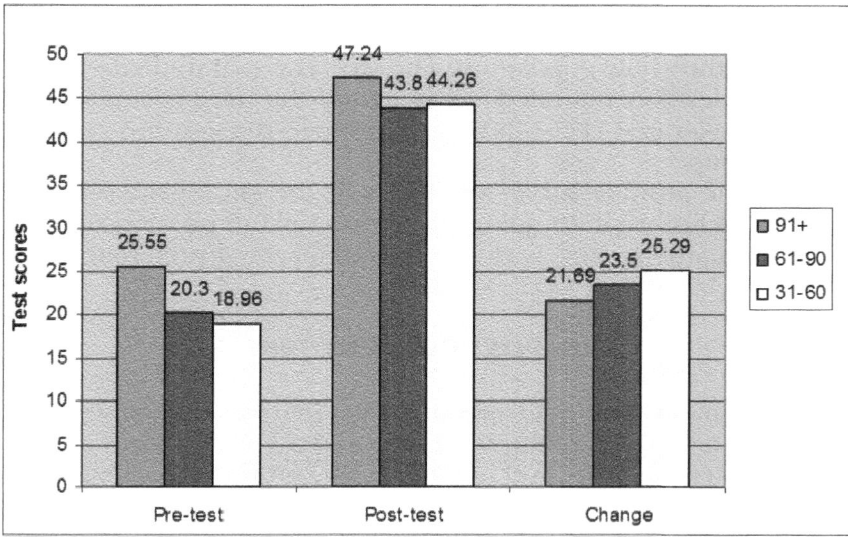


**Figure 8.1.** Mean raw score by number of completed college credits.

The attitude survey used was compiled from surveys found on the National Science Foundation's Online Evaluation Resource Library (OERL, n.d.) website. The majority of the survey was from the Maryland Collaborative for Teaching Preparation (MCTP) Survey Instrument, the *Attitudes and Belief about the Nature of and the Teaching of Mathematics and Science* (McGinnis, 2003). Statements also were gleaned from the *OERL Teacher/Faculty Survey and Student Attitude Survey*. These surveys were tested and found to be reliable. The survey compiled for this study, from those sources mentioned above, included:

- 8 statements on the belief about the nature of science
- 6 questions relating to attitude toward science
- 6 questions related to beliefs about the teaching of science
- 4 questions regarding attitude toward learning to teach science
- 8 questions about attitude toward teaching science

The individual statements were scored on a 5-point Likert scale, ranging from 1 = "Strongly Disagree" to 5 = "Strongly Agree." The data were disaggregated for each question, and the means, standard deviations, and frequency counts recorded. The disaggregated data were then compared by groups and the mean attitude charted. Three separate studies of the total population were made. The first comparison study looked at the numbers of college credits completed (Seniors-Juniors-Sophomore) and the whole population. The second group studied was split by the grades they intended to teach, and the third group studied was the accelerated ITC-LINKS group versus the non-links regular college group.

The same attitude survey was given at the end of the semester with the inclusion of 10 questions asking for information on each undergraduate's science background and teaching preferences.

## PRE- POST-CONTENT TEST

The pre-content test was given during the first week of the semester and consisted of 24 multiple choice and 11 short answer questions to determine the level of science background knowledge that each undergraduate student brought to the class. The test covered six science content areas specific to the 4th–6th grade California Content Standards. The themes were water, matter, magnetism, electricity, plate tectonics, and photosynthesis. Questions on the exam were written at the eighth-grade level. The results from the pretests were used to guide the content discussions and help the instructor focus on misconceptions the undergraduates had in science. The posttest consisted of 36 questions and was given at the end of the semester to assess change in the students' understanding of science content.

## JOURNALS

Each undergraduate kept a journal to record discussions on writing prompts, lab instructions, class lecture, and thoughts on the class activities. Journal prompts were taken from weekly discussions and trainings during the two-hour class. The prompts included questions on pedagogy, problems encountered during teaching, methods of using materials correctly, methods of dealing with parents, working with time constraints and management, use of questioning strategies, addressing misconceptions, and methods of presenting science content. Students kept their weekly responses in a journal that was collected and scored twice during the semester; at week 7 and week 15. The final journal prompt was designed to specifically to encourage students to write about any changes in their attitude toward teaching science that may have taken place throughout the semester. This final prompt read as follows:

1. How has the Hands-on Lab made learning science relevant for you?
2. How do you think it makes science relevant for the elementary school students?

Journals were collected, read, and scored based on the attitude of the written work. A 5-point Likert scale was used to identify negative, slightly negative, neutral, slightly positive, and positive responses to the prompt.

Undergraduates also were required to document their teaching experience by explaining how they used scientific material and the manipulatives for each station they facilitated. They had the opportunity to teach six labs, so they had six lab descriptions, one for each science theme. The lab write-ups were graded on a 5-point rubric, looking for understanding of the science concept and whether materials were used to explain the concept. The labs varied from examples of magnetic fields to concepts such as capillary action. The electricity lab included comparing parallel and series circuits, examining what is in a light bulb and how it works, and exploring what materials serve as good conductors or insulators. Evidence of tectonic plate movement with puzzles and fossils, earthquakes, subduction boundaries, and convection currents were explored in the plate tectonics lab. In all, there were over 36 labs activities that each student was exposed to and of those, they had actual practice for one of the labs in each of the six different topics.

## POST-SURVEY AND JOURNAL TEST

The journal entries and lab write-ups were collected for a final time during the last week of the semester. Posttests and post-surveys were compared to the pretests and pre-surveys for significant changes in knowledge and attitude. The post-attitudinal survey contained additional questions related to the students' science background and desire to teach following graduation. The posttest was administered during final exam week.

## DATA ANALYSIS PROCEDURES

A paired *t*-test was used to determine if there was any significant increase in the science content knowledge of the students. The statistical program Statistical Package for the Social Sciences (SPSS, 1998) was used to form the analysis of the paired *t*-test, and Minitab was used to create charts. The pre/post attitude surveys were tallied and the means, standard deviations, and frequency counts of the statements (1 = Strongly Disagree, 2 = Sort of Disagree, 3 = Not Sure, 4 = Sort of Agree, 5 = Strongly Agree) given to the 70 interns were charted for analysis. Attitude samples from the intern journals were tallied on a 5-point Likert scale and data collected and analyzed qualitatively.

## RESULTS

### Pre- and Post-Content Knowledge Test Results

Students in the class were given a pretest at the beginning of the semester to assess current knowledge in the subject matter that would be covered throughout the semester. They were given a posttest at the end of the semester to determine how much knowledge they gained. In both administrations, the total possible score for the multiple choice section of the test was 24, and for the short answer section it was 32, for an overall total of 56 points possible. All students showed some improvement in their content knowledge, even the few science majors who had had up to eight science courses. Figure 8.2 compares the mean pretest raw scores to the mean posttest raw scores, with 95% confidence interval bars included. The paired samples *t*-test showed that the posttest scores were significantly greater than the pretest scores. The data show that the test was reliable and significant.

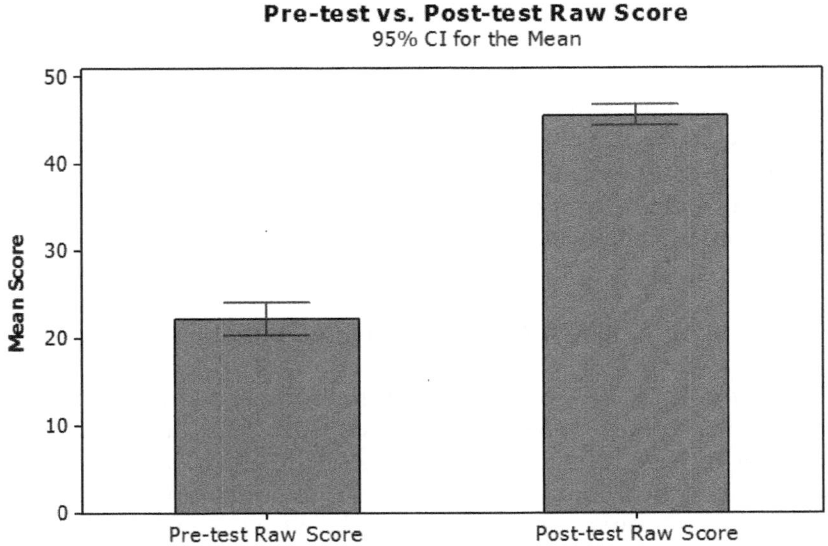


Figure 8.2. Pretest vs. posttest raw score.

## Means

The means data of the pre- and post-testing indicates that students who had completed 31–60 units of college credit (sophomores) had the lowest raw pre-scores but showed the greatest improvement between pre- and posttest scores. The raw scores of the sophomores increased by 25.2% between the pre- and posttest administrations; juniors increased by 23.5%, and seniors by 21.5%.

When investigated in regard to the grade they think they will teach, the students interested in teaching K–3 had the greatest change in total raw score, with 24.5%. Those students who wanted to teach 4th–8th grade showed a 23.5% change, while the scores of those who plan to teach high school improved by 19.2%. Students in the ITC-LINKS group had the greatest change in raw score at 25.77% compared to the average of all others, which was 21.8%.

It was expected that students who have taken fewer science courses would have the most to gain from the content portion of the course. It also was expected that students interested in teaching the lower grades would have less interest in science content, and therefore, have more to learn from the science curriculum portion of the course. Following this same line of reasoning, students thinking about teaching 9th–12th grade science would have come to the lab with a better background in science and should have started with a high pre-score and ended with a high post-score. The pre-post data support these expectations.

## Pre- Post-Survey Results

The results of the survey showed a significant increase in positive attitudes of the undergraduates in the Internship in Science Teaching class. Individual statements were scored on a 5-point Likert scale. The data were disaggregated for each question, and the means, standard deviations, and frequency counts recorded. The disaggregated data were then compared by groups and the mean attitude charted.

The total population was disaggregated into three different groups. The first comparison study looked at attitudes based on class standing (sophomore, junior, senior). The second comparison study looked at attitudes based on the grade level students intended to teach. The third method of disaggregating the data separated the attitudes of the ITC-LINKS group from those of the non-ITC-LINKS students.

The overall pre- post-comparisons of the surveys show significant movement in the students' attitudes. Changes were recognized as being significant when there was a change of 0.5 or more in the mean score between pre- and post-test. Data were reviewed and analyzed as to the amount of change in the pre- post-surveys. The whole group data were plotted and reviewed for large shifts in attitude. The overall pre- post-comparisons of the surveys show movement in many areas of the students' attitude.

One of the largest shifts was seen in the sophomores on question 8. These undergraduates increased their belief that using technology would improve students' understanding of science from a mean score of 3.22 to a post mean of 4.19, with the mean-shift change of 0.97. The greatest pre-post change in the attitudes of juniors was on question 14, believing that students did not learn only by repeating problems (shift of 1.1). The largest gain made by seniors was with their feeling of preparedness to teach science to a variety of grade levels (36, 1.03). The overall population showed a mean change of 0.67 on this question, the highest gain noted for any single question.

Juniors also increased their attitude toward how prepared they feel to teach and explain science (.90 for each of 33, 34, and 35). Seniors increased their attitude about attempting science (.64 mean increase in attitude for question 6). Other significant increases were in the junior group regarding liking science (15), and using technologies for all students (19). Both seniors and juniors specifically, and the group as a whole, showed a marked increase in confidence in science content knowledge (26 with .61 and .50 respectively).

Both seniors and sophomores showed improved attitudes regarding their desire to turn students on to science (question 28). This was the second highest gain for sophomores and total population. Seniors and juniors increased by .75 and .70 respectively in confidence of having the skills to teach science (29).

Results of the attitude survey indicated that there were many areas where attitude was significantly improved after a semester working with children in the Hands-on Lab in the Internship in Science Teaching course. The survey showed the following significant changes:

- belief about the nature of science: two significant changes
- attitudes toward science: five significant changes
- beliefs about the teaching of science: two significant changes
- attitudes toward learning to teach science: one significant change
- attitudes toward teaching science: six significant changes

The largest overall change came in the attitude of feeling prepared to teach science to a variety of grade levels and in knowing what to do to help students like science. This result was very encouraging since the class gave students the opportunity to teach a grade level that they thought they would not enjoy teaching. Finding that teaching science can be enjoyable helped the students feel more positive about taking the content classes they need in order to be well prepared to teach the subject. Table 8.1 shows the individual percentage of responses to the pre- and post-survey questions.

**Table 8.1. Survey Questions % Spread, Group E Attitude Towards Teaching Science**

Group E Attitudes towards teaching science = (8 Statements)

The means, standard deviations, and frequency counts of the statements (1=Strongly Disagree, 2=Sort of Disagree, 3=Not Sure, 4=Sort of Agree, 5=Strongly Agree) given to the 70 undergraduates are presented in the table below.

| *Statement* | *Mean (SD)* | *Sort of Disagree* | *Not Sure* | *Sort of Agree* | *Strongly Agree* |
|---|---|---|---|---|---|
| 26. I am confident that I will be able to answer students' science questions. | | | | | |
| Pre | 3. 14 (.97) | 14 (20%) | 22 (31%) | 28 (40%) | 2 (3%) |
| Post | 3.56 (.83) | 9 (13%) | 15 (21%) | 40 (57%) | 5 (7%) |
| 28. I do not know what to do to turn students on to science. | | | | | |
| Pre | 2.89 (1.15) | 26 (37%) | 14 (20%) | 18 (26%) | 6 (9%) |
| Post | 2.24 (.91) | 41 (59%) | 9 (13%) | 8 (11%) | 1 (1%) |
| 31. I think that I will enjoy answering students' science questions. | | | | | |
| Pre | 3.53 (.93) | 8 (11%) | 22 (31%) | 29 (41%) | 10 (14%) |
| Post | 3.79 (.95) | 6 (9%) | 16 (23%) | 31 (44%) | 16 (23%) |

*(Table continues on next page)*

**Table 8.1. (Continued)**

| Group E Attitudes towards teaching science = (8 Statements) | | | | | |
|---|---|---|---|---|---|
| The means, standard deviations, and frequency counts of the statements (1=Strongly Disagree, 2=Sort of Disagree, 3=Not Sure, 4=Sort of Agree, 5=Strongly Agree) given to the 70 undergraduates are presented in the table below. | | | | | |
| *Statement* | *Mean (SD)* | *Sort of Disagree* | *Not Sure* | *Sort of Agree* | *Strongly Agree* |
| 32. I think that I will prefer to teach mathematics and science emphasizing connections between the two disciplines. | | | | | |
| Pre | 2.86 (1.05) | 17 (24%) | 25 (36%) | 17 (24%) | 3 (4%) |
| Statement | Mean (SD) | Sort of Disagree | Not Sure | Sort of Agree | Strongly Agree |
| 33. The idea of teaching science scares me. | | | | | |
| Pre | 2.96 (1.22) | 20 (29%) | 12 (17%) | 23 (33%) | 6 (9%) |
| Post | 2.59 (1.22) | 30 (43%) | 9 (13%) | 13 (19%) | 6 (9%) |
| 34. When a student has difficulty understanding a science concept, I think that I will be at a loss as how to help the student understand it better. | | | | | |
| Pre | 2.53 (.93) | 32 (46%) | 19 (27%) | 11 (16%) | 1 (1%) |
| Post | 2.31 (1.00) | 39 (56%) | 10 (14%) | 7 (10%) | 3 (4%) |
| 35. I think that I will find it difficult to explain to students why science experiments work. | | | | | |
| Pre | 2.77 (1.04) | 27 (39%) | 16 (23%) | 19 (27%) | 2 (3%) |

*(Table continues on next page)*

**Table 8.1. (Continued)**

| Group E Attitudes towards teaching science = (8 Statements) | | | | | |
|---|---|---|---|---|---|
| The means, standard deviations, and frequency counts of the statements (1=Strongly Disagree, 2=Sort of Disagree, 3=Not Sure, 4=Sort of Agree, 5=Strongly Agree) given to the 70 undergraduates are presented in the table below. | | | | | |
| *Statement* | *Mean (SD)* | *Sort of Disagree* | *Not Sure* | *Sort of Agree* | *Strongly Agree* |
| Post | 2.31 (.96) | 40 (57%) | 10 (14%) | 8 (11%) | 2 (3%) |
| 36. I feel prepared to teach science to a variety of grade levels. | | | | | |
| Pre | 2.60 (.98) | 27 (39%) | 22 (31%) | 11 (16%) | 2 (3%) |
| Post | 3 27 (1 12) | 16 (23%) | 15 (21%) | 27 (39%) | 8 (11%) |

## Qualitative Information From Student Journals

Undergraduate students were responsible for self-evaluating their progress in the HOL through weekly journal entries. It became clear that their interest and skills increased as the semester progressed. Undergraduates were able to adjust their teaching strategies in order to cover the appropriate content. They became more effective in asking questions and learned the importance of wait time before prompting the elementary students toward formulating an answer. They became increasingly comfortable with the use of the lab equipment and discovered that they were more confident in their background knowledge.

During the debriefing meeting following the first lab, the undergraduates discussed their fear of teaching for the first time, their frustration with teaching, their lack of confidence in their content knowledge, and their lack of skills in working with school-age children and equipment. A typical example of a first journal entry is as follows: "I am feeling very nervous. I am going to be uncomfortable with teaching these children. It feels very strange to know that I am jumping right into it. I have no idea what to expect. I am freaking out because it is very scary and new."

During the second week of teaching, the undergraduates repeated the same unit and lab activity. This time at the debriefing meetings, they talked about their successes. They felt less anxious about teaching; they were prepared with content knowledge and felt comfortable about using the material resources. They were practiced with the timing of the labs and able to complete each activity in the allotted time. Most undergraduates reported seeing and recognizing the elementary students' grasp of the content and/or were able to adjust their teaching/facilitating so that all of the children in their small group were able to make the connection between the materials and the content objective. Typical comments include:

> I feel like my teaching was pretty good—I used a visual example that really made the students understand cohesion and surface tension. So, overall my teaching and comfort level really improved.

> I had the Mixture lab and I didn't feel like what I knew was good enough, but when it came down to it, it went great! The lab actually turned out pretty well! I finally relaxed and communicated the material to the students casually and tried to make it as understandable as possible. The kids did great! I was worried about the flow, but it turned out great! After the first (group of kids) I had it figured out.

> Today's teaching experience was much better than last week. I was much more confident and comfortable with teaching.

By the fourth teaching week, the undergraduates felt so comfortable with the science content and lab activities that they were asking to switch labs in order to teach a new topic. Most undergraduates wanted to learn and teach all six of the lab activities by mid-semester. It was evident that they were feeling confident in their abilities to learn and teach each subsequent lab activity more quickly than at the beginning of the semester.

Journal reflections at the end of the semester showed the undergraduate students had increased their belief in their ability to learn and teach science content to elementary school children. In general, the undergraduates had enjoyed the experience, and reported that they intended to teach science in their elementary school classrooms. The majority found that they were excited at the prospect of teaching science and their journal entries reflected this. A selection of journal entries are presented here:

> I did implement a few changes. First, I quickly learned how important it was not to allow the lab material to sit in front of the students until I am ready for them to use them. It's just too tempting to have cool stuff lying around! Also, I asked the students more questions the second time around and waited longer for them to answer. I tried really hard not to let the silences push me to give the answer/explanation. Instead, I rephrased the question or I went back to a previous comment, re-explained it, and then asked my question again.

It has made science relevant to me; it showed me how I could teach science to students. Science teaching is what scares me the most in teaching, this put concepts into teachable and learnable way. I felt like the labs not only helped the students, but also help to reinforce concepts and break misconceptions for me.

Overall, I feel much more confident in my teaching skills and I feel confident that the students that I taught fully grasped the concepts which were introduced.

## Quantitative Data From Journals

The undergraduates' journals were collected during the semester and scored with respect to their attitudes. A 5-point Likert scale was used to chart responses to the journal prompts. A negative statement was given a double dash (--), a slightly negative statement a double dash/0 (--/0), a neutral statement was given (0), a slightly positive statement was given (0/+) and a positive statement received a plus (+). A journal writing prompt was given every week during the entire 15-week semester. The journals were reviewed by the day and hour that undergraduates taught in the labs to determine if there was an attitude trend among a particular group of interns, within a theme, or depending upon which elementary classes were visiting the lab.

Overall attitudes were examined, and results compiled, to measure each undergraduate's response to the prompts. There were 41 responses to the first question regarding previous experience in science assessing their feelings about their elementary or high school science. There was an even spread of responses, with 17 interns having very negative experiences, 10 neutral responses, 6 slightly positive, and 18 positive experiences (Table 8.2).

Some of the most informative journal entries were those written during the first few weeks of the semester before, and just after, the undergraduates' first teaching experiences. The "before" teaching prompt asked the to respond to their preparedness in terms of content knowledge and ability to use the teaching materials. Only 14 of 67 undergraduates responded positively. There were 17 negative, 3 slightly negative, 26 neutral, and 7 slightly positive responses. In contrast, after teaching only one lab, 37 undergraduates replied positively about their teaching experience, 10 slightly positive, 15 neutral, 2 slightly negative, and 3 negative. After the second week of teaching, the positive responses grew to 48, with 8 slightly positive, 5 neutral, and zero negative or slightly negative. It was obvious that a shift in the undergraduates' attitude about science teaching and their ability to teach had already occurred by week 3.

**Table 8.2. Likert Scale of Attitudes From Journal Prompts in Undergraduate Journals**

| Attitudes From Journal Prompt | | | | | |
|---|---|---|---|---|---|
| | *1* | *2* | *3* | *4* | *5* |
| | *(- -)* | *(- -/0 )* | *0* | *( 0 / + )* | *( + )* |
| Experience | 17 | 0 | 10 | 6 | 18 |
| Pre-lab 1 | 17 | 3 | 26 | 7 | 14 |
| Post-lab 1 | 3 | 2 | 15 | 10 | 37 |
| Post-lab 2 | 0 | 0 | 5 | 8 | 48 |

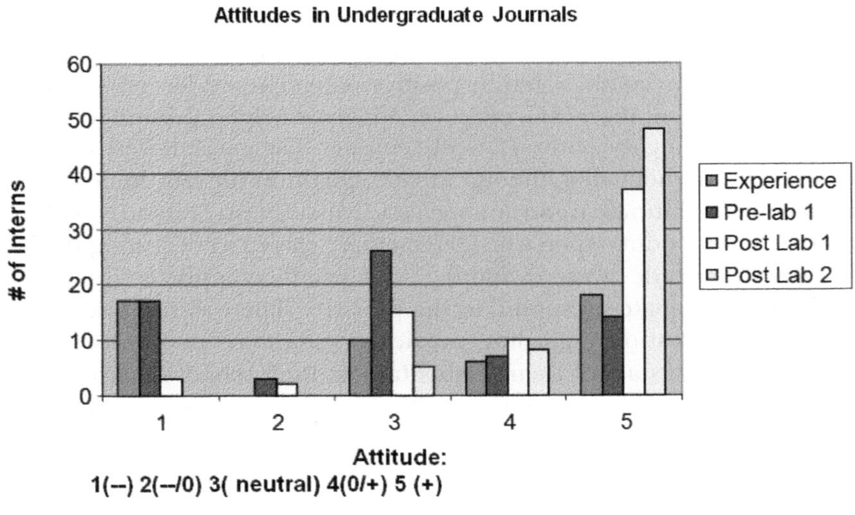


**Figure 8.3.** Change in attitudes as noted in undergraduate journals.

## DISCUSSION

One of the best arguments for having an activities-based class comes from a study done by Uline, Wilson, and Cordry (2004, p. 459):

> Classroom management has long been a topic of serious interest for new teachers. Although pre-service teachers receive well-developed materials addressing classroom management and student behavior, the hands-on

experience in the clinical setting creates an increased awareness of the complexity involved in managing student behavior.

Practicing with real materials and becoming familiar with science content help improve attitude toward science teaching. As Watters and Ginns found, "the direct experience teaching children science in the field enhanced their outcome expectancy for the teaching of science" (2000, p. 572). Watters and Ginns also found that working directly with children allowed preservice teachers to become more in tune with the variations in an individual child's knowledge, understanding and capabilities.

In a previous study, Watters and Ginns (1997) ascertained that these activities should occur in an environment allowing risk-free, meaningful learning to take place. In order to improve or increase their attitude and desire to teach science, preservice teachers need a nonthreatening place to practice with supportive instruction and guidance. The HOL class is organized in such a manner that each undergraduate student and future teacher is assigned to a small group of children, working with them for a short period of time, on a small amount of content. As Anita Woolfolk maintains (Shaughnessy, 2004), the preparation of teachers should be a gradual, continuing process where prospective teachers assume more and more responsibility. Starting with a small, snack-sized, safe, and supportive experience in science teaching in the Hands-on Lab maximizes the likelihood of success which ultimately should translate into classroom implementation.

The first author's experience in the Hand-on Lab from the six previous semesters to the one in which the study was conducted allowed her to study the undergraduates' attitudes toward science in an informal way. Through journal reflections and short surveys it was discovered that the students came into the class with a mixed view of and varied experiences in science. The majority of students had little science in elementary school, and they typically reported having had neutral or negative experiences or only one limited positive experience. These findings were supported by Bell in his *Reflective Journal* studies that showed that preservice teachers' attitudes and beliefs about science teaching are greatly influenced by their personal experiences in learning science (Bell, 2001; Watters & Ginns, 1997).

Other studies also have shown that undergraduates lack experience in science content and in the actual teaching of science. Downing reported that "studies have indicated that elementary pre-service teachers do not feel adequately prepared for teaching science content areas" (Rice & Roychouldhury, 1994, cited in Dowing, Filer, & Chamberlain, 1997, p. 3). Research also shows that undergraduate preservice teachers haven't had sufficient science background, are afraid to teach science, and prefer to teach reading and math. Downing showed a significant positive correlation between elementary preservice teachers' ability to perform science process skills and their attitudes toward science (Dowing et al., 1997).

It is imperative that higher education institutions include teaching strategies designed to increase students' positive attitudes toward science education in required science classes. Journal writing is one such strategy college instructors can easily utilize. Uline et al. (2004, p. 459) found that allowing preservice teachers to reflect on clinical experiences can improve the quality of their learning significantly. Research also supports the idea that preservice teachers who participate in science courses that were taught using constructivist instructional methods developed a positive attitude toward science. It was stated by Gibson et al. (2001) that this may translate into their interest in teaching science in the elementary classroom. Experiences that impact the college student should be as positive as possible. A change from lecture halls to science teaching labs such as the Hands-on Lab and the Internship in Science Teaching class will help motivate prospective teachers to engage in more science teaching. In 1995, Watters and Ginns found that science teaching outcome expectancy increases during experiences that involve children learning science. They found that observation of the successful impact of science teaching on children appears to be necessary to help change the attitudes of preservice teachers. Learning by textbook alone did not result in significant gain in understanding of science or in the confidence of preservice teachers to teach science. Education that includes the exploration of materials and ideas strengthens the knowledge base and the attitude of the learner.

## CONCLUSION

Liberal studies undergraduate students who are already in the teaching pathway demonstrate an increase in positive attitudes toward science and an increase in content knowledge after taking the Internship in Science Teaching class, gaining experience teaching science to elementary school children. Undergraduate students in the Internship in Science Teaching class enter the lab with limited content knowledge, and most of them claim to dislike science. Some students have had positive experiences in high school, and some have had few or no positive experiences in science anywhere in their educational past. Prior experience greatly affects the attitudes and beliefs of students toward subject matter. The HOL experience provides these students with a guided, positive approach to learning and teaching science.

Of the 70 students who took part in this study, 70% now claim that they like science. Seventy five percent (up from 46%) claim they know what to do to help their students like science, and 50% say they are prepared to teach science (up from 18%) and know that they have the skills to do so. In addition, because of the change in attitude toward science, the experience may help the students have a clearer understanding of the relevance that other, more content-based, college science courses have for their future teaching careers.

This study shows that the undergraduates' science content knowledge improved after a semester in the HOL. The pre- and post-tests of science knowledge showed significant increases on both multiple choice and short answer scores with the latter indicating that students improved their deeper understanding of science concepts. The mean improvement in the total raw score was 23.3 points, or about 42%. All students showed improvement in content knowledge, and students with the weakest science backgrounds showed the greatest percent improvement.

Based on the change in attitude toward science and the increased science content knowledge on the part of the undergraduates taking the Internship in Science Teaching class, this study demonstrates that the Hands-On Lab experience contributes to students' positive sense of self-efficacy. This has implications for the future teaching practices of these students since self-efficacy as related to teaching is defined by Gibson and Dembo (1984, p. 571) as a belief that "one's abilities can bring about positive changes in students' behaviors and achievement."

## SUMMARY/EXTENSIONS

The Hands-on Lab is not just an exciting place for the college student, but for the elementary students as well. The enthusiasm shown by the elementary students helps the college students realize the value of teaching science using hands-on activities. Teachers who bring their students to the HOL may use the experience as an introduction, enhancement, or conclusion to their own science units. The HOL provides the elementary class with vocabulary, pre-post tests, and additional standards-based activities. Teachers who escort their students to the HOL see their students being engaged in the learning of science. This is the first time, for some teachers, to view their students in a hands-on science learning environment. The HOL provides the visiting classroom teacher with a professional development opportunity associated with the HOL learning experience. These opportunities contribute to the continuation of the Hands-on Lab experience into the curriculum of the classroom the school site. Additionally, many of the classroom teachers tell the undergraduates that they are impressed at how well the college students understand and present their content. This interactive exchange between inservice and preservice teachers provides positive feedback for the undergraduates as well as a positive science example for the teachers.

Specialized professional development workshops and institutes have been conducted focusing on English language development strategies for the lessons provided in the Hands-on Lab. The lab has been utilized to develop hands-on exercises for grades K–3. Also, middle school teachers have brought their students to the Hands-on Lab to learn the lessons in order to teach them to elementary school students in service learning projects back in their home districts. The HOL represents a positive experience for all participants and represents a productive bridge between the K–12 environment and institutions of higher education.

## REFERENCES

Bell, G. L. (2001, January). *Reflective journal writing paired with inquiry-based science instruction: Effects on elementary pre-service teachers' science and science teaching beliefs*. Paper presented at the Annual Meeting of the Association for the Education of Teachers in Science, Costa Mesa, CA.

California Department of Education. (2000). *Science content standards for California public schools, kindergarten through grade twelve.* Sacramento, CA: CDE Press.

Cochran, K. F. (1997, January 14). *Pedagogical content knowledge: Teachers' integration of subject matter, pedagogy, students, and learning environments.* Research matters—to the science teacher, No.9702. Greely, CO. National Association for Research in Science Teaching. Retrieved from http://www.narst.org/publications/research/pck.cfmv

Czerniak, C. M., & Chiarelott, L. (1990).Teacher education for effective science instruction—A social cognitive perspective.*Journal of Teacher Education, 41*(1), 49–58.

Dowing, J. E., Filer, J. D., & Chamberlain, R. A. (1997, November). *Science process skills and attitudes of pre-service elementary teachers.* Paper presented at the Annual Meeting of the Mid-South Educational Research Association, Eastern Kentucky University, Memphis, TN.

Gibson, H. L., Bernhard, J., Kropf, A., & Van Strat, G. A. (2001, January). *The impact of constructivist instructional methods on pre-service teachers' attitudes toward teaching and learning science.* Proceedings of the Annual International Conference of the Association for the Education of Teachers in Science. Costa Mesa, CA.

Gibson, S., & Dembo, M. H.(1984). Teacher efficacy: A construct validation.*Journal of Educational Psychology, 76,* 569–582.

Jarrett, O. (1999, Spring). Science interest and confidence among preservice elementary teachers.*Journal of Elementary Science Education, 11*(1), 49–59

Johnston, J. D. (2003, November). *Active learning and pre-service teacher attitudinal change.* Paper presented at the Annual Meeting of the Mid-South Educational Research Association, Biloxi, MS.

Jones, M., & Carter, G. (2008). Science teacher attitudes and beliefs. In S. Abell & N. Lederman (Eds.), *Handbook of research on science education* (pp. 1067–1104). New York, NY Routledge.

Kelly, J. (2000). Rethinking the elementary science methods course: A case for content, pedagogy, and informal science education. *International Journal of Science Education, 22*(7), 755–777.

Lederman, N. (2008). Nature of science: Past, present, and future. In S. Abell & N. Lederman (Eds.), *Handbook of research on science education* (pp. 831–879). New York, NY: Routledge.

Marzano, R. J., Pickering, D. J., & Pollock, J. E. (2001). *Classroom instruction that works.* Alexandria, VA: Association for Supervision and Curriculum Development.

McGinnis, R. J. (1998, April). *A longitudinal assessment of teacher candidates' attitudes and beliefs in a reform-based mathematics and science teacher preparation program.* A poster session presented at the annual meeting of the meeting of the National Association for Research in Science Teaching, San Diego, California.

McMurrer, J. (2008). Instructional time in elementary schools: A closer look at specific subjects. Center for Educational Policy. Retrieved from www.cep-dc.orghttp://www.cep-dc.org/displayDocument.cfm?DocumentID=309

Murphy, C., Beggs, J , & Carlisle, K. (2003, March). *The impact of co-teaching between science student teachers and primary classroom teachers on children's enjoyment and learning of science and student teacher confidence.* Paper presented at the Annual Meeting of the National Association for Research in Science Teaching, Philadelphia, PA.

National Research Council. (1996). *National Science Education Standards.* Washington, DC: The National Academies Press.

National Research Council. (2007).*Taking science to school: Learning and teaching science in grades K–8* (Committee on Science Learning, Kindergarten Through Eighth Grade. Richard Duschl, Heidi A. Schweingruber, and Andrew W. Shouse, Editors. Board on Science Education, Center for Education Division of Behavioral and Social Sciences and Education). Washington, DC: The National Academies Press.

Online Evaluation Resource Library (OERL). (n.d.). Teacher Education Instruments. Retrieved from http://oerl.sri.com/instruments/te/teachsurv/instr49.html

Ritter, J. M., Boone, W. J., & Rubba, P. A. (2001, January). *An extension analysis on the self-efficacy beliefs about equitable science teaching and learning instrument for prospective elementary teachers.* Paper presented at the Annual Meeting of the Association for the Education of Teachers in Science, Costa Mesa, CA.

Shaughnessy, M. F. (2004). An interview with Anita Woolfolk: The educational psychology of teacher efficacy. *Educational Psychology Review, 16*(2), 153–176.

Statistical Package for the Social Sciences. (1998). *SPSS Base 8.0 for Windows User's Guide.* Chicago, IL: SPSS.

Tosun, T. (2000).The beliefs of pre-service elementary teachers toward science and science teaching. *School Science and Mathematics, 100*(7), 374–379.

Uline, C., Wilson, D., & Cordry, S. (2004). Reflective journals: A valuable tool for teacher preparation, *Education, 124*(3), 0013–1172.

Watters, J. J., & Ginns, I. S. (1995). *Origins of and changes in pre-service teachers' science teaching self efficacy.* Retrieved from http://eprints.qut.edu.au/archive/00001761/01/1761.pdf

Watters, J. J., & Ginns, I. S. (1997, March). *Impact of course and program design features on the preparation of pre-service elementary science teachers.* Paper presented at the Annual Meeting of the National Association of Research in Science Teaching, Chicago, IL.

Watters, J. J., & Ginns, I. S. (2000).Developing motivation to teach elementary science: Effect of collaborative and authentic learning practices in pre-service education. *Journal of Science Teacher Education, 11*(4), 301–322.

CHAPTER 9

# REFRAMING NON-SCIENCE MAJORS' FUNDAMENTAL UNDERSTANDINGS ABOUT SCIENTIFIC INQUIRY AND SCIENTISTS

**Gayle Buck, Pazit Koren, Xinying Yin, and Varda Bar**

## ABSTRACT

Our knowledge of people's naïve beliefs about inquiry has grown, and a greater attention to scientists and the work that they do has been achieved. Research, however continues to document naïve understandings. The non-science majors entering our undergraduate science course were no exception. Although they initially responded to our query that anyone can be a scientist, when their beliefs in this regard were explored, stereotypical understandings of scientists and scientific inquiry emerged. Thus, we sought to reframe our practice. The questions guiding our research were: To what extent did course reconceptualization efforts lead to broader and more specific understandings of scientific inquiry and scientists? and In what ways did the strategies enacted in the reconceptualization course foster or hinder students' understandings of scientific inquiry and scientists? In this chapter, we provide a description of our intervention, the findings associated with implementation, and subsequent revisions.

*Research Based Undergraduate Science Teaching*, pp. 275–300

Over the past several decades, our understanding of people's naïve beliefs about the nature of scientific inquiry has grown, and a greater attention to scientists and the work they do has been achieved. Despite the more diverse, and often glamorized, image of scientific inquiry and scientists in venues such as television shows, movies, and video games, research continues to document naïve understandings about science and scientists (Buldu, 2006; Finson, 2002; Lyons & Quinn, 2010; Rubin & Cohen, 2003; Unver, 2010). The non-science majors entering our undergraduate science course were no exception. Although they initially responded to a query that anyone can be a scientist, when their belief systems in this regard were explored, real scientists were limited to Newton and Einstein, and real scientific work followed a narrowly defined scientific method in a room full of beakers, test tubes, and chemicals. Thus, we sought to foster a reframing of their fundamental understandings about scientists and scientific inquiry.

Our study was rooted in the belief that frameworks for fostering undergraduate students' understandings of scientists and the nature of their inquiries emerge through systematic examination, reflection, and learning about our own teaching at the university level (Loughran, 2006). The purpose of this study, one of several guiding our revisions, was to enhance our efforts in adjusting our course by monitoring the impact of our intervention on undergraduates' fundamental understandings. The undergraduates who enroll in this course are freshmen and sophomores from a variety of non-science majors such as elementary education, early childhood education, psychology, folklore and ethnomusicology, family studies, and communication and culture. The course goals for these majors include (1) examine the nature of science, (2) enhance understandings of scientists and scientific careers, (3) experience creating scientific understandings from inquiry-based interactions, (4) develop/refine the lab skills necessary to be successfully engaged in scientifically oriented questions, (5) develop a thorough understanding of scientific inquiry, (6) experience science inquiry through the theme of environmental science, and (7) refine skills at making relevant life decisions based on evaluating credible scientific evidence.

This study specifically focused on our teaching efforts in regards to goals 2 and 5 and was guided by two research questions:

1. To what extent did course reconceptualization efforts lead to broader and more specific understandings of scientific inquiry and scientists?
2. In what ways did the strategies enacted in the reconceptualized course foster or hinder students' understandings of scientific inquiry and scientists?

Our initial inquiries into our students' understandings revealed that their perceptions of science and scientists were highly stereotypical. Although this image may be accurate in regards to some scientists, it fails to capture the diversity of people, interests, and approaches involved in science. Thus, we sought to foster a broader understanding; not seeking one particular perception but a degree of diversity within the perceptions. In addition, their initial responses lacked examples and elaborations. We also adjusted our practice to foster a more specific understanding, seeking responses that included actual examples of scientists and scientific inquiry.

## GUIDING LITERATURE

The theoretical underpinning of our undergraduate science course for non-science majors is scientific inquiry. Scientists "share certain basic beliefs and attitudes concerning this process: regarding what they do and how they view their work, [and that] scientific inquiry is not easily described apart from the context of particular investigations" (American Association for the Advancement of Science [AAAS], 1990, pp. 2–4). Research has shown that stereotypical understandings of scientists and their work are associated with limitations in students' understandings of various aspects of the nature of science, such as the practices and procedures of a scientist's work (van Eijck & Roth, 2008), and may constrain undergraduates' research experiences (Ryder & Leach, 1999). Negative perceptions of scientists can translate into negative perceptions of science (Finson, Beaver, & Cramond, 1995). The stereotypical perception, for example, of a scientist wearing the white lab coat and spending all day doing chemical experiments with beakers and test tubes is related to a lack of understanding about the diversity of scientific research fields. Thus, addressing students' understandings of scientific inquiry is intimately related to their understandings of scientists. Many individuals from various age groups, socioeconomic conditions, and ethnicities perceive a scientist as middle-aged or elderly western, White/Caucasian male wearing a white lab coat and eyeglasses, having facial hair, doing experiments in a lab with test tubes and beakers in an atmosphere of secrecy and danger (Chambers, 1983; Eisenhart, Finkel & Marion, 1996; Finson et al., 1995; Mead & Metraux, 1957; Rosenthal, 1993; Rubin & Cohen, 2003; Thomas, Henley, & Snell, 2006; Unver, 2010). We sought to address this stereotypical image by exploring the many intersections of diversity that contribute to scientific endeavors.

In the spring of 2010, we revised our course to both broaden and deepen the students' understandings of scientific inquiry and scientists. This reconceptualization effort was supported and guided by empirical work in these areas as explored below.

## University Students' Conceptions of Scientists and Scientific Inquiry

Many college students have been shown to have persistent stereotypical understandings of scientists and scientific inquiry (Bovina & Dragul'Skaia, 2008; Quita, 2003; Rubin & Cohen, 2003; Thomas et al., 2006; Unver, 2010). A study conducted by Thomas et al. (2006) revealed American college students' images of scientists. The college students in this study, from a variety of majors, demonstrated a stereotypical image of a scientist. Their mean number of stereotypical indicators was similar to that of the fourth and fifth graders in Chamber's study conducted in 1983. Bovina and Dragul'Skaia (2008) found that Russian college students perceive scientists to be intelligent and old. The humanities students had more negative representations of a scientist than natural science students, such as being poor, mad, and solitary. Education majors, from the U.S. and other countries, are among the adults demonstrating stereotypical images of scientists and scientific inquiry (e.g., Kahle, 1987; McDuffie, 2001; Moseley & Norris, 1999; Rampal, 1992; Rosenthal, 1993; Rubin & Cohen, 2003). Research indicates that education majors tend to be fearful of teaching science (Rosenthal, 1991), and their stereotypical image of scientists and scientific inquiry is related to their images of themselves as future science teachers, which impacts their teaching self-efficacy (Finson, 2001).

The previous research on students' images of scientists almost always finds accompanying stereotypical understandings of scientific inquiry. The image of the person, however, was ultimately emphasized: the now common picture of a white male with crazy hair and lab coat. The lack of emphasis on the stereotypical view of scientific inquiry is unfortunate given its importance. An understanding of the nature of science, and science inquiry, is important for all fields, including the education majors who serve as the majority enrollment in our course. Research supports the need for education majors to develop understandings about the nature of scientific inquiry in order to ultimately teach their students about the elements of scientific inquiry (Haefner & Zembal-Saul, 2004). In a study of education majors conducting independent inquiry on an ecology topic, Roth (1999) found they had considerable trouble creating research questions, assumed cause-and-effect relationships for correlation questions, had difficulties in manipulating operational variables, and constructed claims that either did not relate to the original question or did not logically extend from the data collected. In a later study, Windschitl (2004) examined how education majors reference folk theories of "inquiry" within independent investigative experiences. Findings by Windschitl indicated that some facets of the participants' conceptual models of scientific inquiry were congruent with authentic inquiry, but the misrepresentations of fundamental aspects of science were prominent. A hypothesis, for example, functions as a guess about an outcome, and theory is merely an optional tool one might use at the end of a study to help explain results.

## Strategies for Enhancing Students' Understandings of Scientists and Scientific Inquiry

Research has shown that effective strategies to enhance students' conceptions of scientists include inviting scientists to classrooms, providing role models, visiting scientists' workplaces, interviewing scientists, engaging in inquiry-based learning contexts, participating in authentic research, and engaging in explicit reflections (Buck, Leslie-Pelecky & Kirby, 2002; Cakmakci, Tosun, Turgut, Orenler, Sengul, & Top, 2011; Painter, Jones, Tretter, & Kubasko, 2006; Rahm & Downey, 2002; Scherz & Oren, 2006). Compared to the studies on precollege students, however, very few studies have been conducted to address these concerns at the university level.

Learning cycle lessons were designed by Reap et al. (1994) involving inquiries in physical and life science contexts combined with activities emphasizing a multicultural, gender-neutral view of science in an elementary science methods course, and these obtained a positive effect of reducing university students' stereotypical images of scientists. The researchers concluded that students' prevalent stereotypical images of scientists are affected by the way science is taught in their university programs. In another study, Crawford, Zembal-Saul, Munford and Friedrichsen (2005) reported that a group of prospective secondary teachers in an advanced science methods course did not have initial overall understandings of scientific inquiry, especially a working knowledge of the diverse ways scientists use evidence and construct explanations. After using technology and inquiry-based tasks on evolution, however, some of the sample demonstrated an enhancement of understandings of the nature of scientific work. But the effects of involving college students in inquiry experiences on developing their conceptions of inquiry are not always clear. Windschitl (2001, 2002), for example, found that the experience refined the inquiry conceptions of those who already had more sophisticated understandings of scientific investigations while did little more than confirming the simplistic notions of inquiry for those who held an image of a one-dimensional scientific method.

Even fewer studies were found that addressed students' understandings of scientists and scientific inquiry in university science courses for non-science majors. Scientific inquiry has not been a prominent feature of the undergraduate science courses that non-science majors' take (Zembal-Saul, Blumenfeld, & Krajcik, 2000). Typical freshmen-level college science classes are generally confirmatory laboratories (Trumbull & Kerr, 1993; Windschitl, 2004). Most college students have never been engaged in learning science as inquiry or been exposed to effective, inquiry-based instruction (Roth, McGinnis, & Bowen, 1998). University science courses could serve as an arena in which to facilitate genuine inquiry experiences in a manner that would ultimately support students in constructing more positive and accurate understandings of science (Painter et al., 2006). For example, students engaging in scientific inquiry during an inquiry-based innovative life science course were found by Haefner and Zembal-Saul (2004) to have developed more appropriate understandings of scientific inquiry. The students, unfortunately, still failed to grasp the complexity found in the development and evaluation of scientific explanations.

In order to further enhance non-science majors' understandings of scientific inquiry, research apprenticeships have been suggested (Sadler, Burgin, McKinney, & Ponjuan, 2010). Through the apprenticeship process, it is anticipated that inquiry skills and understandings will be transferred from the expert scientist to the student. The apprenticeship model is supported by research on situated cognition (Brown, Collins, & Duguid, 1989) in that it recognizes that the novice and expert are from different worlds. The novice becomes an expert "through the mechanism of acculturation into the world of the expert" as implicit, tacit knowledge is conveyed (p. 466). When, in the process, the novices' personal theories are challenged, they are forced to revise their current thinking; hence, creating a more informed understanding (Farnham-Diggory, 1994). After a 10-week apprenticeship, Varelas, House, and Wenzel (2005) found that students came to appreciate certain aspects of science practice such as the nonlinearity, messiness, and risk taking and the collaboration, autonomy, and mentoring within a community. Participation in authentic scientific research with scientists for an extended period is viewed as a valuable learning experience through which students can develop the ability to do scientific inquiry and understandings about scientific inquiry (Barab & Hay, 2001; Ritchie & Rigano, 1996). It has been found, however, that students do not necessarily develop adequate conceptions of scientific inquiry in apprenticeship programs, with a possible reason being that scientists often do not explicitly articulate the nature of scientific inquiry (Aydeniz, Baksa, & Skinner, 2011; Bell, Blair, Crawford, & Lederman, 2003; Ryder, Leach, & Driver, 1999; Sadler & McKinney, 2010). Such opportunities often are only available for science majors and very limited for most non-science major college students (Bhattacharyya, 2009).

## INTERVENTION

Based on this literature, we reconceptualized our science course for non-science majors to lead to broader and more specific student understandings of scientists and scientific inquiry. We made adjustments, explained below, that specifically addressed understandings of scientists and scientific inquiry. The intervention was to foster a continual process of change in which the instructor acts as a guide in a dialog between the former and desired understandings. This process of guidance consists of four essential features: eliciting prior knowledge, creating cognitive dissonance, application of new knowledge with feedback, and reflection on learning (Baviskar, 2009). The approach to fostering this process was guided by the research noted above. Since much of that research was completed at the precollege level or in an education preservice teachers' methods course, we were utilizing work from other contexts and participants. We structured this study to monitor and guide our efforts with this specific context and population. The study associated with our efforts was conducted in one of the freshmen/sophmore-level science courses at our university. Table 9.1 lists those aspects of the course that address the specific goals focusing this study.

### Explore Existing Beliefs and Biases

College students need learning experiences that encourage them to articulate and examine their prior ideas about sciences and scientists. We, therefore, set out to surface and foster self-reflection on the students' views of scientific inquiry and scientists. The students were asked to draw a scientist at work and describe the properties inherent in the picture (Draw-a-Scientist Test [DAST], Finson et al., 1995).

The students also were asked to give names of scientists. They then reviewed all of the lists to determine (1) the average number of scientists' names appearing in the list, (2) the scientists' gender, (3) common names/specializations given in the scientists' name lists, (4) the most common description of a typical day of the scientist or that it changes from one scientist to another. Many of these strategies have been used in research in which outsiders explore such understandings. We utilized these strategies, however, to foster self-discovery and self-reflection on the part of the students. Thus, their responses were analyzed and reported back to them in cumulative form. The students were asked to look across all of the findings to determine typical images of scientists and scientific inquiry in their group of peers. These were then discussed in groups of 5–6 students. The discussions were guided by questions such as, "What do these results tell us about our knowledge of and attitudes toward scientists?"

**Table 9.1. Course Activities**

| *Emphasis* | *Activities* |
|---|---|
| Exploring Existing Beliefs and Biases | Pre-survey on scientists & scientific inquiry |
| | Reflection and discussion of survey results in groups |
| Exploring Real Scientists and Authentic Scientific Inquiry | Exploring known scientists and their research on computers |
| | Webquest on multicultural scientists |
| | Videos and discussions on Fossey/Goodall/Galdikas |
| | Inquiry-based labs such as predicting the timing and location of the next Hawaiian volcano (Russo, Mattox, & Kildau, 2010). Research scientists studying same phenomena as in inquiry labs. For example, identify volcanologists and how they conduct their work. |
| Getting to Know Local Scientists and Scientific Inquiry | Interview scientist from field of interest |
| | Visit scientists lab and go through mini inquiry experience |
| | Share and discuss scientists |

## Exploring Real Scientists and Authentic Scientific Inquiry

It has been shown that people who lack experiences with authentic scientific inquiry have static conceptions of science and an authoritative picture of how the world works (Anderson, 2003). In order to provide our students with opportunities to get to know scientists and engage in inquiry-based experiences, we replaced/restructured several of our inquiry labs to explicitly reflect authentic inquiry experiences by real scientists. We incorporated inquiry-based labs on global warming, solar energy, and the Hawaiian Islands (see Russo et al., 2010). These labs were enhanced with Internet research that revealed the many faces of the scientists who study the phenomenon being investigated and the ways in which they do so. The studies were purposely selected to demonstrate various methods of scientific inquiry. For example, the examples noted above include two experiments and one descriptive study.

We extended this work by incorporating "Familiarity with Scientists and Their Inquiries." In this activity, students became acquainted with scientists from various backgrounds through history. They did this by completing a WebQuest to find a diverse group of scientists in regards to gender, race, and socioeconomic backgrounds. The search involved characteristics and methods of research in addition to personal characteristics of the scientists. The WebQuest was followed by decontextualized lessons on three female scientists: Jane Goodall, Dian Fossey, and Birute Gladikas. These scientists were selected in response to the earlier results from the "Exploring Existing Beliefs about Scientists and Scientific Inquiry" activity noted above. The students viewed videos featuring each of these scientists and completed follow-up questionnaires. The questionnaire included human characteristics, "What was her motivation to devote herself to research?" and research approach, "Which aspects of her research are similar or different from your knowledge of scientific research?"

### Getting to Know Local Scientists

Apprenticeships have been shown to improve people's understanding of and skills in doing scientific inquiry. Our responsibility for covering many course objectives in one semester and the sheer number of students (~240 semester), however, impeded us from incorporating a full apprenticeship into each student's experiences. We implemented, instead, a new course component, "Familiarity with Local Scientists and Their Inquiries," and coupled it with a revised version of the previously required inquiry project. The students met local scientists at their places of work and interviewed them about their research field, data collection and analysis methods, motivation of doing the research, and life of being a scientist. Although many of these scientists were found by reviewing the work in various areas of research at our university, several were found once the students contacted local businesses (e.g., pharmacy). In the 6th week of the 15-week semester, we had the students, working in pairs, select a scientist to visit and interview based on their area of interests. For the following four weeks, the student pairs contacted the scientists and conducted their interviews and lab visits outside class time. These were then shared in class.

## METHOD

This research was rooted in our desire to improve our efforts at enhancing non-science majors' understandings of scientific inquiry by introducing them to real individuals who are contributing to science. We completed an action research study focused on our efforts to reconceptualize one science course. Action research is pragmatic social science conducted in collaboration among researchers and practitioners in efforts to enact solutions to problems deemed important by the local community (Levin & Greenwood, 1998). As pragmatic research, it involves knowledge generation through action and experimentation. Our research incorporated the necessary characteristics of pragmatic action research as described by Levin and Greenwood (1998, 2002). The process was context-bound and addressed real-life problems. It involved collaborative communicative processes in which all researchers'/practitioners' contributions were taken seriously; the diversity of experiences and capacities within our group was considered an opportunity for enrichment of the action research process.

## Participants

We were active participants in the research. Our action research team included one science education faculty member assigned to coordinate the efforts associated with the course. This faculty member also served as the instructor for the course throughout the study. A second team member taught another section of this course as a doctoral student instructor; this team member, however, did not have teaching responsibilities associated with this particular course section. Also part of the team were two science educators from a different university. These educators also taught science courses for non-science majors at their home institutions and had conducted previous studies on enhancing female students' understandings of scientists.

All students in one section of the course ($n = 22$) were given the opportunity to take part in the study. Of these, 18 students (82%, 15 females and 3 males) elected to participate. All were Caucasian. These students were freshmen and sophomores declaring non-science majors including elementary education, special education, speech and hearing, psychology, family studies, and exploratory. The instructor-of-record was not present when the students were invited to participate. As a result, she was unaware of who had elected to participate. A second member of the research team had access to the names and tagged the data.

## Data Collection

Students' views of scientific inquiry were assessed pre- and post-course through a modified version of the Views of Scientific Inquiry-2 (VOSI-2) (Schwartz, Lederman, & Lederman, 2008). The modified questionnaire of eight items was developed from questions in the pool of VOSI items (Schwartz et al., 2008) that specifically addressed our aims for the course. The students completed the modified VOSI-2 in partners in order to stimulate discussion and search for a consensus. The modified questionnaire included the following items.

1. What types of activities do scientists do to learn about the natural world? Be specific about how they go about their work.

2. What scientists choose to study and how they learn about the natural world may be influenced by a variety of factors. How do scientists decide what and how to investigate? Describe all the factors you think influence the work of scientists. Be as specific as possible.
3. Write a definition of a scientific experiment. Give an example from something you have done or heard about in science that illustrates your definition of a scientific experiment. Explain why you consider your example to be a scientific experiment.
4. A person interested in birds looked at hundreds of different types of birds who eat different types of food. He noticed that birds that eat similar types of food tended to have similar shaped beaks. For example, birds that eat hard shelled nuts have short, strong beaks, and birds that eat insects from tide pools have long, slim beaks. He concluded that there is a relationship between beak shape and the type of food birds eat. Do you consider this person's investigation to be scientific? Please explain why or why not. Do you consider this person's investigation to be an experiment? Please explain why or why not.
5. Some people have claimed that all scientific investigations must follow the same general set of steps or method to be considered science. Others have claimed there are different general methods that scientific investigations can follow. What do you think? Is there one scientific method or set of steps that all investigations must follow to be considered science?
6. If several scientists, working independently, ask the *same question*, will they necessarily come to the *same conclusions*? Explain why or why not. Does your response to (a) change if the scientists are working together? Explain.
7. What does the word "data" mean in science? Is "data" the same or different from "evidence"? Explain.
8. What is "data analysis"? What is involved in doing data analysis?

In addition to the modified VOSI-2, the research team analyzed several written documents administered on the first two days of the course. These documents were completed again near the end of the semester (after all noted interventions had been completed).

1. Draw-a-Scientist Test (DAST) (Finson et al., 1995).
2. List five names of known scientists and mention their specialization (Koren & Bar, 2009).
3. Declare the level of agreement to short statements regarding the characteristics of the scientist (Rubin, Bar, & Cohen, 2003).

4. Compose an essay about the "day in the life of a scientist" (Rubin et al., 2003).

To further track the development of student understandings of scientists and scientific inquiry, we also captured their course work. The work was designed to have them reflect on examples of scientists and scientific inquiry. The data included students' audiotaped reflections on their own responses to the documents noted above (e.g., Draw-a-Scientist Test), as well as their written reflections on the videos and internet research. In addition, we collected their written reports associated with the apprentice experiences. Copies of all lab instruction and course activities were collected to track our instruction over the semester. The data were used to connect the student findings with actual course instruction. We also kept an instructor/researcher log in which we recorded our reflections of the experience.

## Analysis

Data analysis was an iterative process occurring concurrently with data collection, during group planning sessions, and at the completion of the study. Reflection by students of their thoughts and experiences provided us with opportunities to evaluate the efficacy of current strategies and identify approaches that would improve learning outcomes. We discussed emergent themes, similarities and differences among our interpretations, and future directions for practice. This continuous analysis became part of the data, which, in turn, served as additional data for analysis at the completion of the study.

The written documents used in this study were analyzed through the use of both qualitative and quantitative approaches. The quantitative data were analyzed by recording the percentages of answers (for example: percentages of participants who gave each name in the "name five scientists' tools"). The qualitative data, such as written essays and discussion group reflections, were analyzed by categorical analysis. The data were read multiple times, categories were developed, the data were coded based on these categories, and statements were made about the categories.

The students' responses to the modified VOSI-2 were evaluated based on their epistemic value related to the different aspects of scientific inquiry by the authors over a sustained period of time (Aydeniz et al., 2011). These aspects included the activities involved in scientific investigations (Q1); factors influencing scientists' investigations (Q2); the conception of experiment (Q3); scientific investigation, observation, and experimentation in scientific inquiry (Q4); the singularity and multiplicity of scientific inquiry methods (Q5); the social nature of scientific inquiry and subjectivity and objectivity of scientific inquiry (Q6 and Q7); the difference between data and evidence (Q8); and what data analysis means in scientific inquiry (Q9). In analyzing students' responses, we coded all the responses in the two questionnaires and constructed interpretations of each pair of students' conceptions using the codes. We then compared each pair's pre- and post-questionnaire to identify their learning gains (or lack thereof) regarding to scientific inquiry over this course. In the findings, we use "appropriate" to indicate students' conceptions aligned with the contemporary views of scientific inquiry in the science education community. But we did not only evaluate students' responses as "appropriate" or "inappropriate" since, very often, students have more complete or elaborate conceptions shown in their post-questionnaires, which did not mean that they had inappropriate conceptions on the pre-questionnaires.

This analysis process was validated by peer review and repeated readings of the raw data and was the consensus of the authors. To ensure we used instruction as intended, we monitored and reviewed our instruction. We made notes of when we explicitly addressed the concepts and compared them to the analysis of outcomes.

## FINDINGS

### To What Extent did Course Reconceptualization Efforts Lead to Broader and More Specific Conceptions of Scientific Inquiry and Scientists?

Data were collected using the DAST, list of scientists' names, characteristics of a scientist checklist, essay on the life of a scientist, and the modified VOSI-2 to provide insight into the students' developing understandings of scientists and scientific inquiry. We reviewed the findings and developed collective statements on the major categories that emerged. These statements are provided below, using quantitative and qualitative findings as supporting evidence.

#### *Scientists*

In the beginning of the semester, the familiar scientists listed by the students portrayed many of the stereotypical characteristics of a scientist. The most frequently named scientists were Einstein and Newton. Of the known scientists, 90% were males. Most of the students gave a very general field (biology, geology). Only a few were able to give a brief description of the work of a named scientist. For example, Newton is related to gravity but not to astronomy. Einstein is related to relativity, but not to the photo effect or Brownian motion. The most frequently noted pieces of equipment were beakers and test tubes. The equipment was mostly related to chemistry; there was no hint in the drawings of out-of-lab activity, and such activity was rarely noted in any of the writings. Only two computers were shown or mentioned, and no other more sophisticated equipment (e.g., camera, oscilloscope, electron microscope) was shown or mentioned. Scientists were generally described as lonely. The noted characteristics of scientists revealed that the students doubted that they were creative; however, the students indicated that they were intellectual and diligent. Scientists' daily activities were rather ambiguous, indicating that students were not really acquainted with what scientists did in a typical day.

At the end of the intervention, the percentage of women in the drawings increased to 55% from 10% on the pretest. One group of pictures contained both a man and a woman with notes that there are all kinds of scientists, working in all kinds of fields (including fieldwork). Field work appeared in the drawings; however, out-of-lab activity was still underestimated in the drawings compared to the number of scientists students were exposed to who did fieldwork. Posttest drawings contained specific equipment related to the scientists' work; both special tools and samples were drawn. Some of this equipment was high technology, such as a microscope attached to the computer and analysis machines (RNA, DNA). The number of computers increased, appearing in 33% of the pictures as compared to 8% on the pretest. The dress became more relaxed (jeans and T-shirts). Lab coats and safety accessories appeared in a minority of the cases (17%) with captions noting that these are used when needed. In the field they were dressed in clothes fit for fieldwork. By the end of the semester, a majority of the familiar names of scientists given were those of contemporary scientists (57%) who worked locally with 60% of them female names. The description of the work of the named scientist was specific and detailed; for example, evolution of insects and immunology. The students wrote that scientists were very creative, enthusiastic, and motivated, in addition to intellectual and diligent. They also noted that scientists ask a lot of questions. The typical day was described as being rather flexible, changing according to the type of activity that needed to be done. Those activities were described as experimenting, analyzing results, reading, writing, applying for grants, and teaching. Some work was noted to be very long and the scientist had to be ready to go to the laboratory and check it at any time, day and night.

### Scientific Inquiry

***Methods of inquiry.*** When asked at the beginning of course about what types of activities scientists do to understand the natural world, all pairs of students noted that scientists observe and collect data, and half of the questionnaires (50%) included references to the fact that scientists conduct experiments. Other activities mentioned included applying science to real-world situations (10%), using prior knowledge (20%), and supporting the well-being of people (10%). Overall, students did not demonstrate a coherent notion of methods of inquiry. Their vague conceptions of scientific activities were limited to the physical activities involved in data collection. On the post-questionnaires, more types of activities involved in scientific inquiry were noted, and student descriptions were much more elaborate and specific. For example, all pairs of students described experiments, and one pair also mentioned "using experiment for testing hypothesis." In addition, asking questions appeared on 30% of the post-questionnaires, whereas no reference was made on the pre-questionnaires. Of the questionnaires, 40% noted that scientists need to research previous studies before they can begin their own projects or to use that research to further explain experiment results. On making observations and collecting data, three pairs of students elaborated on their earlier understanding: One pair added references to quantitative and qualitative data, another added references to the five senses, and the third pair added the role of inferences in explaining the observation. Other knowledge the pairs had gained about the activities of scientists included data analysis and evaluation (10%), making recordings, making data tables and graphs based on results of experiment (10%), and publishing (10%).

When asked about their conceptions of a scientific experiment in the beginning of the semester, most students did not have a clear and coherent definition. The definitions provided could be summarized as a hands-on investigation that is trying to answer a question or to prove something. The post-questionnaires revealed more elaborate conceptions of scientific experiments. For example, one pair noted that an experiment is "a method of investigating causal relationships among variables, using manipulated, responding, and controlled variables to justify their examples of experiment." The examples given within the various definitions specifically reflected their learning from the lab visits or the class activities. For example, one pair of students referenced a particular scientist who was studying genes while defining scientific experiment. Another pair of students referenced a descriptive study in which a molecular biologist was studying the stages of ovary development using fruit flies, noting that "she did not follow one specific scientific method, [but] let her research progress in a course of discovery." Another pair noted the different approaches the two different scientists they visited took, noting "they both study insect evolution of dung beetles, but they had different procedures and followed different steps in order for them to get a conclusion."

By the end of the course, students understood more about the factors that influenced a scientist's decision to study a particular phenomenon. On the pre-questionnaire, 100% of the students noted that scientists conducted investigations based on interest, which remained on the post-questionnaire. A few pairs also mentioned other factors such as issues going on in the world (10%), overall well-being (10%), funding (10%), or administration (10%). On the post-questionnaire, 100% of the students came to know that other than personal interests, scientific research was influenced by the previously noted factors in addition to social issues, place and culture, and public policy. One pair of students, for example, noted that a scientist needed to consider significance of the research to society.

***Data and data analysis.*** When asked about scientific data on the pre- and post-questionnaires, 100% of the students demonstrated an adequate notion of data in science, which was generally described as the information collected from observation and experiments. Of the questionnaires, however, 70% contained more informed understandings after the course. One of the pairs, for example, noted that "some data can be counted as evidence if it is supportive data. However, if the data does not support the claim or inquiry question, then it isn't evidence."

When asked about data analysis, 63% (2 did not provide a response) of the students demonstrated a naive idea on the pre-questionnaire. On the post-questionnaire, 60% of the pairs of students demonstrated more informed understandings of data analysis, adding references to certain aspects such as interpretation or approaches to analysis. One pair of students, for example, originally noted on their pre-questionnaire that data analysis was "evaluating information." They elaborated on this description on the post-questionnaire noting that, "data analysis is evaluating the information collected during the experiment process. The evaluation is done in order to derive meaning from the data so that it can be used as evidence or not for the proposed answer to the investigable question." Students generally began the semester with some appropriate understandings of the specific aspects of data analysis, such as looking for trends and relationships between the data, comparing different sets of data, making tables and graphs, checking validity of experiments, and explaining meanings. It was evident that students had developed more specific and elaborated descriptions on their post-questionnaire, reflecting their learning from lab visits. One pair, for example, had only "studying the data and deciding what it means" on their pre-survey, while at the end of the semester they added, "This could mean to make different graphs or tables to look at the data more clearly or even running it through different machines."

***Subjectivity and objectivity.*** When asked whether several scientists, working independently, asking the same question would come to the same conclusion, 100% of the student pairs initially responded "no." Two pairs of students elaborated on this response with the first pair noting that scientists "have their own opinions and previous views that influence what they believe and how they conduct their experiments" and the second pair noting "there isn't a great likelihood of several scientists setting up the exact same experiment because individuality, creativity, interests, and results can be very individual to each inquiry." The post-questionnaire showed the students retained their initial beliefs. However, 40% of the questionnaires contained more elaborated explanations with supporting details, such as noting different sources of information used, different ways to interpret the data, different ways of collecting the data, and ways of compiling their findings noting that each scientist is unique in the way they think and handle situations.

When asked whether their response would change if the scientists are working together, 40% of the questionnaires revealed that the students believed that even if scientists worked together, they would not necessarily come to the same conclusions because of different ways of thinking. Of the responses, 10% could be categorized as a maybe; indicating that it is more likely they will come up with the same answers when working together. The rest of the students insisted on the pre-questionnaire that the group of scientists would come to the same conclusion. These students elaborated with comments such as, "if the scientists are working together then they are following the same procedures and methods and working toward one conclusive answer," or noting that the scientists would eventually come to a consensus even though they may still have different methods or interpretations and the collaboration would have broader and more accurate results. Overall, 90% of the students retained their initial beliefs about subjectivity and objectivity within collaborative work throughout the course. However, 40% of the questionnaires contained more specific justifications for those beliefs. One pair stated, for example, that "We don't think our response would change. We both interviewed scientists that study the dung beetle in the same lab, but after looking at our different results, [we realize] the two scientists came to different conclusions about the dung beetles." Of the students, 10% did change their ideas.

## In What Ways Did the Strategies Enacted in the Course Foster or Hinder Students' Understandings of Scientific Inquiry and Scientists?

Data were collected using instructor notes, audiotaped reflections from the discussion groups, written reflections, and course documents to provide insight into the instructional strategies that fostered a broader understanding of scientific inquiry and scientists. We reviewed the findings and developed collective statements on the major categories that emerged. These major categories are discussed in the following sections, using student quotes as supporting evidence.

### *Exploring Existing Beliefs and Biases*

The first strategy we explored was having the students reflect on personal biases and beliefs in regard to scientists and scientific inquiry. For this reflection, the students were provided with the accumulated student data from the DAST, the list of known scientists, a checklist of characteristics of a scientists, and essays on a typical day in the life of a scientist. The data clearly indicated (1) the average number of scientists' names appearing in the list, (2) the scientists' gender, (3) common names/specializations given in the scientists' name lists, (4) the most common description of a typical day of the scientist or how it changes from one scientist to another. In discussion groups of 5–6, they explored and discussed those findings. The subsequent discussions were audiotaped and transcribed.

The students were very surprised to find an overwhelming consistency in their own images: The scientist was a stereotypical experimentalist male with goggles. They attributed this image to the media and school. The students explained: "Obviously when someone thinks of science, they are going to think of like test tubes and goggles, like most of the experiments and lab settings we have [in school]." The students noted their thinking showed no evidence of creativity or humanity in science, instead describing scientists as logical, accurate, methodical, hard-working, and lonely, thereby filling out the stereotype. They noted that the scientists they described seemed to be mechanical, just following the rules. They noticed that outdoor research was hardly mentioned as they put science in the laboratory. The students also noted their own lack of attention to technology. They concluded that they were not sure about the role of technology in the development of science, nor did they know how to draw or write about sophisticated equipment. They also noted the lack of references to female scientists. In one group discussion, they started to make references to Jane Goodall but, upon reflection, noted their limited knowledge about her work, so they were hesitant to discuss her work further and the thread of discussion quickly ended. One student concluded that the lack of references to female scientists was a direct reflection of their actual knowledge about women in science. Another student noted that the scientists they were taught about were old and from ancient times when women and minority scientists did not exist.

To summarize, students were surprised about the extent of the stereotypical image revealed by the group overall and defensive about their own responses. Most group reflections ended with them declaring that the findings from the pre-questionnaires were more of a reflection of their lack of knowledge about what to replace the stereotypical image with than it is of their existing image.

### *Exploring Real Scientists and Authentic Scientific Inquiry*

Students were further exposed to scientists and scientific inquiry through a WebQuest, lessons on scientists studying the phenomenon being explored in class, interviews of Jane Goodall and Birute Gladikas, and viewing the movie *Gorillas in the Mist.*

No evidence was found to show an impact from the exploration of scientists studying the same scientific topics the students were studying (e g., global climate change) and only a minimal amount of evidence was found that showed an impact from the WebQuest. The WebQuest provided a more diverse list of names of scientists in one in-class activity completed halfway through the semester. These names, however, did not carry over to the final data collection on names of scientists.

The data suggested that the students' ideas of scientists and scientific inquiry were influenced by watching interviews of scientists and watching the movie. Jane Goodall, Beruite Galdikas, and Dian Fossey were all on the list of known scientists completed at the end of the course, and their work was mentioned quite extensively in the post-questionnaires and course documents. In the reflections on the movie, one student noted, "I thought that scientists wear lab coats (and) work in the laboratory with test tubes and beakers—this is a different kind of research, done in the field in a jungle on a mountain side." Reading the students' reflections on the movie, it can be concluded that the students encountered a research experience very different from that which they described in the beginning of the course. The movie, for example, described science done by a woman and carried out in a mountain jungle. The students noted how Fossey's research was based mainly on observations. One student noted that "by being there she was able to learn the family structure, and other things that we otherwise wouldn't know; and even though she introduced them to things such as cameras, I still think she mostly adapted to their environment and way of doing things while she interacted with the gorillas." The students noted interventions that were done in order to gain the confidence of the gorillas. Some interventions they identified were mimicking the gorillas' motions and gestures until they began to imitate her and got closer to her. The students further noted that her motivation was heavily based in empathy. One student indicated that "Fossey was enchanted by the gorillas put her heart to them, protected them from harm."

To summarize, students were given a lot of exposure to different scientists throughout the course. In the final analysis, little evidence was found in the course documents to show an impact from our efforts to have the students explore the many faces of scientists who study the phenomenon we studied in class or through a WebQuest. Data, however, suggested that an impact was realized from the decontextualized lesson on three female scientists, Goodall, Fossey, and Galdikas.

### *Getting to Know a Local Scientist*

With this project, our students contacted local scientists researching topics of interest to them and arranged a meeting at their places of work. During the meeting, students interviewed the scientist about their research field, data collection and analysis methods, motivation to do their research, and life as a scientist. The students then shared their experiences with the other members of the class.

A review of the reports showed that the group of scientists who were ultimately interviewed were relatively young and came from several scientific domains. Male (42%) and female (58%) scientists were both well represented. About two-fifths did field research, or a combination of field and laboratory research. The students found that even within this small representative group of scientists, research was being done in many different fields including cutting-edge research in microbiology dealing with special neurological molecules, plate tectonic research, pollution management (for the city or government), pharmacologic work, insect evolution studies (development and mutations), petrol oils, environmental research related to geology and biochemistry (mammals), and toxicology. All of the of research projects the students encountered needed high-tech equipment. Such equipment included DNA and RNA analyzers, microscopes, wave measuring machines for plate tectonic research, and observation tools like electronic looking glasses for animal watching.

Our final data analysis revealed that, as a result of this project, our students learned that scientific inquiry is not limited to one method. In reflections about the activity, they noted, "I never really knew that there were that many types of research and that they all go about their research in different ways, and there are no specific steps to doing science." They also stated, "The scientific method, more often than not, does not describe the scientific process," and "there are so many different aspects of the job that I had never really associated with scientists." They reported that many of the scientists they came to know did not fit the stereotypical image of a scientist, noting that "the diversity of scientists really showed as well in regards to gender, ethnicity, age and field; scientists are not the stereotypical, lab-coat wearing nerds with no life—they are real people." As a result of this activity, the students "realized that scientists are much different than working professionals in other career fields. Their occupational pressures are different (experiment results, securing funding), but they are dedicated to love what they do."

The students indicated that scientists' motivation was derived from good experiences with science at school, with one saying, "Most of the scientists that were shared, have wanted to do something in science, or were good at the subjects of math and science, since a young age." They noted that the scientists' motivation was spurred by interest, challenges to overcome, success, and a sense of being a part of an important and useful project like taking care of the environment. They were also exposed to the difficulties experienced by scientists such as failed experiments, patient treatments that failed, grants, and financial insecurity. They saw and stated that scientific work needs many hours to accomplish, so scientists, both male and female, have to balance work, social life, and family.

This project appeared to help these students develop more positive attitudes about scientists and science. Students comments included: "My view on the importance of scientific work changed—amazing research and could potentially change the lives of those who are suffered from the disease" and "This project made me look at scientist and science in respectful ways. I have a greater appreciation for their work." Our final analysis revealed that this project greatly impacted our students. Some aspects of this experience could be seen in all of our post-intervention data. References to these experiences were found in their understandings of scientists and scientific inquiry.

## CONCLUSION AND IMPLICATIONS

Our findings, in general, show that our actions produced broader and more specific conceptions of scientific inquiry and scientists among our students. These findings provided us with a deeper understanding of the impact of the specific aspects of our intervention. Although we acknowledge that, at this point, it is hard to pull out a particular component without realizing an impact on others, our findings revealed some components that seemed to have greater impacts than others. These findings and their implications for our teaching are discussed below.

We intentionally structured our intervention to foster a continual process of change from existing stereotypical understandings to broader and more specific understandings of scientific inquiry and scientists. To start this process, we had the students explore their own initial biases and beliefs. The expressions of dismay and discomfort on the part of our students indicated that this strategy had an impact. The students concluded that they did not have an accurate understanding of scientists and scientific inquiry. This conclusion set the stage for a reconstruction effort. We plan to maintain this aspect of the intervention.

In contrast to the findings of Mbajiorgu and Iloputaife (2001), our findings revealed that providing students with general information on scientists' lives does not change substantially their image of scientists or scientific inquiry. A short-term impact on the students' understandings was realized as a result of the WebQuest and general online readings; however, these components of our intervention did not seem to impact their images over time. The exception was the information presented on Jane Goodall, Beruit Galdikas, and Dian Fossey. Looking back over our intervention efforts, we can see that our presentation of these three women was much more personal than that on the other scientists. The interviews of Jane Goodall and Beruit Galdikas featured the women themselves talking about their lives and their work. The movie viewed was very personal and involved an emotional component. In contrast, the WebQuest featured written text about scientists that was, for the most part, impersonal. Another difference in our approach was the extent of exposure. The information on the scientists included in the WebQuest, which also involved many women, was in the form of short biographies for these scientists, which the students glanced through seeking answers to specific questions. Thus, it is possible that the difference in effect may just be that amount of exposure for which we allowed, but we have no evidence for this conclusion at this point. We do not plan to maintain our approach to providing the students with general information on scientists' lives, with the exception of the lessons on the Jane Goodall, Beruit Galdikas, and Dian Fossey. In the future, we would like to expand our approach by adding more scientists introduced in a similar manner to how we introduced these three scientists.

For us, little doubt remains as to the value of the lab visits and interviews. After interviewing scientists and visiting their workplaces, our students' descriptions of scientists became localized and realistic, and they expressed broader and deeper understandings about scientific inquiry. Interviewing a scientist with specific guiding questions and visiting his/her workplace were found to assist students in broadening their understanding of scientific research and scientists. Through such an approach, our students saw what is currently happening in the real world of science. We acknowledge that in such a brief interaction, however, with little background knowledge in the specific science area, our students' understandings about the scientists' research may have been limited. Participation in authentic scientific research in the form of apprenticeships for an extended time has been found to enhance students' conceptions about scientists, understandings about scientific inquiry, and abilities to do scientific inquiry (Sadler & McKinney, 2010). Most of the students explored in this previous research, however, were from science and engineering disciplines while non-science major college students usually have limited science learning experience in college and little opportunity to experience authentic science (Bhattacharyya, 2009). Multi-section courses for these non-science majors, such as ours, have a very limited time allotment for such internship experiences. We plan to keep this component of our intervention, but would like to explore possible approaches that would allow us to extend the level of interactions closer to that of long-term internships explored in the previous research without sacrificing the other goals for the course overall.

Through this action research study, we came to better understand the conceptual understandings of scientific inquiry and scientists that are shaping the learning experiences of our students. By turning a critical eye back on our intervention efforts, we are now better positioned to reframe our students' fundamental understandings about scientific inquiry and scientists.

## REFERENCES

American Association for the Advancement of Science (AAAS). (1990). *Science for all Americans.* New York, NY: Oxford University Press.

Anderson, A. (2003, March). *Difficulties with inquiry in science classrooms.* Paper presented at the meeting of the National Association for Research in Science Teaching, Philadelphia, PA.

Aydeniz, M., Baksa, K., & Skinner, J. (2011). Understanding the impact of an apprenticeship-based scientific research program on high school students' understanding of scientific inquiry. *Journal of Science Education and Technology, 20,* 403–421.

Barab, S. A., & Hay, K. E. (2001). Doing science at the elbows of experts: Issues related to the science apprenticeship camp. *Journal of Research in Science Teaching, 38,* 70–102.

Baviskar, H. (2009). Essential criteria to characterize constructivist teaching: Derived from a review of the literature and applied to five constructivist-teaching methods articles. *International Journal of Science Education, 31,* 541–550.

Bhattacharyya, P. (2009). Incorporating authentic scientific research in an introductory general-education course for nonmajors. *Journal of College Science Teaching, 39,* 43–53.

Bell, R. L., Blair, L. M., Crawford, B. A., & Lederman, N. G. (2003). Just do it? Impact of a science apprenticeship program on high school students' understandings of the nature of science and scientific inquiry. *Journal of Research in Science Teaching, 40,* 487–509.

Bovina, I. B., & Dragul'Skaia, I. U. (2008). College students' representations of science and the scientist *Russian Education and Society, 50*(1), 44–64.

Brown, J. S., Collins, A., & Duguid, P. (1989). Situated cognition and the culture of learning. *Educational Researcher, 18*(1), 32–42.

Buck, G. A., Leslie-Pelecky, D., & Kirby, S. K. (2002). Bring female scientists into the elementary classroom: Confronting the strength of elementary students' stereotypical images of scientists. *Journal of Elementary Science Education, 14*(2), 1–9.

Buldu, M. (2006) Young children's perceptions of scientists: a preliminary study. *Educational Research, 48*(1), 121–132.

Cakmakci, G., Tosun, O., Turgut, S., Orenler, S., Sengul, K., & Top, G. (2011). Promoting an inclusive image of scientists among students: Towards research evidence-based practice. *International Journal of Science and Mathematics Education, 9,* 627–655.

Chambers, D. W. (1983). Stereotypic images of the scientist: The Draw-a-Scientist Test. *Science Education, 67*, 255–265.

Crawford, B. A., Zembal-Saul, C., Munford, D., & Friedrichsen, P. (2005). Confronting prospective teachers' ideas of evolution and scientific inquiry using technology and inquiry-based tasks. *Journal of Research in Science Teaching, 42*, 613–637.

Eisenhart, M., Finkel, E., & Marion, S. (1996). Creating the conditions for scientific literacy: A re-examination. *American Educational Research Journal, 33*, 261–295.

Farnham-Diggory, S. (1994). Paradigms of knowledge and instruction. *Review of Educational Research, 64*, 463–477.

Finson, K. D. (2001). Investigating preservice elementary teachers' self-efficacy relative to self-image as a science teacher. *Journal of Elementary Science Education, 13*(1), 31–42.

Finson, K. D. (2002). Drawing a scientist: What we do and do not know after fifty years of drawing. *School Science and Mathematics, 102*(7), 335–345.

Finson, K. D., Beaver J. B., & Cramond, B. L. (1995). Development and field test of a checklist for the draw-a-scientist test. *School Science and Mathematics, 95*(4), 195–205.

Haefner, L. A., & Zembal-Saul, C. (2004). Learning by doing? Prospective elementary teachers' developing understandings of scientific inquiry and science teaching and learning. *International Journal of Science Education, 26*, 1653–1674.

Kahle, J. (1987). SCORES: A project for change? *International Journal of Science Education, 9*, 325–333.

Koren, P., & Bar, V. (2009). Perception of the image of the scientist by Israeli student teachers from two distinct communities in Israel: Arabs and Jews. *The Eurasia Journal of Mathematics, Science and Technology Education, 5*(4), 347–356.

Levin, M., & Greenwood, D. J. (1998). *Introduction to action research: Social research for social change*. Thousand Oaks, CA: Sage.

Levin, M., & Greenwood, D. J. (2002). Pragmatic action research and the struggle to transform universities into learning communities. In P. Reason & H. Bradbury (Eds.), *Handbook of action research* (pp. 103–113). Thousand Oaks, CA: Sage.

Loughran, J. (2006). *Developing a pedagogy of teacher education: Understanding teaching and learning about teaching*. New York, NY: Routledge.

Lyons, T., & Quinn, F. (2010). *Choosing science. Understanding the declines in senior high school science enrolments*. Armidale, NSW: University of New England.

Mbajiorgu, N. M., & Iloputaife, E. C. (2001) Combating stereotypes of the scientist amongpre-service science teachers in Nigeria. *Research in Science and Technological Education, 19*(1), 55–67.

McDuffie, T. (2001). Scientists—Geeks and nerds? *Science and Children, 38*(8), 16–19.

Mead, M., & Metraux, R. (1957). Image of the scientist among high school students: A pilot study. *Science, 126*, 384–390.

Moseley, C., & Norris, D. (1999). Preservice teachers' views of science, *Science and Children, 37*(6), 50–53.

Painter, J., Jones, M. G., Tretter, T. R., & Kubasko, D. (2006). Pulling back the curtain: Uncovering and changing students' perceptions of scientist. *School Science and Mathematics, 106*(4), 181–190.

Quita, I. N. (2003). What is a scientist? Perspectives from teachers of color. *Multicultural Education, 11*(1). 29–31.

Rahm, J., & Downey, J. (2002). "A scientist can be anyone!" Oral histories of scientists can make "real science" accessible to youth. *The Clearing House: A Journal of Educational Strategies, Issues and Ideas, 75*(5), 253–257.

Rampal, A. (1992). Images of science and scientists: A study of school teachers' views of the characteristics of scientists. *Science Education, 76*, 415–436.

Ritchie, S., & Rigano, D. (1996). Laboratory apprenticeship through a student research project. *Journal of Research in Science Teaching, 33*, 799–815.

Rosenthal, D. (1991). A reflective approach to science methods courses for preservice elementary teachers. *Journal of Science Teacher Education, 2*, 1–6.

Rosenthal, D. B. (1993). Images of scientists: A comparison of biology and liberal arts studies majors. *School Science and Mathematics, 93*(4), 212–216.

Roth, W. M. (1999, April). *Scientific research expertise from middle school to professional practice*. Paper presented at the Annual Meeting of the American Educational Research Association, Montreal, Quebec.

Roth, W. M., McGinnis, M. K., & Bowen, G. M. (1998). How prepared are preservice teachers to teach scientific inquiry? Levels of performance in scientific representation practices. *Journal of Science Teacher Education, 9*, 25–48.

Rubin, E., Bar, V., & Cohen, A. (2003). The images of scientists and science among Hebrew and Arabic-speaking pre-service teachers in Israel. *International Journal of Science Education, 25*(7), 821–846.

Russo, J., Mattox, S., & Kildau, N. (2010). Predicting the timing and location of the next Hawaiian volcano. *Science Scope, 33*(5), 32–38.

Ryder, J., & Leach, J. (1999). University science students' experiences of investigative project work and their images of science. *International Journal of Science Education, 21*, 945–956.

Ryder, J., Leach, J., & Driver, R. (1999). Undergraduate science students' images of science. *Journal of Research in Science Teaching, 36*, 201–219.

Sadler, T. D., Burgin S., McKinney L., & Ponjuan L. (2010). Learning science through research apprenticeships: A critical review of the literature. *Journal of Research in Science Teaching, 47*, 235–256.

Sadler, T. D., & McKinney, L. (2010). Scientific research for undergraduate students: A review of the literature. *Journal of College Science Teaching, 39*(5), 43–49.

Scherz, Z., & Oren, M. (2006). How to change students' images of science and technology. *Science Education, 90*, 965–985.

Schwartz, R. S., Lederman, N., & Lederman, J. (2008, March). *An instrument to assess views of scientific inquiry: The VOSI questionnaire*. Paper presented at the international conference of the National Association for Research in Science Teaching. Baltimore, MD.

Thomas, M, D., Henley, T. B., & Snell, C. M. (2006). The draw a scientist test: A different population and a somewhat different story. *College Student Journal, 40*(1), 140–148.

Trumbull, D. & Kerr, P. (1993). University researchers' inchoate critiques of science teaching: Implications for the content of preservice science teacher education. *Science Education, 77,* 301–317.

Unver, A. O. (2010). Perceptions of scientists: A comparative study of fifth graders and fourth year student teachers. *Necatibey Faculty of Education Electronic Journal of Science and Mathematics Education,4*(1), 11–28.

van Eijck, M., & Roth, W.-M. (2008). Representations of scientists in Canadian high school and college textbooks. *Journal of Research in Science Teaching, 45,* 1059–1082.

Varelas, M., House, R., & Wenzel, S. (2005). Beginning teachers immersed into science: Scientist and science teacher identities. *Science Education, 89,* 492–516.

Windschitl, M. (2001, April). *An analysis of preservice science teachers' independent inquiry experiences.* Paper presented at American Educational Research Association Conference. Seattle, WA.

Windschitl, M. (2002). Inquiry projects in science teacher education: What can investigative experiences reveal about teacher thinking and eventual classroom practice? *Science Education, 87,* 112–143.

Windschitl, M. (2004). Folk theories of "inquiry": How preservice teachers reproduce the discourse and practices of an atheoretical scientific method. *Journal of Research in Science Teaching, 41,* 481–512.

Zembal-Saul, C., Blumenfeld, P. C., & Krajcik, J. S. (2000). The influence of guided cycles of planning, teaching and reflection on prospective elementary teachers' content representations. *Journal of Research in Science Teaching, 37,* 318–339.

CHAPTER 10

# TEACHING UNDERGRADUATE INTRODUCTORY PHYSICS WITH AN INNOVATIVE RESEARCH-BASED CLICKER METHODOLOGY

**Lin Ding and Neville W. Reay**


## ABSTRACT

Personal response systems, also known as clickers, have been broadly used in undergraduate physics classrooms to promote student active learning. Typical physics clicker materials are individual concept questions created and used as discrete items. Although useful, these materials are not directly aimed at mitigating the issues of contextualized learning. Since learning is context sensitive, students often have trouble applying learned knowledge and skills across different situations. To address this issue, we have developed an innovative clicker methodology framework, in which clicker questions are systemically crafted into coherent sequences. Within each sequence, 3–4 individual questions are devised to contain the same underlying concept but are embedded in different contextual surface features. By consistently implementing each sequence as a cohesive microsystem in class, we afford students with increased opportunities to apply a key concept across diverse contexts,


*Research Based Undergraduate Science Teaching*, pp. 301–330

therefore promoting flexible learning. This chapter reports creation, validation, implementation, and evaluation studies of our clicker question sequences. Successful classroom practices and student learning outcomes are also reported.

With the burgeoning popularity of clickers in higher education, a reoccurring question among many educators and researchers has been how to most effectively use this tool to teach undergraduate sciences. Clickers, often also known as student response systems, are handheld wireless electronic devices, much like a TV remote control, that students can use in class to anonymously transmit their responses to an instructor's computer when answering questions posted on a screen in front of a classroom. By instantaneously tallying and analyzing student responses, the clicker technology can provide immediate feedback to both instructors and students. Various benefits of clicker usage have been documented (Caldwell, 2007; Crossgrove & Curran, 2008; Martyn, 2007; Preszler, Dawe, Shuster, & Shuster, 2007; Woelk, 2008; Wood, 2004). Perhaps the most important advantage of this technology is that it can improve science learning through student active engagement and cooperative participation (Duncan, 2005; Knight & Wood, 2005; Mazur, 1997).

In science learning and particularly in physics learning, much attention is justifiably focused on helping students rectify incorrect preconceptions and construct scientifically sound and robust understanding, so that students can flexibly utilize acquired knowledge to solve problems in novel situations. However, learning is sensitive to the context in which it is embedded (Bransford, Brown, & Cocking, 2000; Spiro, Feltovich, Jacobson, & Coulson, 1991). Oftentimes, learners have difficulty in applying what they have acquired in one context to others, despite the fact that the same knowledge is applicable (diSessa & Wagner, 2005; Hammer, 2000; Sabella & Redish, 2007). This context dependence of learning has been extensively reported in science education (see below for details) and, if not properly handled, can be a serious barrier to students' conceptual understanding and application.

We seek to address the context dependence of learning by proposing and applying a new methodology framework for clicker use in college-level introductory physics classes. In this framework, sequences of three or four individual questions are systemically designed. Within each sequence, questions share the same underlying concept but have different contexts. We refer to contexts as specific situations through which a question is framed and presented. Different contexts, for example, can involve distinct narrative scenarios, entities, states and conditions, and representational formats. They are primarily nonessential surface features and are separate from the core ideas needed to answer the questions. In this regard, the application of a basic physics concept or principle can be embedded into a range of contexts and hence can be transformed into various isomorphic questions. We use these isomorphic clicker questions in temporal proximity as a coherent sequence. As such, students can be explicitly instructed to pay particular attention to the underlying core concepts and learn not to be distracted by superficial features. Based on this framework, we have developed, validated, implemented, and evaluated 167 clicker question sequences, or approximately 500 individual questions. The sequences are extensive enough in number to populate a typical year of college-level introductory physics. Our goal is to use this clicker methodology to enhance student conceptual learning of introductory physics.

Centered on the two themes of clicker use in science teaching and the context dependence of learning, this chapter is organized into four sections. In the first section, we provide a major review of relevant literature. In the second section we introduce, in detail, our new clicker methodology framework as a linkage to connect current research with classroom teaching. The third section reports our original studies of developing, validating, implementing, and evaluating clicker question sequences for use in introductory physics classes, followed by a general discussion in the next. We close this chapter with a brief summary.

## LITERATURE REVIEW

In what follows, we review relevant literature from four perspectives: context dependence of learning, theoretical frameworks thereof, teaching strategies, and clicker technology in science teaching and learning.

### Context Dependence of Learning

An ultimate goal of learning is that a learner must be able to make transfer, or simply put, to apply what she or he has learned in one context to new contexts in order to solve a variety of emerging problems (Bransford et al., 2000; Broudy, 1977). However, an overwhelming body of evidence from cognitive and educational research has shown that learning can be highly sensitive to contextual cues, and therefore transfer across different contexts is not an easy task for students.

Perhaps one of the most well-known stances in cognitive psychology that specifically addresses the context dependency of learning is "situated learning." It emphasizes the idea that learning bears a deep imprint of the specific environment in which it is embedded (Anderson, Reder, & Simon,

1996; Clement, 1982; Koedinger & Nathan, 2004). For example, dairy workers and grocery shoppers, who are skilled at nonschool mathematics in performing rather demanding daily-life tasks, may not be able to do the same type of computations to solve problems presented in the standard school format (Lave, 1988; Scribner, 1984). Similarly, for students who rely on environmental settings as external cues to learn a given material, an approach called imagery technique, their subsequent understanding of the material can be noticeably decreased when the cues change (Eich, 1985). Cognitive studies in the area of analogical reasoning also demonstrate that automatic transfer rarely happens. Even after spending sufficient time studying a solved problem, students still have difficulty in carrying out the same solution to tackle another isomorphic problem, one that is presented in a different case scenario but can be solved by using the same solution scheme (Gick & Holyoak, 1980, 1983).

In science education, contextualized learning is ubiquitous. Transfer across different subject domains has shown to be very challenging for students (Bassok, 1990; Bassok & Holyoak, 1989; Rebello, Cui, Bennett, Zollman, & Ozimek, 2007). This is particularly true when a set of knowledge learned in a context-rich domain (e.g., physics) needs to be applied to a more general and abstract domain (e.g., mathematics). As Bassok and Holyoak claimed, this challenge is mainly derived from the contextualization of the knowledge that students have acquired (Bassok, 1990; Bassok & Holyoak, 1989). A high level of contextualization without abstract schemas poses a barrier for students to recognize the relevance between similar problems and to retrieve previously learned knowledge.

Even within the same subject domain, the context dependence of learning is still frequently observed and is broadly documented among novice students in learning sciences. When using analogical examples (e.g., a magnetic pendulum and a chaos bowl) to help students understand the nondeterministic nature of chaotic systems in physics, Duit and his colleagues found that if no guidance was provided, students could be easily distracted by some unintended superficial features (e.g., geometric appearances) and hence lose sight of the key ideas (Duit, Roth, Komorek, & Wilbers, 2001). Similarly, in problem sorting tasks, typical students often tended to categorize physics problems into groups that were defined by surface features such as entities contained in the problem statements, rather than into groups characterized by major physics principles (Chi, Feltovich, & Glaser, 1981; Hardiman, Dufresne, & Mestre, 1989).

Moreover, even fine-grained representational change also can induce significant fluctuations in student performance on isomorphic problems. In studying how the framing of physics problems may affect student responses, Steinberg and Sabella (1997) discovered a persistent mismatch between students' performance on qualitative questions in the multiple-choice format and their performance on similar exam problems presented in the quantitative show-work format. Oftentimes students could successfully solve quantitative problems through mathematical manipulation but failed to answer equivalent conceptually based qualitative questions. Along the same research line, Meltzer (2005) compared student responses to a set of isomorphic physics questions framed in four different representation formats: verbal, graphical, diagrammatic, and mathematical/symbolic. Results confirmed the significant influences of representational context on student knowledge and reasoning. Recent research further suggested that novice students typically lack flexibility in switching between multiple representations in solving physics problems (Kohl & Finkelstein, 2006; Kohl, Rosengrant, & Finkelstein, 2007). Therefore, representation formats can greatly affect students' choice of problem solving strategies, which in turn affect their performance (Kohl & Finkelstein, 2005).

## Theoretical Framework for Context Dependence of Learning

In light of the above observations, researchers have proposed theoretical accounts to elucidate the underlying causes for the context dependence of learning. In this section, we review four relevant frameworks: cognitive flexibility theory, phenomenological primitive model, knowledge resources model, and coordination class theory.

The cognitive flexibility theory (CFT) emphasizes that knowledge needs to be represented in different conceptual and case viewpoints so as to be useful for flexible understanding in future use (Spiro et al., 1992). According to the CFT, knowledge domains can be roughly divided into two large families—ill-structured and well-structured domains, and learning in the former can be highly sensitive to context. This is because ill-structured domains, like physics, typically require learners to simultaneously consider multiple schemas in each incident of knowledge application (known as complexity), and each of these application incidents often varies considerably across cases nominally of the same type (known as irregularity). In other words, knowledge in ill-structured domains "has to be used in too many different ways for them all to be anticipated in advance" (Spiro et al., 1992, p. 64). As a result, there is no precompiled "prescription" that can be readily brought to bear in all cases through simple memory retrieval. Therefore, learners must improvise suitable schemas to solve each problem on a case-by-case basis. It is this cross-case diversity that leads to the context dependency of learning in many ill-structured domains.

Another framework that can explain the context dependency of learning is the phenomenological primitives model, which was initially developed to account for student intuitive conceptions. In this model, student intuitive ideas are described as a set of small-sized, loosely interconnected cognitive structures (diSessa, 1988, 1993). DiSessa called these cognitive structures phenomenological primitives, or *p-prims*. Some examples of p-prims include "closer means stronger," "maintaining agency," "dying away," and "Ohm's p-prim" (diSessa, 1993). Ohm's p-prim describes the relationship between agency, impediment, and effect. Agency is positively associated with effect, whereas impediment is negatively associated with effect. These p-prims are not by themselves right or wrong; instead, how they are used can be correct or incorrect. For instance, one cannot judge whether or not the notion of "closer means stronger" is correct without knowing the specific context in which it is used. Since p-prims exist in students' minds at an abstract level and are not tied to any specific situation, when to use which p-prim can largely be determined by external cues, such as the entities contained in a problem statement or the representational format of a question. This provides some preliminary explanations from the learners' perspective for why science learning and reasoning can be sensitive to context.

Based on the p-prim model, Hammer and colleagues and Redish and Sabella further proposed a "knowledge resource" model, in which students' knowledge is conceptualized as a network of nodes interconnected by links (Hammer, 1996a, 1996b, 2000; Hammer, Elby, Scherr, & Redish, 2005; Sabella & Redish, 2007). The nodes represent basic declarative and procedural knowledge elements, such as p-prims and some common sense interpretations of the physical world, which are obvious and irreducible to students. The links represent the interrelations among the nodes. When a student encounters a particular problem, part of her or his knowledge network is accessed and activated, thus becoming available as "resources" for tackling the problem. Notably, the access and activation of the "knowledge resources" is highly dynamic, depending on the contextual cues perceived by the problem solver in the following mechanisms: (1) In some cases, one activated resource can further activate more related resources, much like a spreading effect; (2) in other cases, while some resources are activated, others may be blocked; (3) yet in other cases, multiple conflicting resources can be simultaneously activated in response to a particular contextual cue, causing the problem solver to be in a conflicting mental state, called "cognitive conflict." According to Sabella and Redish, how the activation of a resource leads to the activation or blocking of another (pattern of association) is crucial for problem solving (Sabella & Redish, 2007). The fact that many novice students are easily influenced by surface features when answering questions is perhaps due largely to their lacking principle-oriented patterns of association.

Recently, DiSessa and Wagner (2005) proposed a coordination class theory, in which they claimed that a particular type of concept, namely coordination class, is responsible for knowledge transfer. The coordination class has two functions: allowing a learner to "read out" relevant information from various situations, and invoking a "causal net" to determine the relevance of the readout information and make inferences. According to diSessa and Wagner, both the information readout and causal net processes are dependent on learners' prior knowledge. If a learner lacks well-prepared prior knowledge, both processes can go awry. Consequently, the learner either cannot recognize the applicability of learned knowledge in the case of a different but similar situation due to causal net failures, or cannot determine relevant information due to the readout deficiencies (diSessa & Wagner, 2005). This is why novice students frequently try out their bits and pieces of old intuitive ideas in an inconsistent manner when encountering the same problem embedded in new contexts.

## Teaching Strategies to Overcome Difficulties Arising From Contextualized Learning

The above theoretical frameworks provide useful insight into the effective design of teaching methods to alleviate difficulties arising from the context dependence of learning. One strategy is to present students a variety of first-hand experiences that share similarities in fundamental concepts. The goal of this method is to rely on inductive reasoning to help students recognize underlying connections among apparently diverse examples. Minstrell, for example, successfully used this approach to teach students how to analyze forces for objects that are "at rest" (Minstrell, 1982).

Another strategy is to use analogical reasoning to facilitate application of basic concepts across different contexts. For instance, Clement, Brown, and Zietsman implemented "bridging cases" in introductory physics classes (Clement, 1988, 1991; Clement, Brown, & Zietsman, 1989). They first identified students' correct understandings in specific examples as anchors, and then attempted to connect these anchors to target examples by progressively changing certain contextual surface features. Since, for students, the anchors and targets often bear seemingly little analogical resemblance, the intermediate examples can serve as a bridge to lead students across the conceptual gap. The rationale of this approach is that not all student preconceptions are misconceptions, and that transfer across different contexts can be effectively promoted if connected sequences of analogical examples are used with student proper conceptions as a starting point (Clement, 1982; Clement et al., 1989).

Focusing on multiple knowledge representations is also a useful method. For example, in teaching introductory mechanics, Dufresne, Gerace, and Leonard (1997) deliberately asked students to reconstruct each given physics problem into several representational formats, such as strobe diagrams, equations, and graphs, different from the one in which it was originally presented, before solving it. As such, students could better understand the problem in diverse contexts and be more flexible in pursuing solutions. Similarly, Van Heuvelen and Zou (2001) taught students to use verbal, pictorial, bar chart and mathematical representations to learn work-energy processes. A post-intervention assessment indicated that students using this strategy had an increased understanding of the work-energy topics. Several other studies also demonstrated that teaching with multiple representations can enhance student understanding of physics concepts (Lasry & Aulls, 2007; Savinainen, Scott, & Viiri, 2005).

It is worth noting that in order for the above teaching strategies to be effective, explicit guidance for comparison among multiple cases is needed. Traditional teaching by examples without directing student attention to the underlying connections among them is ineffective (Brown, 1992). Conversely, it is more successful if multiple source examples are used together with explicit instruction for comparing among them (Novick & Holyoak, 1991). The purpose of such instruction is to help students extract useful schema at an abstract level optimal for identifying and mapping the common relational structures among individual cases (i.e., the underlying deep structure) (Gentner, Loewenstein, & Thompson, 2003; Gick & Holyoak, 1983; Kurtz, Miao, & Gentner, 2001). According to diSessa, Brown, and Clement, both abstract schema and concrete examples are crucial for successful transfer. The former facilitate recognition and mapping of deep structures across different contexts, whereas the latter provide opportunities to enrich and solidify the shared structures at a concrete level (Brown, 1993; Brown & Clement, 1989; diSessa & Wagner, 2005).

## Clicker Technology in Science Teaching and Learning

Various teaching strategies including those mentioned above can be further augmented if used in conjunction with suitable educational technologies, given that they are carefully designed and properly implemented. Student response systems (or clickers) are a popular technological tool that has been shown effective in increasing student active engagement and learning performance in science classes.

This technology provides a number of benefits for effective learning and teaching. It facilitates instructors' acquisition of student information, such as student attendance, preparedness, and affective opinions (Caldwell, 2007; Knight & Wood, 2005; Martyn, 2007; Paschal, 2002; Woelk, 2008). More importantly, it offers an opportunity for conducting in-class formative assessment, which allows both the instructor and students to be informed of their progress in real time so that subsequent actions can be purposely focused on student learning difficulties (Caldwell, 2007; Woelk, 2008). Perhaps most importantly, when properly used, clickers can spark lively discussions among students and between students and instructor, thereby creating a desirable learning environment in which students are actively engaged and collaboratively work on the same tasks (Duncan, 2005, 2006; Mazur, 2009; Wood, 2004). An example is Mazur's Peer Instruction (Mazur, 1997). In this method, students are periodically asked conceptual questions in a lecture. Each time, they first individually answer a question using clickers and then verbalize their reasoning in discussion with neighboring peers before answering the same question again. As such, students can clarify their thoughts and better improve their conceptual learning (Crouch & Mazur, 2001; Duncan, 2005; Fagen, Crouch, & Mazur, 2002; Smith et al., 2009).

A number of empirical studies have documented the positive effects of clicker use on student learning of sciences. Deslauriers, Schelew and Wieman (2011) compared two sections of a lower-division introductory physics course. One was a traditional lecture delivered by an experienced instructor; the other was taught by trained novice instructors that used clickers as a primary tool to promote student engagement and cooperative learning. Results showed that the latter class performed significantly better on a post concept test than did the former class. At the upper-division level, Knight and Wood (2005) partially changed a traditionally taught biology course into a more interactive format by reducing lecture time to compensate for active student participation in frequent in-class clicker assessment. This change subsequently resulted in higher student learning gains and improved conceptual understanding. Recently, Crossgrove and Curran (2008) studied the effect of clicker use on student long-term knowledge retention and revealed that using clickers helped students in a non-major biology class to better retain learned knowledge up to four months after the course ended. Furthermore, Preszler et al. (2007) systematically studied a broad range of biology classes and divided them into three levels of frequency in clicker usage (low, medium, and high) and found that that an increased clicker use in lecture positively affected student performance on exam questions.

Despite these reported successes, researchers have unanimously cautioned us that clickers are only a pedagogical tool. The rationale behind clicker question design and the pedagogy of implementation are far more important (Caldwell, 2007). Some general guidelines for designing and using clicker questions have been recommended. For example, Beatty et al. suggested a three-prong consideration for learning goals (covering content, cognition, and metacognition) and offered some general tactics for writing good clicker questions to address these goals (Beatty, Gerace,

Leonard, & Dufresne, 2006). Along the same line, Sullivan (2008) used Bloom's taxonomy of cognition levels to call for good clicker questions that target higher-level of reasoning, such as application, analysis, synthesis and evaluation. Also, Mestre, Gerace, Dufresne, and Leonard (1997) proposed a cyclic method to effectively implement clickers in classroom teaching, which includes seven steps of question generation, presentation, student group work, answer collection, histogram display, class discussion, and closure.

While these general tips are useful for individual question design and implementation at a broader level, they are not set forth to uniquely address the context dependence of learning. It is well recognized as one of the design tips that clicker questions need to present learners with materials in different contexts (Caldwell, 2007). This chapter introduces a new clicker methodology framework to effectively fulfill this need through clicker use.

## FROM RESEARCH TO PRACTICE: A NEW METHODOLOGY FRAMEWORK OF CLICKER QUESTION SEQUENCES

We have designed and assessed a new clicker methodology framework to specifically address learning difficulties originating from the context dependence of learning. In our framework, clicker questions are not written or used individually to help students learn isolated concepts, as conventionally practiced. Instead, sequences of three or four individual clicker questions are systemically created; within each sequence, questions address the same concept but have different contextual features. When used in classroom teaching, each sequence is implemented as a cohesive microsystem to direct students' attention onto the shared underlying structure and to stimulate their analytical thinking about a basic concept in diverse contexts.

This clicker sequence framework is built on the aforementioned theoretical studies of learning and transfer, and its intrinsic benefits can afford a practically effective means to connect research with real classroom teaching. From the perspective of cognitive flexibility theory, a single concept or principle in ill-structured domains (e.g., physics) can take on many forms of usage across different instances (Spiro et al., 1992). Therefore, consistently showing learners multiple conceptual applications close together in time can be more favorable than presenting them sporadically. Similarly, from the stance of the knowledge resources model, principled knowledge associations are fragile for novice students (Hammer, 2000; Hammer et al., 2005;

Sabella & Redish, 2007), so target training with interconnected sequences of isomorphic clicker questions can effectively strengthen association links centered on basic concepts and weaken those connected by superficial contexts. Also, in view of the p-prims model, students' primitive ideas are rather abstract (diSessa, 1988, 1993). Hence, providing multiple concrete cases temporally together in a sequence can beneficially lead students to solidify and enrich the application of their abstract ideas. Additionally, from the coordination class standpoint, information readout and causal net failures directly impede knowledge application in new contexts (diSessa & Wagner, 2005). This means that helping students learn to extract and extend the key information across a unified sequence of diverse situations is necessary.

Moreover, a range of isomorphic cases within each clicker question sequence are presented to students in the problem-solving format, similar to what students will likely encounter in their future knowledge learning and skill acquisition. Bransford et al. emphasized that preparing students with experiences that resemble future use can most effectively facilitate their knowledge application in novel situations, and that the line of similarity in fact lies in the problem-oriented experiences (Bransford, Franks, Vye, & Sherwood, 1989).

Furthermore, the connectedness between the individual clicker questions within each sequence can catalyze more in-depth discussions in class to increase learners' cognition and metacognition skills. Student attentions are not only focused on fulfilling the cognitive requirements needed to answer each clicker question, but also directed to evaluating their own understanding of the relational structure shared by the individual questions. This usage of clicker question sequences can enrich student peer discussions and entire-class dialogues.

This clicker methodology encompasses an iterative process of four major stages: clicker sequence creation, validation, classroom implementation, and evaluation (see Figure 10.1). Following these four stages, we have generated a repository of 167 clicker question sequences, approximately 500 individual questions. The sequences are extensive enough in number to populate a typical year of college-level introductory physics. Our studies in this area include both the conceptualization of sequence design and actual teaching in real classrooms; they serve as a linking bridge to connect research with practice. In what follows, we report our original studies in college-level introductory physics from four viewpoints, each corresponding to one of the four stages in Figure 10.1.

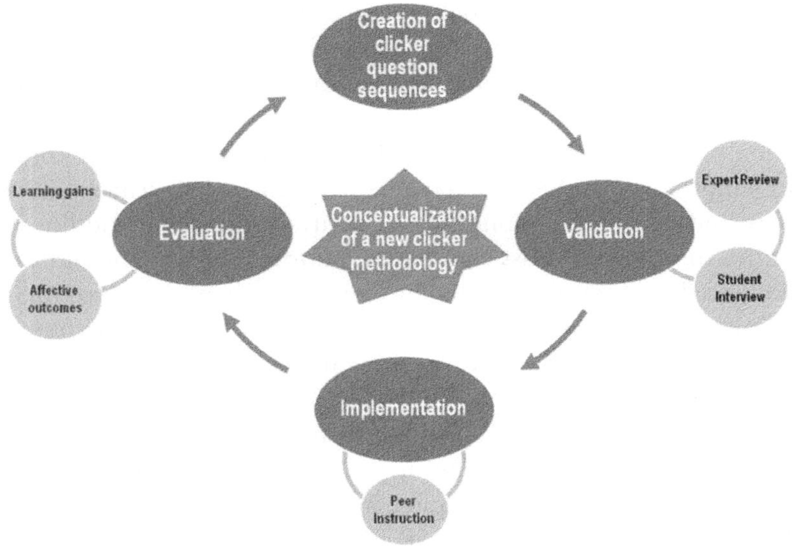


**Figure 10.1.** Interactive approach to developing clicker question sequences.

## CREATION, VALIDATION, IMPLEMENTATION, AND EVALUATION OF CLICKER QUESTION SEQUENCES

### Creation of Clicker Question Sequences

We created clicker question sequences with an intention to cover major concepts typically discussed in a calculus-based introductory physics course. These concepts were extracted from commonly used textbooks (Halliday, Resnick, & Walker, 2008; Knight, 2008) and were consulted on with faculty from institutions of different types to ensure what is covered in our clicker material is valued by the broad physics community (Lehrer & Shauble, 2000). For each concept, three or four isomorphic multiple-choice questions were designed into a sequence with varying features. They were purposely made to be vastly distinct in contexts to provoke an enriched knowledge base for a shared underlying concept (diSessa & Wagner, 2005). The techniques we used to diversify these contextual features include creating dissimilar story lines, providing different entities or different states/conditions thereof, changing perspectives or viewpoints, and switching between several distinct representational modes, including verbal, pictorial, diagrammatical, graphical, symbolic, and mathematical (Lee, Ding, Reay, & Bao, 2011; Reay, Bao, Li, Warnakulasooriya, & Baugh, 2005; Reay, Li, & Bao, 2008).

Particular attention also was given to the cognitive levels involved in each clicker question sequence. Specifically, we carefully avoided questions that target lower levels of cognition, such as recall of facts or execution of routine procedures. Instead, we strived to create questions requiring higher-level skills such as knowledge differentiation and integration, comparing and contrasting, inductive/deductive reasoning, causal analysis, summarizing, categorizing, estimation, trouble-shooting, and evaluation. These questions were also designed to include distracters that target student misconceptions and intuitive ideas. These distracters were drawn from the authors' years of observations on student responses to homework and exam questions as well as from the rich resource in relevant studies of physics education research (e.g., Heron, 2004; McDermott & Redish, 1999).

Moreover, we deliberately arranged the questions of each sequence in the order of increasing difficulty to progressively raise student cognitive and metacognitive attention. It was our hope that by the end of each sequence, students would benefit from thinking of their own thoughts in answering the individual questions and come to realize the connective thread running through the entire sequence. However, we tried not to make the last question too challenging, thereby enabling students to gain a sense of achievement after completing a sequence.

During this creation stage, all question sequences were constantly reviewed and revised internally among the authors. More details for the external review and validation processes are reported in the following subsection. In Figure 10.2, we show two sample sequences to illustrate the typical structure and rationale of question sequence design.

Figure 10.2 shows a mechanics question sequence on the topic of force analysis. It provides three cases with different perspectives. The first question asks students to diagrammatically represent all individual forces on a block sitting at rest on a slide. The second question then shifts the perspective from the individual forces to the net force on the block. Students need to consider not only the net effect of all forces on the block, but also the net effect due to the slide. If one still dwells on the individual components, it is unlikely that she or he will be able to correctly determine the answer. The third question further changes the focus from the block to the slide, yet another perspective to look at the same event. It now requires students to decide how the interaction between the block and the slide causally affects the motion of the slide. It is a challenging question, because other than the need of changing the view angle to analyze the situation, it also integrates force and the effect of force—two separate elements—into one. Collectively, these three questions form a coherent sequence, offering students an opportunity to consider the same topic from multiple viewpoints. This is beneficial for helping students attain the goals of flexible knowledge application (Spiro et al., 1992).

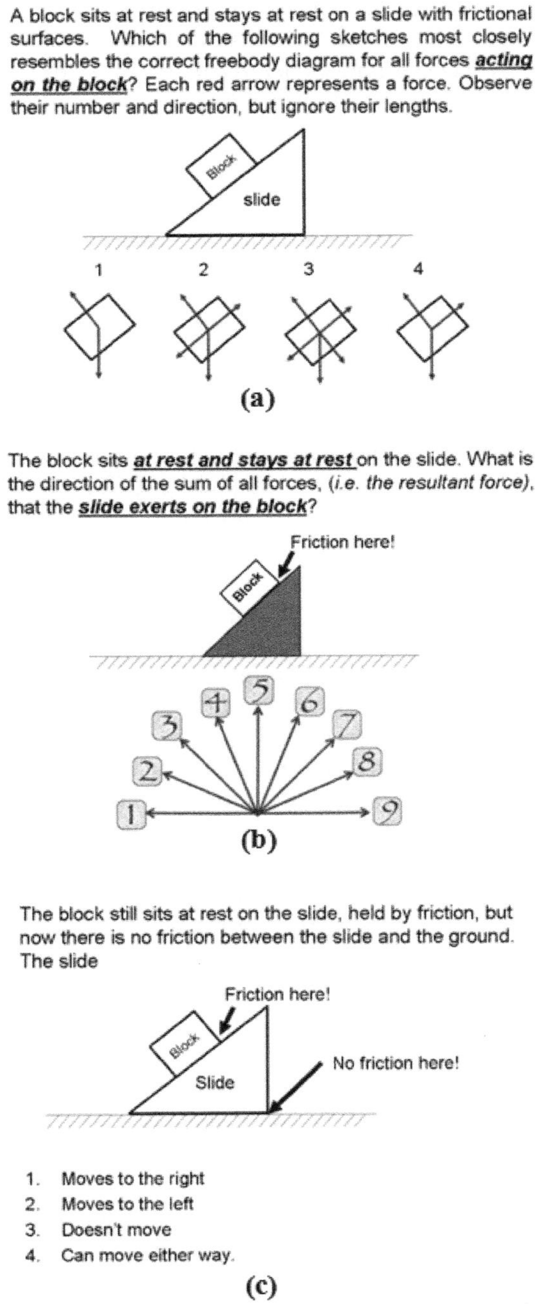


**Figure 10.2.** A mechanic's clicker question sequence.

Figure 10.3 illustrates a sequence in electricity and magnetism. It targets the concept of Faraday's law and consists of three questions that employ diverse entities and representational modes. The first question requires students to compare and contrast the induced EMF of two loops with different sizes as they enter a uniform magnetic field. Since both loops are square, the larger one generates a greater EMF. If used alone, this question would be less instructive, as many false positives may occur. Students can use the "larger means more" p-prim to get a correct answer. So, this question is combined with the next one that involves four loops of different dimensions. This second question is purposely designed to call for the knowledge differentiation skill—the delineation between magnetic flux and the rate of change in magnetic flux. Those with false positives in the first question now can be differentiated from the true positives. The third question further asks students to extend their knowledge to an unfamiliar case—a circular loop, which is not regularly encountered in typical textbooks. Additionally, it requires students to make a representational translation between diagrammatic and graphical modes in answering the question. This sequence highlights the use of contrasting cases by presenting multiple entities with perceptually different dimensions. This can lead to students' increased conceptual skills and better prepare them for future knowledge transfer (Bransford & Schwartz, 1999; Schwartz & Bransford, 1998).

## Validation of Clicker Question Sequences

The clicker question sequences were then validated, first with experts. Six physics faculty members, two postdoctoral fellows, and 45 graduate teaching instructors from four universities and colleges examined our clicker question sequences and provided open-response feedback. Their task was to ensure the clicker questions covered important topics, were technically correct, and addressed common student incorrect conceptions. Each question was judged against these three criteria and ranked on a five-point Likert scale with 1 being a poor question and 5 being a good one. All questions received an average above 4, and experts' open-response feedback typically contained minor suggestions for the improvement of language and representation. Based on these suggestions, we revised the questions. Revisions were then cross-examined among the authors to ascertain that the revised questions answered the concerns raised by the experts.

Using these expert-validated clicker questions, we conducted student interviews. Traditionally, student interviews focus on uncovering learners' naïve ideas or learning difficulties. A common assumption is that students are beginners who must possess misconceptions. However, we viewed student interviewees as "expert consultants." Though they were

The figure shows two wire loops, with edge lengths of $L$ and $2L$, respectively. Both loops will move through a region of uniform magnetic field $\boldsymbol{B}$ at the *same constant velocity*. Rank them according to the emf induced ***just as their front edges enter the B field region***.

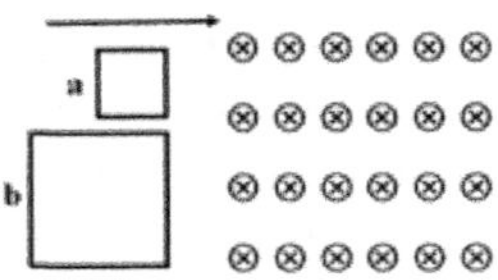


1. a > b
2. a = b
3. a < b
4. Depends on the magnitude of their common velocity
5. Depends on the magnitude of the B field

**(a)**

The figure shows four wire loops, with edge lengths of either $L$ or $2L$. All four loops will move through a region of uniform magnetic field $\boldsymbol{B}$ at the same constant velocity. Rank them according to the emf induced ***just as they enter the B field region***.

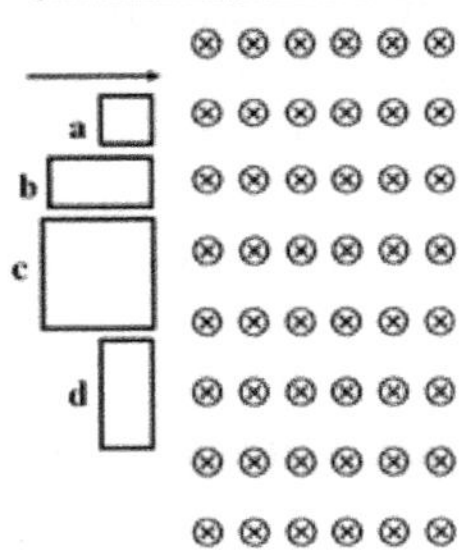


1. a < b < d < c
2. a < b = d < c
3. a < b < c < d
4. a = b < c = d
5. a = b < d < c

**(b)**

A circular wire loop moving at constant velocity enters a long region of uniform magnetic field B. Which one of the graphs describes the emf $\varepsilon$ in the loop as a function of time $t$?

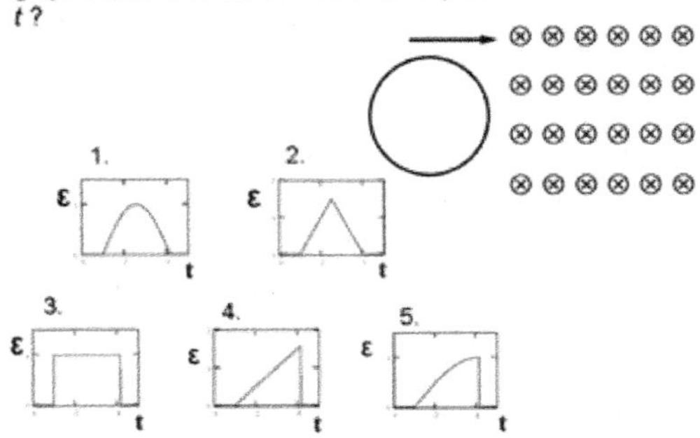


**(c)**

**Figure 10.3.** An E&M clicker question sequence.

not content gurus, they offered valuable new perspectives showing how our questions were interpreted by students (Ding, Reay, Lee, & Bao, 2009). In our interviews, students were specifically asked to rephrase each question without answering it, evaluate the accuracy of its language, and report any problems in graphs or pictures. Then they were required to provide solutions while "talking aloud." We started with extensive interviews with dozens of students, from which we selected 16 who were verbal and detail-oriented for repeated weekly interviews. By conducting these interviews, we validated the entire set of the 167 clicker question sequences, with each sequence reviewed by at least three different students. All these students were attending a then ongoing physics course and represented diverse achievement levels measured by their final course grades (ranging from A to D).

Our original intention of the validation study was to broadly conduct interviews so that statistics could be established to describe and infer the validity of each sequence (Lee, 2009). To our surprise, many of these clicker questions, which had already been validated by physics experts, were misleading for students. Oftentimes, students interpreted questions in an unintended (but not necessarily incorrect) way, and hence ended up with an incorrect response. For instance, when answering the question shown in Figure 10.4(a), a student mistakenly selected "zero net force." A close examination of the interview transcript showed that the student correctly drew a series of inward arrows along the circle to represent the net force on the car at each time instant. But he thought the question asked for a "time average" of the net force, an unintended but not incorrect interpretation. We therefore revised the question to specifically ask for the net force on the car at the time instant shown in the diagram [see Figure 10.4(b)].

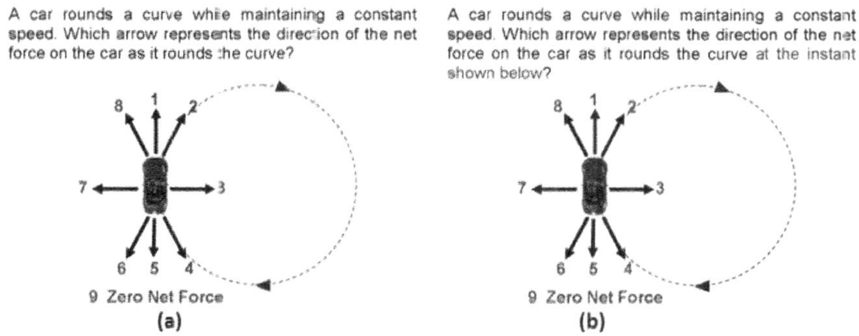


**Figure 10.4.** A clicker question before and after student interviews: (a) An expert-validated question and (b) Revised version as a result of student interviews.

The student interviews led to additional changes to 49% of the clicker question sequences already validated by experts, and a quarter of these changes were significant. Here, significant changes are those without which an able student will deviate from the intended way of reasoning and/or answering the question. Table 10.1 shows the numbers and percentages of changes made to the clicker question sequences as a result of both the expert review and student interviews.

**Table 10.1. Revised Clicker Question Sequences Due to Expert Review and Student Interviews**

| | *Revised Question Sequences* | *Question Sequences Revised Significantly* |
|---|---|---|
| Expert Review | 55 (33%) | 9 (5%) |
| Student Interviews* | 81 (49%) | 21 (13%) |

*Note:* The asterisk (*) is to highlight the fact that the questions used in student interviews were already validated with experts.

It is also notable that during the interviews we were particularly vigilant to any possible questions that students could answer through simple recall or blind mathematical manipulations. We did not detect such cases; instead we observed that students all tried (successfully or unsuccessfully) to invoke cognitive skills at the level of application or above to answer questions. This provides some evidence that our clicker materials target higher-level of cognition.

Furthermore, in the interviews we always used questions of each sequence together as a cluster and asked students at the end to verbalize their thoughts regarding the connections among them. In most cases, students could relate these questions in terms of a key concept, although they did not always successfully carry it out in answering all questions. We included this step in our interviews to verify whether or not sequenced questions could, at least to some extent, induce student recognition of the underlying concept. Our observations revealed promising signs.

## Implementation of Clicker Question Sequences

We have implemented all of our clicker question sequences in a number of introductory physics classes, spanning mechanics, electricity and magnetism (E&M), waves, optics, and modern physics (including quantum physics, particle physics, and relativity). In a typical class of 48 minutes, 2–4 clicker question sequences are used, interspersed with mini-lectures or demonstrations. Each sequence normally takes 7–9 minutes to complete and is carried out using variations of the following procedure.

Students individually answer each question and are then shown a response histogram. But the correct answer is not revealed. After discussions with their neighbors or a whole-class discussion initiated by the instructor, students either answer the same question again or then move on to the next one if the original number of correct answers is high (typically above 90%) (Reay et al., 2005; Reay et al., 2008).

The first question is relatively easy and can be used as a warm-up exercise. Unless a significant portion of the class misses the question, discussion often is brief. However, the second question is more difficult, and student responses may split among several alternative choices. The instructor calls for peer discussions while circulating around the classroom to closely observe student conversation. Peer discussion often is not sufficient if the number of correct answers is low (below 40%). In this case, a class-wide dialogue may be initiated and moderated by the instructor before students revote on the same question. Student responses usually converge to the correct choice after this process. The third question, which is slightly more challenging than the second one, is used to check if students have assimilated the key concept underlying the sequence. Typically the majority of a class at this stage can apply the concept to correctly answer the questions, and a follow-up peer discussion further solidifies students' understanding and application of this key concept. However, if a large percentage of the class misses the question, the instructor may choose to repeat the discussion-revote process as is done in the second question. After completing the sequence, it is important that the instructor direct students' attention to the interconnections among all questions and heighten students' self-evaluative awareness by asking them to revisit their own thoughts in answering the questions.

We use clicker question sequences as a facilitator for classroom formative assessment. Student performance on the clicker questions is not graded; instead we use this opportunity to gather evidence for the purpose of improving learning. Three layers of evidence are collected in our own teaching: students' comprehension of a key concept, application across multiple contexts, and analysis of the connectivity among the questions. The formative results then help us determine the subsequent direction of a lecture at any time—a method known as contingent teaching. The remainder of the lecture can be modified in real time to focus on revisiting difficult concepts, opening a student-centered discussion before revoting, or moving on. Since it is important to motivate students to answer questions seriously, we offer a small amount of extra credit worth 3–5% of the total course grade to those who can correctly answer at least 20% (a near random guessing rate) of all the questions used in an academic term. When offering this small incentive, student in-class participation rate increases from approximately 90% to 98%.

Also, we use our clicker materials to enhance student active engagement and cooperative learning. As described above, peer discussion and whole-class dialogues are an integral component in implementing our clicker question sequences. Students are given time to collaboratively work in small groups, articulating their own thoughts and critically evaluating each other's ideas. In our classes, not only do we instigate student discussions between the two polls of the same question, as typically practiced in peer instruction, but we also use them as a mediating transition between isomorphic questions in each sequence. Students' attention therefore is guided to answering the individual questions locally as well as to thinking their own thoughts globally for finding the commonalities at the end. With some added comments from the instructor during the final round of discussion, the idea of using the same concept across different contexts is further reinforced and extended to other possible cases not shown in the sequence. We use this opportunity to alert students that physics learning is not about rote memorization of facts or formulas, but rather is to flexibly apply basic concepts in new situations.

## Evaluation of the Effectiveness of Clicker Question Sequences

To evaluate how the use of our clicker questions sequences affected student learning, we looked into two aspects: students' conceptual learning gains and affective outcomes. In addition, we conducted a series of studies using clicker question sequences as part of an intervention to promote students' solving quantitative synthesis problems. However, due to the different nature of research goals, the results of these studies are reported elsewhere (see Ding, Reay, Lee, & Bao, 2011).

### *Learning Gains*

Conceptual learning gains were measured in terms of student performance on concept inventories. Here, we used Hake's normalized gain, as shown below, to quantify student improvement from pre to post performance. The advantage of this measure resides in the fact that it takes into consideration absolute gain in relation to potentially maximal gain (Hake, 1998):

$$\text{Normalized Gains} = (\text{Posttest} - \text{Pretest}) \,/\, (100\% - \text{Pretest}).$$

In each term, student learning gains were compared between two parallel classes of the same physics course, one adopting our clicker materials and the other not. For convenience, we call them clicker and non-clicker classes, respectively. These two classes shared the same course syllabus, labs, recitations, homework assignments, quizzes, midterms, and final exam, but were taught by two different instructors. Nonetheless, historical data indicated that the difference in instructors caused little divergence in student learning gains (Ding, Reay, Lee, & Bao, 2008). For instance, despite the fact that there were nine different instructors teaching the same traditionally delivered calculus-based E&M classes in the three-year period between 2003 and 2005, student normalized gains on the CSEM Conceptual Survey of Electricity and Magnetism (Maloney, O'Kuma, Hieggelke, & Van Heuvelen, 2001) were statistically similar (Ding, Reay, Lee, & Bao, 2008).

However, our historical data also revealed that test timing and conditions had a significant impact on test results (Ding et al., 2008). When CSEM was initially used in our traditional E&M classes, it was administered as both a pretest after several days into a term and a posttest in the last or second-to-last week, both without incentives (Condition 1). Student normalized gains consistently remained around 18.5%. Then we moved the pretest to the first day of class and administered the posttest with credit for participation, while keeping other conditions the same (Condition 2). Students' normalized gains increased to 25.5%. After we further changed the posttest incentive to only offering credit to those who scored 90% or higher (Condition 3), normalized gains increased again to 37.3%. Finally, when we made the posttest a part of the final examination (Condition 4), students' normalized gains jumped to 48.2% (see Figure 10.5). In short, by adjusting test timings and incentives, results for normalized gains fluctuated dramatically from 18.5% to 48.2%. It is worth noting that all classes, from which the data were collected, were traditionally delivered calculus-based introductory E&M courses that did not use clickers.

Informed by these results, we carefully monitored test conditions when making comparisons between the clicker and non-clicker classes. To be consistent, we always gave the pretest on the first day of class, before students received any instruction in class, and the posttest on the final examination. Following this method, we carried out evaluation studies in three academic terms of a calculus-based introductory physics course, including mechanics, E&M, and modern physics.

In the mechanics term, we used a concept test of 26 multiple-choice items selected from the Force Concept Inventory (Hestenes, Wells, & Swackhamer, 1992), Mechanics Baseline Test (Hestenes & Wells, 1992), and those designed by the authors. By doing so, we were able to cover all the major key concepts that were taught in the mechanics course. In the E&M term, the Conceptual Survey of Electricity and Magnetism (Maloney et al., 2001) was used. As for the modern physics term, we complied a test of 17 multiple-choice items to cover major topics in waves, optics, quantum physics, particle physics, and theory of relativity. Questions of these inventories were similar to the clicker questions used in class in terms of key concepts but differed in context.

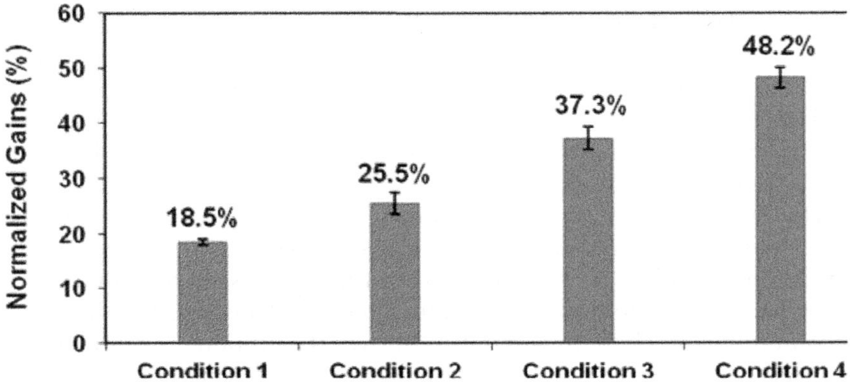


**Figure 10.5.** Test conditions affect student normalized gains on CSEM. Condition 1: pretest after several days into a term, posttest in last week or 2nd-to-last week, no incentives. Condition 2: pretest on first day, posttest with credit for participation. Condition 3: posttest with credit for performing well. Condition 4: posttest on final examination.

Comparison results are displayed in Table 10.2. As shown, the clicker class consistently outperformed the non-clicker class, and its better performance was found to be statistically significant for both mechanics and modern physics, but not for E&M. In the E&M term, students were only shown the clicker question sequences once during class and did not have another chance to review them. Retrospectively, we do not consider this approach optimal, because large learning effects normally should not be expected of students from a one-shot encounter with training materials in a limited time period even in the best possible learning situation (Bransford & Schwartz, 1999). The fact that one-time exposure to our clicker question sequences yielded limited student learning gains does not negate the effectiveness of our clicker materials; instead it illustrates the importance of proper practices of implementing research-based instructional materials in real classrooms. Therefore, in the mechanics and modern physics terms, we uploaded all clicker question sequences used in class to a secured course website without providing any answers. Students who were then taking the clicker class had full access to the materials for after-class review. This change (with other conditions being the same as before) upgraded students' learning gains in the clicker class to a level significantly better than that in the non-clicker class, with medium to high effect sizes.

**Table 10.2. Comparison in Student Conceptual Learning Gains Between the Clicker and Non-Clicker Classes**

| *Term* | *Normalized Gains (%)* | | *p value (t-test)* | *Effect Size* |
|---|---|---|---|---|
| | *Clicker* | *Non-Clicker* | | |
| Mechanics (Fall 2007) | 28.5 ($N$ = 184) | 22.7 ($N$ = 171) | $2.6 \times 10^{-5}$ | 0.44 |
| E&M (Spring 2007) | 52.0 ($N$ = 106) | 48.4 ($N$ = 122) | 0.25 | 0.15 |
| Modern Physics (Spring 2008) | 60.1 ($N$ = 156) | 44.4 ($N$ = 140) | $8.6 \times 10^{-8}$ | 0.64 |

### *Affective Outcomes*

In addition to the conceptual learning gains, we also measured student affective opinions about clicker use (Lee, Ding, Reay, & Bao, 2011). A written survey of 17 statements was constructed and tested with diverse students at the end of each term. It targeted student views about how using clickers affected their self-portrayed learning outcomes and classroom engagement. Some sample statements are: "I liked using clickers"; "Clickers helped me understand lectures better"; "Clickers made me feel involved in the course"; and "I would recommend using clickers in all future introductory physics courses." Students were asked to rate their levels of agreement with these statements by using a 5-point Likert scale, ranging from +2 for strong agreement to –2 for strong disagreement.

Average scores for these questions varied from +1.3 to +1.8, with typical histograms shown in Figure 10.6 (Lee et al., 2011). To reduce the bias of using only positively phrased statements, we also included occasional inverse questions such as "I would not recommend using clicker in future physics courses." In all such cases, student ranking was highly negative, a consistent result regardless of the question phrasing.

## GENERAL DISCUSSION

Context dependence of learning is a commonly observed phenomenon that often impedes student understanding and application of basic science concepts. The results of our studies have demonstrated that the new clicker methodology discussed herein can be a useful framework to help enhance student conceptual learning in introductory physics. However, several aspects of our studies are worth stressing.

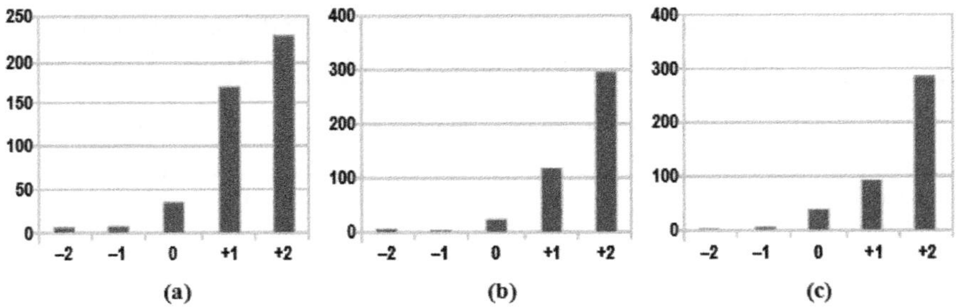


**Figure 10.6.** Student affective opinions on clicker use. (a) Student response to the statement "clicker helped me understand lectures better." (b) Student response to the statement "clicker made me feel involved in the course." (c) Student response to the statement "I would recommend using clickers in all future introductory physics courses."

From the conceptualization to the enactment of our clicker methodology framework, we conducted a series of studies, focusing on the various stages of an effortful process of developing effective course materials. Although we report the four components of our research on creation, validation, implementation, and evaluation of clicker materials in separate sections, they collectively form a dynamic iterative whole in reality. In other words, each time an issue in our clicker materials emerged, regardless of during which stage, an immediate action would always be taken to either revise it or redesign it from scratch, followed by another round of validation and implementation. This cyclic iteration was consistently executed throughout the entire span of our research.

We also re-underscore the importance of the proper design and implementation of our clicker question sequences if optimal learning gains are to be achieved. As previously reported, our clicker questions in each sequence were systemically crafted with diverse contexts. This should not imply that the creation of each sequence is just a simple assembly of several isomorphic questions with different appearances. These questions, albeit sharing the same underlying key concept, differ in a way that each requires the invocation of some unique cognition that is specifically linked to its particular context. For instance, the whole-and-part relation between net force and individual forces in the second question of the force diagram sequence (Figure 10.2) and the cause-effect relation between force and motion in the third question are distinct features that make them unique other than their obvious appearances. If they were different only in their exterior contexts but not in the cognition skills they called for, it is less likely that the student experience with the same key concept would be enriched and solidified (diSessa & Wagner, 2005).

Moreover, the clicker question sequences need to be properly carried out in class, which means that focusing on student active engagement is essential. By doing so, the foundational purpose of increasing flexible application of physics concepts can be maximized through student articulation and evaluation of their own thoughts. Additionally, allowing students a full access to the clicker material after class is equally critical. Since students need time to digest and internalize the key ideas contained in each question sequence, limited exposure within a short time period may not afford them sufficient experience for future use of what they have learned (Bransford & Schwartz, 1999).

Although the above studies were carried out in the context of college-level introductory physics, our clicker methodology framework can be, upon future testing across diverse courses, equally useful for teaching and learning of other sciences and engineering. Derived from cognitive and educational theories, this framework specifically seeks to address the context dependence of learning, which is a difficulty frequently observed in a broad range of disciplines, so it is conceivable that by following this framework, similarly effective course materials can be developed for other subject domains. Future research along this line would be valuable. Moreover, we enacted this framework based on using clickers, a popular technology in many classes. This fact notwithstanding, the core of this framework can be used with other technologies, such as student laptops, iPads, and smartphones, as well as low-tech methods such as flash cards. In general, the new clicker approach introduced in our framework has broader implications for teaching and learning in various classrooms.

## SUMMARY

This chapter introduces a new clicker methodology framework to promote flexible student application of conceptual physics knowledge. Based on this framework, we have created, validated, implemented, and evaluated 167 clicker question sequences, approximately 500 individual clicker questions. The sequences are extensive enough in number to populate a typical year of college-level introductory physics. Each sequence was systemically designed to contain 3–4 questions that share the same underlying key concept but are embedded in different contexts. These sequences were validated through both expert review and student interviews, and were implemented in real classrooms using peer instruction and collaboration to actively engage students. Results from student performance on concept inventories showed consistently increased learning gains for those who used the clicker materials. Students also reported high opinions for clicker use in introductory physics classes. These clicker question sequences may be adopted or adapted for use in many college-level physics classes. Interested readers can contact the authors to access our clicker materials.

## REFERENCES

Anderson, J. R., Reder, L. M., & Simon, H. A. (1996). Situated learning and education. *Educational Researcher, 25*(4), 5–11.

Bassok, M. (1990). Transfer of domain-specific problem-solving procedures. *Journal of Experimental Psychology, 16*(3), 522–533.

Bassok, M., & Holyoak, K. J. (1989). Interdomain transfer between isomorphic topics in algebra and physics. *Journal of Experimental Psychology, 15*(1), 153–166.

Beatty, I. D., Gerace, W. J., Leonard, W. J., & Dufresne, R. J. (2006). Designing effective questions for classroom response system teaching. *American Journal of Physics, 74*(1), 31–39.

Bransford, J. D., Brown, A. L., & Cocking, R. R. (2000). *How people learn: Brain, mind, experience, and school* (expanded ed.). Washington, DC: National Academy Press.

Bransford, J. D., Franks, J. J., Vye, N. J., & Sherwood, R. D. (1989). New approaches to instruction: Because wisdom can't be told. In S. Vosniadou & A. Ortony (Eds.), *Similarity and analogical reasoning* (pp. 470–497). Cambridge: Cambridge University Press.

Bransford, J. D., & Schwartz, D. L. (1999). Rethinking transfer: A simple proposal with multiple implications. *Review of Research in Education, 24*(61–100).

Broudy, H. S. (1977). Types of knowledge and purposes of education. In R. C. Anderson, R. J. Spiro, & W. E. Montague (Eds.), *Schooling and the acquisition of knowledge* (pp. 1–17). Hillsdale, NJ: Erlbaum.

Brown, D. E. (1992). Using examples and analogies to remediate misconceptions in physics: Factors influencing conceptual change. *Journal of Research in Science Teaching, 29*(1), 17–34.

Brown, D. E. (1993). Refocusing core intuitions: A concretizing role for analogy in conceptual change. *Journal of Research in Science Teaching, 30*(10), 1273–1290.

Brown, D. E., & Clement, J. (1989). Overcoming misconceptions via analogical reasoning: abstract transfer versus explanantory model construction. *Instructional Science, 18*, 237–261.

Caldwell, J. E. (2007). Clickers in the large classroom: Current research and best-practice tips. *CBE-Life Sciences Education, 6*(1), 9–20.

Chi, M. T. H., Feltovich, P. J., & Glaser, R. (1981). Categorization and representation of physics problems by experts and novices. *Cognitive science, 5*(2), 121–152.

Clement, J. (1982). Algebra word problem solutions: Thought processes underlying a common misconception. *Journal for Research in Mathematics Education, 13*(1), 16–30.

Clement, J. (1988). Observed methods for generating anaologies in scientific problem solving. *Cognitive Science, 12*, 563-586.

Clement, J. (1991). Non-formal reasoning in experts and science students: The use of analogies, extreme cases and physical intuition. In J. F. Voss, D. N. Perkins, & J. W. Segal (Eds.), *Informal Reasoning and Education* (pp. 345–362). Hillsdale, NJ: Lawrence Erlbaum.

Clement, J., Brown, D. E., & Zietsman, A. (1989). Not all preconceptions are misconceptions: finding "anchoring conceptions" for grounding instruction on students' intuitions. *International Journal of Science Education, 11*(5), 554–565.

Crossgrove, K., & Curran, K. L. (2008). Using clickers in nonmajors- and majors-level biology courses: student opinions, learning, and long-term retention of course material. *CBE-Life Sciences Education, 7*, 146–154.

Crouch, C. H., & Mazur, E. (2001). Peer instruction: Ten years of experience and results. *American Journal of Physics, 69*(9), 970–977.

Deslauriers, L., Schelew, E., & Wieman, C. (2011). Improved learning in a large-enrollment physics class. *Science, 332*, 862–864.

Ding, L., Reay, N. W., Lee, A., & Bao, L. (2008). Effects of testing conditions on conceptual survey results. *Physical Review Special Topics—Physics Education Research, 4*(010112), 1–6.

Ding, L., Reay, N., Lee, A., & Bao, L. (2009). Are we asking the right questions? Validating clicker question sequences by student interviews. *American Journal of Physics, 77*(7), 643–650.

Ding, L., Reay, N. W., Lee, A., & Bao, L. (2011). Exploring the role of conceptual scaffolding in solving synthesis problems. *Physical Review Special Topics—Physics Education Research*, 7(020109), 1–11.

diSessa, A. (1988). Knowledge in pieces. In G. Forman & P. Pufall (Eds.), *Constructivism in the computer age* (pp. 49–70). Hillsadale, NJ: Erlbaum.

diSessa, A. (1993). Towards an epistemology of physics. *Cognition and Instruction, 10*(2–3), 105–225.

diSessa, A., & Wagner, J. F. (2005). What coordination has to say about transfer. In J. Mestre (Ed.), *Transfer of learning: Research and perspectives* (pp. 121–154). Greenwich, CT: Information Age Publishing.

Dufresne, R. J., Gerace, W. J., & Leonard, W. J. (1997). Solving physics problems with multiple representations. *The Physics Teacher, 35*, 270–275.

Duit, R., Roth, W., Komorek, M., & Wilbers, J. (2001). Fostering conceptual change by analogies—between Scylla and Charybdis. *Learning and Instruction, 11*, 283–303.

Duncan, D. (2005). *Clickers in the classroom: How to enhance science teaching using classroom response systems*. Boston, MA: Addison-Wesley.

Duncan, D. (2006). Clickers: A new teaching aid with exceptional promise. *Astronomy Education Review, 5*(1), 70–88.

Eich, E. (1985). Context, memory, and integrated item/context imagery. *Journal of Experimental Psychology, 11*(4), 764–770.

Fagen, A. P., Crouch, C. H., & Mazur, E. (2002). Peer instruction: Results from a range of classrooms. *The Physics Teacher, 40*(4), 206–209.

Gentner, D., Loewenstein, J., & Thompson, L. (2003). Learning and transfer: A general role for analogical encoding. *Journal of Educational Psychology, 95*(2), 393–408.

Gick, M. L., & Holyoak, K. J. (1980). Analogical problem solving. *Cognitive Psychology, 12*, 306–355.

Gick, M. L., & Holyoak, K. J. (1983). Schema induction and analogical transfer. *Cognitive Psychology, 15*, 1–38.

Hake, R. R. (1998). Interactive-engagement versus traditional methods: A six-thousand-student survey of mechanics test data for introductory physics courses. *American Journal of Physics, 66*(1), 64–74.

Halliday, D., Resnick, R., & Walker, J. (2008). *Fundamentals of physics* (8th ed., Extended ed.). Hoboken, NJ: John Wiley & Sons.

Hammer, D. (1996a). Misconceptions or p-prims: How may alternative perspectives of cognitive structure influence instructional perceptions and intentions? *Journal of The Learning Sciences, 5*(2), 97–127.

Hammer, D. (1996b). More than misconceptions: Multiple perspectives on student knowledge and reasoning, and an appropriate role for education research. *American Journal of Physics, 64*(10), 1316–1325.

Hammer, D. (2000). Student resources for learning introductory physics. *American Journal of Physics, 68*(S1), S52–S59.

Hammer, D., Elby, A., Scherr, R. E., & Redish, E. F. (2005). Resource, Framing, and Transfer. In J. P. Mestre (Ed.), *Transfer of Learning from a Modern Multidisciplinary Perspective* (pp. 89-120). Greenwich, CT: Information Age Publishing.

Hardiman, P. T., Dufresne, R., & Mestre, J. P. (1989). The relation between problem categorization and problem solving among experts and novices. *Memory & Cognition, 17*(5), 627–638.

Heron, P. R. L. (2004). *Empirical investigations of learning and teaching, Part I: Examining and interpreting student learning*. In E. F. Redish & M. Vicentini (Eds.), *International School of Physics "Enrico Fermi" Course CLVI* (pp. 341–350). Varenna, Italy: IOS Press, Amsterdam..

Hestenes, D., & Wells, M. (1992). A mechanics baseline test. *The Physics Teacher, 30*(3), 159–166.

Hestenes, D., Wells, M., & Swackhamer, G. (1992). Force concept inventory. *The Physics Teacher, 30*(3), 141–158.

Knight, J. K., & Wood, W. B. (2005). Teaching more by lecturing less. *Cell Biology Education, 4*, 298–310.

Knight, R. D. (2008). *Physics for scientists and engineers: a strategic approach* (2nd ed.). San Francisco, CA: Pearson Addison-Wesley.

Koedinger, K. R., & Nathan, M. J. (2004). The real story behind story problems: Effects of representations on quantitative reasoning. *Journal of The Learning Sciences, 13*(2), 129–164.

Kohl, P. B., & Finkelstein, N. D. (2005). Student representational competence and self-assessment when solving physics problems. *Physical Review Special Topics—Physics Education Research, 1*(1), 010104.

Kohl, P. B., & Finkelstein, N. D. (2006). Effects of representation on students solving physics problems: A fine-grained characterization. *Physical Review Special Topics—Physics Education Research, 2*(1), 010106.

Kohl, P. B., Rosengrant, D., & Finkelstein, N. D. (2007). Strongly and weakly directed approaches to teaching multiple representation use in physics. *Physical Review Special Topics—Physics Education Research, 3*(1), 010108.

Kurtz, K. J., Miao, C., & Gentner, D. (2001). Learning by analogical bootstrapping. *The Journal of the Learning Sciences, 10*(4), 417–446.

Lasry, N., & Aulls, M. W. (2007). The effect of multiple internal representations on context-rich instruction. *American Journal of Physics, 75*(11), 1030–1037.

Lave, J. (1988). *Cognition in practice: Mind, mathematics, and culture in everyday life.* Cambridge, UK: Cambridge University Press.

Lee, A. (2009). *Development and evaluation of clicker methodology for introductory physics courses.* Columbus, OH: The Ohio State University.

Lee, A., Ding, L., Reay, N. W., & Bao, L. (2011). Single-concept clicker question sequences. *The Physics Teacher, 49*(6), 385–389.

Lehrer, R., & Shauble, L. (2000). Developing model-based reasoning in mathematics and science. *Journal of Applied Developmental Psychology, 21*(1), 39–48.

Maloney, D. P., O'Kuma, T. L., Hieggelke, C. J., & Van Heuvelen, A. (2001). Surveying students' conceptual knowledge of electricity and magnetism. *American Journal of Physics, 69*(S1), S12–S23.

Martyn, M. (2007). Clickers in the classroom: An active learning approach. *Educause Quarterly, 30*(2), 71–74.

Mazur, E. (1997). *Peer instruction: A user's manual.* Upper Saddle River, NJ: Prentice-Hall.

Mazur, E. (2009). Farewell, lecture? *Science, 323*, 50–51.

McDermott, L. C., & Redish, E. F. (1999). Resource Letter: PER-1: Physics Education Research. *American Journal of Physics, 67*(9), 755–767.

Meltzer, D. E. (2005). Relation between students' problem-solving performance and representational format. *American Journal of Physics, 73*(5), 463–478.

Mestre, J. P., Gerace, W. J., Dufresne, R. J., & Leonard, W. J. (1997). *Promoting active learning in large classes using a classroom communication system.* In E. F. Redish & J. S. Rigden (Eds.), *The changing role of physics departments in modern universities: Proceedings of the International Conference on Undergraduate Physics Education, Part two: sample classes* (pp. 1019–1036). Woodbury, NY: American Institute of Physics.

Minstrell, J. (1982). Explaining the "at rest" condition of an object. *The Physics Teacher, 20*(1), 10–14.

Novick, L. R., & Holyoak, K. J. (1991). Mathematical problem solving by analogy. *Journal of Experimental Psychology, 17*(3), 398–415.

Paschal, C. B. (2002). Formative assessment in physiology teaching using a wireless classroom communication system. *Adavances in Physiology Education, 26*(4), 299–308.

Preszler, R. W., Dawe, A., Shuster, C. B., & Shuster, M. (2007). Assessment of the effects of student response systems on student learning and attitudes over a broad range of biology courses. *CBE-Life Sciences Education, 6*, 29–41.

Reay, N. W., Bao, L., Li, P., Warnakulasooriya, R., & Baugh, G. (2005). Toward the effective use of voting machines in physics lectures. *American Journal of Physics, 73*(6), 554–558.

Reay, N. W., Li, P., & Bao, L. (2008). Testing a new voting machine question methodology. *American Journal of Physics, 76*(2), 171–178.

Rebello, N. S., Cui, L., Bennett, A. G., Zollman, D. A., & Ozimek, D. J. (2007). Transfer of learning in problem solving in the context of mathematics and physics. In D. H. Jonassen (Ed.), *Learning to solve complex scientific problems* (pp. 223–246). Mahwah, NJ: Lawrence Earlbaum.

Sabella, M. S., & Redish, E. F. (2007). Knowledge organization and activation in physics problem solving. *American Journal of Physics, 75*(11), 1017–1029.

Savinainen, A., Scott, P., & Viiri, J. (2005). Using a bridging representation and social interactions to foster conceptual change: Designing and evaluating an instructional sequence for Newton's third law. *Science Education, 89*(2), 175–195.

Schwartz, D. L., & Bransford, J. D. (1998). A time for telling. *Cognition and Instruction, 16*(4), 475–522.

Scribner, S. (1984). Studying working intelligence. In B. Rogoff & J. Lave (Eds.), *Everyday cognition: Its development in social context* (pp. 9–40). Cambridge, MA: Harvard University Press.

Smith, M. K., Wood, W. B., Adams, W. K., Wieman, C., Knight, J. K., Guild, N., & Su, T. T. (2009). Why peer discussion improves student performance on inlcass concept questions. *Science, 323*, 122–124.

Spiro, R. J., Feltovich, P. J., Jacobson, M. J., & Coulson, R. L. (1992). Cognitive flexibility, constructivism, and Hypertext: Random access instruction for advanced knowledge acquisition in ill-structured domains. In T. M. Duffy & D. H. Jonassen (Eds.), *Constructivism and the technology of instruction: A conversation* (pp. 57–75). Hillsdale, NJ: Lawrence Erlbaum Associates.

Steinberg, R. N., & Sabella, M. S. (1997). Performance on multiple-choice diagnostics and complementary exam problems. *The Physics Teacher, 35*(3), 150–155.

Sullivan, R. (2008). Principles for constructing good clicker questions: Going beyond rote learning and stimulating active engagement with course content. *Journal of Educational Technology, 37*(3), 335–347.

Van Heuvelen, A., & Zou, X. (2001). Multiple representations of work—energy processes. *American Journal of Physics, 69*(2), 184–194.

Woelk, K. (2008). Optimizing the use of personal response devices (clickers) in large-enrollment introductory courses. *Journal of Chemical Education, 85*(10), 1400–1405.

Wood, W. B. (2004). Clickers: A teaching gimmick that works. *Developmental Cell, 7*(6), 796–798.

# PART III

## STUDENT PROBLEM AND RESEARCH STRATEGIES

CHAPTER 11

# INTRINSIC MOTIVATION OF STUDENTS UTILIZING A PROJECT-BASED ORGANIC CHEMISTRY LABORATORY CURRICULUM

**Gail Horowitz**

## ABSTRACT

This action research investigation utilized a qualitative approach to study the intrinsic motivation of students who participated in a project-based organic chemistry laboratory curriculum. The curriculum's goal was to provide students with a motivating environment of enhanced autonomy, authenticity, and suspense by allowing them to adapt and test out a synthetic experiment of their choosing. The study also explored the potentially moderating roles of students' achievement goal orientations, as well as their academic and professional interests. Results showed that an overwhelming majority of students viewed the autonomous aspects of the curriculum as positive. Students' perceptions of the authentic and suspenseful aspects of the curriculum varied, however, with students more positive in affect if they perceived of the degree of challenge of their experiment as moderate (not too difficult, but not too

*Research Based Undergraduate Science Teaching*, pp. 333–378

easy). Mastery-oriented students were more likely to report the curriculum as motivating as compared to more extrinsically oriented students.

A number of project-based science curricula have been reported (Barab, Hay, Barnett, & Keating, 2000; Barron et al., 1998; Edelson, 2001; Edelson, Gordin, & Pea, 1999; Kanter & Reiser, 2007; Kanter & Schreck, 2006; Linn, Clark, & Slotta, 2003; Marx, Blumenfeld, Krajcik, & Soloway, 1997; Means, 1998; Polman, 2000; Sherin, Kanter, Schwarz, Stieff, Herman, & MacKenzie, 2002; Songer, 1998). In some ways, each is unique, having different design features and focusing on different student populations and different content areas. The *Kids as Global Scientists* curriculum, for example (Songer, 1998), focused on earth science at the middle school level and relied heavily on Internet communication, whereas the *Virtual Solar System* curriculum (Barab et al., 2000) focused on astronomy at the college level and utilized three-dimensional modeling tools. It has been argued, however, that, in spite of these differences, all project-based science curricula share one common curricular feature (Blumenfeld, Soloway, Marx, Krajcik, Guzdial, & Palincsar, 1991; Pitts, 2006). These curricula attempt to incorporate elements of authenticity by giving students the opportunity to wrestle with authentic, relevant, real-world problems.

Project-based science curricula both improve student learning (Marx et al., 2004; Rivet & Krajcik, 2004; Williams & Linn, 2003) and enhance student motivation (Higginbotham, Anderson, Rosca, Jencunas, Copman, & Zinkowski, 2006; Kempler, 2006; Mistler-Jackson & Songer, 2000; Pitts, 2006; Songer, 1996). Theorists have argued that project-based approaches are highly motivating to students because they combine a number of key curricular "ingredients" (Blumenfeld et al., 1991; Lepper & Henderlong, 2000; Paris & Turner, 1994) that act in concert to enhance intrinsic motivation. These ingredients, or curricular features, include not only authenticity, but also autonomy, suspense, challenge, and novelty (Blumenfeld et al., 1991; Kempler, 2006; Lepper & Henderlong, 2000; Paris & Turner, 1994; Pintrich, 1991).

## PROBLEMS IN THE ORGANIC CHEMISTRY LABORATORY

Studies of student motivation and project-based science have focused mainly on middle and high school classrooms. College-level science classrooms, however, are a fruitful place to study project-based curricula. The organic chemistry teaching laboratory, in particular, offers opportunities to study student motivation and project-based curricula. It is a setting that is in urgent need of reform. Its curriculum has been characterized as deterministic, inauthentic, and even a waste of time (Gallet, 1998; Glagovich & Swierczynski, 2004; Mohrig, 2004; Pickering, 1988; Wade, 1979). Curricular problems are compounded by a student population that is not interested in chemistry (Gay, 2004; Katchadourian & Boli, 1985), but who simply take organic chemistry to fulfill professional school entrance requirements. Student interest is further diminished by a tense and competitive classroom. Not only does organic chemistry have the reputation of being extremely difficult (Gay, 2004), but the harsh admission requirements of health science graduate schools further foster a pressured and oppressive classroom atmosphere (Tobias, 1990). Thus, not only do many students feel coerced into taking organic chemistry, but many also dread it and wish it over as soon as possible.

## PRIOR RESEARCH

The net result of an inauthentic, deterministic curriculum and a disinterested student population is an organic chemistry laboratory environment in which students' intrinsic motivation is low. Both because of low student motivation and curricular shortcomings, the organic laboratory is in urgent need of reform. A review by this author Horowitz (2007) chronicles recent reform efforts that have taken place within the organic chemistry laboratory. These reform efforts can be categorized into one of three types: discovery-based, inquiry-based, or project-based. What all three approaches seem to have in common (in this author's view) is that they each attempt to foster and encourage student thinking and engagement by incorporating one or more of the following curricular elements: autonomy (allowing for student choice), authenticity (a mirroring of research practice), or suspense (creating uncertainty about experimental outcomes).

So-called "project-based" reforms in the organic laboratory are of particular interest in terms of student motivation because they represent the most broadly based efforts to simultaneously introduce all three motivational elements of autonomy, authenticity, and suspense into the organic chemistry laboratory. Project-based approaches require students to select (autonomy) and work on multiweek research-like projects, in which they must modify existing procedures to solve new synthetic problems (authenticity.) Students conduct a primitive form of research in which their experimental outcomes are unknown (suspense.)

A number of project-based organic chemistry laboratory curricula have been reported in the chemical education literature (see Horowitz, 2007 for a recent review). Most provide only vague descriptions of key curricular elements (Chisholm, 1975; Doyle, 1973; Kovacic, 1972; Kroll, 1985) and only limited anecdotal information regarding student responses (Dobrev,

1996; Gallet, 1998; Hiegel & Belloli, 1971; Kovacic, 1972; Pienta, Regitz, Richards, & Sorrell, 1993; Poon et al., 1999). In the few instances in which survey evaluations have been conducted, only the barest of findings are described (Davis, Hargrove, & Hugdahl, 1999; Kharas, 1997; Kroll, 1985; Ruttledge, 1998). One study by Cooper and Kerns (2006) found that students felt project-based work to be more interesting than traditional experiments. Survey data (Davis et al., 1999; Kroll, 1985; Ruttledge, 1998) and anecdotal evidence (Hiegel & Belloli, 1971; Kharas, 1997; Newton, Tracy, & Prudente, 2006; Pienta et al., 1993) also indicate that organic chemistry students find project-based work to be stimulating and engaging. No published studies exist that focus specifically on the intrinsic motivation of students.

Within the broader field of science education, a number of studies have been conducted on the motivation of students participating in project-based science curricula (Herman & Gomez, 2006; Higginbotham et al., 2006; Kempler, 2006; Krajcik, Blumenfeld, Marx, Bass, Fredricks, & Soloway, 1998; Mistler-Jackson & Songer, 2000; Pitts, 2006; Pitts & Edelson, 2004; Songer, 1996). Both quantitative and qualitative approaches have been utilized with methods including classroom observation, student interviews, and student surveys. Measures of motivation have included self-efficacy, achievement goal orientation, and situational interest. Research findings from these studies (Higginbotham et al., 2006; Kempler, 2006; Mistler-Jackson & Songer, 2000; Pitts, 2006; Songer, 1996) demonstrate that project-based approaches are motivating to students. For example, Mistler-Jackson and Songer (2000) found that students responded enthusiastically and with increased interest and effort to the *Kids as Global Scientists* earth science curriculum. Pitts (2006) found that the scenarios and role-playing present in the *What Will Survive* life sciences curriculum enhanced the motivation and engagement of students.

## THEORETICAL FRAMEWORK

The theoretical framework for this study is depicted in Figure 11.1. In this study, three topics were examined: (1) student perceptions of the curricular goals (column I), (2) student achievement goal orientations and interests (column III), and (3) student intrinsic motivation (column V). Two connections were explored: the connection between students' perceptions of the curriculum and their motivation (connection of column I to column V), and the connection between students' goal orientations, personal interests, and their motivation (connection of column III to column V).

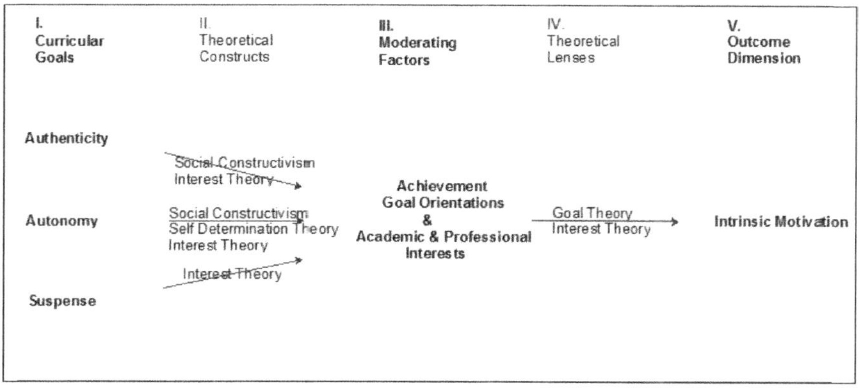


**Figure 11.1.** Theoretical framework illustrating theoretical connections between the curricular goals, moderating factors and the outcome dimension, intrinsic motivation.

## Theoretical Linkages Between Curricular Goals and Motivation

Column II of Figure 11.1 shows the theoretical constructs that link the curricular goals of column I to the outcome dimension, intrinsic motivation (column V). The theoretical constructs that link autonomy, authenticity and suspense to intrinsic motivation are social constructivism (Hickey, 1997; Sivan, 1986), self determination theory (Ryan & Deci, 2000), and interest theory (Krapp, Hidi, & Renninger, 1992).

Social constructivism focuses on how social context facilitates learning, whereby learners are able to engage in social discourse and activities that build upon their individual and collective prior experiences and interests (Driver, 1994; Hickey, 1997). To social constructivist motivational theorists, authenticity is a key motivational element. They argue that student motivation is enhanced when students engage in problems that they find realistic and relevant to their own lives (Blumenfeld, 1992; Blumenfeld, Kempler, & Krajcik, 2006; Blumenfeld, Marx, Patrick, Krajcik, & Soloway, 1997; Cordova & Lepper, 1996; Hennessy, 1993; Lepper & Henderlong, 2000).

Social constructivist motivational theorists also view autonomy as an important motivational factor (Hickey, 1997; Paris & Turner, 1994; Sivan, 1986). The zone of proximal development (Vygotsky, 1978) is characterized by Sivan as inherently motivating because of its non-controlling and autonomous nature, in which responsibility for learning is placed in the hands of the learner. Thus, socioconstructivist motivation theorists see autonomy, in addition to authenticity, as an important and necessary ingredient for fostering student motivation (Paris & Turner, 1994; Sivan, 1986).

Self-determination theory argues that all people are driven to seek autonomy along with relatedness and competence and that "autonomy-supportive" environments, for example, those providing students with freedom of choice, enhance intrinsic motivation (Ryan & Deci, 2000). Interest theorists maintain that intrinsic motivation is enhanced when situational interest is aroused through curricular features such as novelty, surprise, and suspense (Renninger, Bachrach, & Posey, 2008). Interest theory also predicts that autonomy of choice and an authentic learning environment lead to increased personal interest (Lipstein & Renninger, 2007; Paris & Turner, 1994). Autonomy of choice allows students to choose topics of study that are of personal interest to them and thus likely increases students' motivation to engage in the activity. An authentic learning environment in which topics studied are relevant to the lives of students also can arouse the personal interest of students, and through the relevance to them personally, motivate students to sustained involvement and activity.

## Moderating Factors

It was hypothesized that student traits such as professional interests, academic interests, and achievement goal orientations (Figure 11.1, column III) might moderate their responses to the curricular goals of autonomy, authenticity, and suspense (column I). Two motivational theories, goal theory and interest theory (column IV), link student achievement goal orientations and interests to intrinsic motivation. Interest theory predicts that the personal interests of students will moderate their motivational responses. For example, a student who is interested in a research career may be motivated by a curriculum that he or she perceives to be reminiscent of how scientists conduct research (authenticity.) Goal theory predicts that more mastery-oriented students (who desire to develop competence) may respond differently than more performance- or extrinsically oriented students (who wish to outperform others and/or demonstrate competence) (Meece, Blumenfeld, & Hoyle, 1988; Pintrich & Schunk, 2002). An extrinsically oriented student, for example, may be frustrated by a curriculum in which the standards for success are different from those of past experiments, or where the uncertainty (suspense) associated with an open-ended inquiry is not perceived of as expediting his or her goal of performance.

## Intrinsic Motivation

This study explored the relationship between students' perceptions of the curricular goals (column I) and their intrinsic motivation (column V). Intrinsic motivation refers to the desire or drive to "engage in an activity for its own sake" (Pintrich & Schunk, 2002). Intrinsic motivation was identified as the key outcome variable of this study, because unlike extrinsic motivation, which has been linked to superficial learning, intrinsic motivation has been linked to deep information processing and learning (Deci & Flaste, 1995; Nolen, 1988, 1996).

## PURPOSE OF CURRENT STUDY

Project-based organic laboratory approaches appear to possess three key curricular elements (autonomy, authenticity, and suspense) theoretically linked to enhanced student motivation. Prior research in middle and high school settings has demonstrated that project-based science curricula can be motivating to students Given the low motivational level of organic chemistry laboratory students (Gay, 2004), it is crucial that the motivational impact of project-based organic chemistry laboratory curricula be examined in detail The present action research study attempts to address this need. In this study, a project-based organic chemistry laboratory curriculum was implemented. The goals of the curriculum, described further below, were to provide students with an environment of enhanced autonomy, authenticity, and suspense. Research was undertaken to explore the intrinsic motivation of students who experienced this curriculum. The study also explored possible connections between students' motivational responses and two hypothesized moderating variables: student achievement goal orientations and student professional interests.

The research questions of this study were: (1) To what degree do students' behaviors and/or affect indicate that the curriculum is intrinsically motivating to them? (2) What is the connection between students' perceptions of the curriculum's goals and their motivational responses? (3) To what other factors do students attribute their motivation? (4) Do students' achievement goal orientations and/or professional and academic interests play a role in, or moderate, their motivational responses? If so, how?

## METHOD

### Curriculum

In the spring of 2007, a three-week-long project-based curriculum was implemented in the Organic Chemistry II laboratory course. The intent of this curriculum was to create an environment of enhanced autonomy (Blumenfeld et al., 1991; Cordova & Lepper, 1996; Schraw, Flowerday,

& Lehman, 2001), authenticity (Shaffer & Resnick, 1999), and suspense (Blumenfeld et al., 1991; Malone, 1981; Malone & Lepper, 1987).

The intended environment was accomplished as follows: During the first laboratory period of the semester, students selected a synthetic organic reaction from a multi-volume reference text (*Organic Syntheses*) containing thousands of synthetic organic reactions. The selection process provided students with freedom of choice, but also afforded students personal authenticity by allowing them to pick a reaction that was of personal interest or relevance to them (Shaffer & Resnick, 1999). Students subsequently conducted a number of tasks authentic to the practices of synthetic organic chemists. They modified, adapted, and scaled down their *Organic Syntheses* procedures from industrial scales to a microscale, and they completed a purchase order form requesting their needed chemicals and equipment. These tasks were all activities students had never done before but are ones routinely carried out by synthetic organic chemists.

Students were then given two laboratory periods (each four hours in duration) during weeks 11 and 12 of the semester in which to test out their reactions and try to get them to work. Trying to modify and utilize a literature procedure to synthesize a desired compound is a common authentic task of organic chemists, but it is not one typically done by undergraduate organic students. Ordinarily, students such as those in this study conduct reactions that have been pretested by the chemistry department and are known to work. With this project-based curriculum, however, students carried out procedures and conducted reactions they knew had not been pretested at a laboratory scale. The lack of predictability introduced the element of suspense into the curriculum as both the instructors and the students did not know if the reactions would work or what would happen when these modified procedures were conducted. Being given a two-week opportunity to test out their procedures was also something new for students. The second laboratory period provided students with an opportunity to repeat or modify their procedures and thereby improve upon their results. This type of second chance is not something typical for organic students, but it is something commonplace in synthetic research.

One additional element was added to the curriculum to provide students with authenticity or personal relevance: Students were told that if their reaction was successful, it might be adopted by the chemistry department for use with future classes of students. Because the students in this study's population were very connected to one another, where upperclassman often put a lot of effort into helping more junior students, it was expected that the study participants would be highly invested in the opportunity to design an experiment for future students.

A byproduct or somewhat unintentional feature of the curriculum was group work. In order to simplify the logistics of having students run lots of different reactions, it was decided that students would be required to work in groups of three. Each student group was required to jointly select a reaction to test out and to jointly submit a preliminary laboratory report containing a procedural plan, discussion of hazards, and a purchase order request form. Group work was, thus, an additional curricular feature unrelated to authenticity, autonomy, or suspense, which had the potential to enhance the motivation of students (Hidi & Renninger, 2006).

## Participants

This study was conducted at "Institute of Judaism" (IJ), a small, all-male liberal arts college located in the northeastern U.S. It involved the spring, 2007 Organic Chemistry II laboratory class, which consisted of four laboratory sections, each containing about fifteen students. Of the 57 students who carried out the project-based curriculum, 31 ultimately agreed to serve as study participants.

Participants were overwhelmingly pre-health students (planning to attend health science professional schools, with the vast majority [87%] describing themselves as pre-medical [pre-med]). Most were also science majors, with 84% majoring in biology or chemistry. The overwhelming majority also were upperclassmen, with 84% having started organic chemistry as lower or upper juniors. Thirteen percent were honors students, enrolled in IJ's honors program.

The participants in this study all had completed the organic chemistry I laboratory during the fall, 2006 semester. They were familiar with basic organic purification techniques and had been exposed to a number of discovery-based organic experiments (Horowitz, 2003, 2007). They had no prior experience, however, with project-based chemistry experiments.

## Data Collection

### *Overview*

This study was an action research study (Anderson, Herr, & Hinlen, 1994; Carr, 2006; Cochran-Smith & Lytle, 1992). The project-based curriculum was implemented by this author who oversaw the chemistry laboratory program at IJ. The author also was the instructor for two of the four laboratory sections. Data collection was conducted by the author during the spring and summer of 2007. Four qualitative data collection methods were utilized, with different information gleaned from each:

1. Participant observation, captured as field notes and journal entries, was used to examine the behaviors (degree of persistence) of students
2. Artifact collection of student lab reports was used to learn about both the behaviors (degree of persistence) as well as the affective responses, reported by students. Artifact collection also was used to learn about what aspects of the curriculum students reported to be motivating.
3. Semi-structured interviews were used to learn about (a) students' perceptions of the curricular goals, (b) their affective response to the curriculum, (c) their achievement goal orientations, and (d) what aspects of the curriculum they found motivating.
4. A short survey using an open ended, fill-in format was used to gather information about the academic majors and career goals of students.

Because this was an action research study, there was concern that students might feel coerced into participating and/or that their responses might be biased due to a desire to please the researcher. A number of steps were incorporated into the study design in order to address these issues:

1. Grades for the project (as well as all other experiments conducted that semester were assigned solely based on the quality of each individual's laboratory reports and not based upon the degree of effort in the laboratory.
2. Student consent forms and surveys were kept sealed and were not opened by the researcher until after the course was over.
3. All interviews were conducted after the course was over.
4. Interview questions were designed to be open-ended and non-leading. (See Appendix B for the interview protocol.)
5. During the interviews, follow-up questions were asked of students in order to make sure that students could give evidence or specific examples to backup their statements and assertions.
6. Data regarding students' degrees of intrinsic motivation were collected from multiple sources so that self-report data and observational data could be compared and triangulated.

### *Intrinsic Motivation*

In this study, two indexes—student persistence and student affect—were used to describe and characterize student motivation (Pintrich & Schunk, 2002). Student persistence, defined as the ability to persevere in the face of difficulty or failure (Lepper, 1988; Nolen, 2001; Turner, 1995), was operationalized as the quantity of work completed by students relative to the degree of challenge of their experiment (where degree of challenge was defined as the difficulty in obtaining the desired product in high yield and with high purity). Three data sources, participant observation, artifact collection, and semi-structured interviews, were utilized to obtain information regarding students' degrees of persistence. These data sources were used to make comparisons between students' oral and written descriptions, and between my observations and the students' descriptions.

A second index of students' motivation was affect. Information regarding students' affective responses was obtained mainly from their oral descriptions in the semi-structured interviews, but also from the affective responses they wrote about in their laboratory reports. The interview protocol utilized to assess student affect was designed based on the interview protocol of Mistler-Jackson and Songer (2000), a protocol that was utilized to examine the intrinsic motivation of students participating in the *Kids as Global Scientists* curriculum.

### *Perceptions of Curricular Goals*

The main source of information regarding students' perceptions of the curriculum's goals was the semi-structured interview in which students were asked to describe their experiences with the curriculum and to discuss its pros and cons. Students first were asked about their overall experiences with and reactions to the curriculum. Follow-up questions probed more specifically, asking students to describe their experiences with particular curricular features, including challenge, uncertainty, the choosing process, and similarities to scientific research.

### *Achievement Goal Orientations*

Students' mastery and extrinsic achievement goal orientations were examined through the use of semi-structured interviews. Specifically, students were asked to describe how they went about choosing their courses in different subject areas: When did they choose courses based on a desire to learn, and when did they choose courses based on a desire to get an easy A? Students were asked to back up their responses with specific examples in each subject area of which they spoke and were asked to give examples of exceptions to their overall patterns.

## Data Analysis

Interviews were tape recorded and transcribed. A narrative inquiry approach (Maltbia, 2001) was used to organize and condense the large volume of raw data obtained from field notes, interview transcripts, and laboratory reports. These data sources were used to construct three narrative descriptions for each student: (1) motivational response (behavior and affect), (2) the curricular features found motivating or a-motivating, and (3) achievement goal orientation. Following the construction of the three narratives, they were used to assign summary codes or labels (see Table 11.1) to each student.

**Table 11.1. Coding of Curricular Features**

| *Code* | *Type* | *Description* |
|---|---|---|
| Autonomy: Choice | Theoretical | Getting to choose one's own reaction or experiment. |
| Autonomy: Independence | Emergent | Feeling grownup, in charge, entrusted, on one's own, free |
| Autonomy: Ownership | Emergent | Feeling that the experiment belongs to oneself, that the chemicals, the reaction are unique and belong to oneself. |
| Autonomy: Responsibility | Emergent | Feeling that one is responsible for the consequences of one's actions. |
| Autonomy: Problem solving | Emergent | Getting to respond to problems. Getting to implement changes in order to solve problems. |
| Autonomy: Decision making | Emergent | Making decisions about what to do, what changes to make. |
| Autonomy: Procedural design & modification | Emergent | Getting to design one's own procedure or modify a literature procedure. |
| Autonomy: Applying prior knowledge | Emergent | Getting to use prior knowledge to solve problems or make decisions. |
| Autonomy: Thinking | Emergent | Getting to think for oneself. |
| Student authenticity: Of personal interest | Theoretical | Feeling that the experimental topic is something one is personally interested in, even interested in outside of class. |

*(Table continues on next page)*

**Table 11.1. (Continued)**

| *Code* | *Type* | *Description* |
|---|---|---|
| Student authenticity: Relevant to IJ | Theoretical | Thinking about future IJ students using this experiment (scenario). |
| Student authenticity: Relevant to organic lecture | Emergent | Feeling that the experimental topic is something discussed in, covered in, and/or important to lecture; that lab helps with mastering and learning the organic lecture material. |
| Student authenticity: Biological relevance | Emergent | Feeling that one's experiment is related to biological topics. |
| Student authenticity: Getting to think/make decisions | Emergent | Thinking for oneself and making decisions on one's own are skills necessary for future doctors. |
| Student authenticity: Independence | Emergent | An opportunity to act independently is important and valued because it is rarely encountered in college. |
| Scientist authenticity: Modifying a literature procedure | Theoretical | Adapting existing procedures is something research scientists do and know how to do. |
| Scientist authenticity: Uncertainty | Theoretical | In the real world of research, the outcomes of experiments are truly not known. Research is about finding answers to things that are not known. |
| Scientist authenticity: Multiweek | Theoretical | The fact that the experiment ran over more than one week and that there was the opportunity to learn from prior work and make changes and plans accordingly—this is more like how real research is done. |
| Scientist authenticity: Getting to think/make decisions/have choices | Emergent | Getting to make decisions, choose among options, having to solve problems are all things that real research scientists do and know how to do. |

*(Table continues on next page)*

**Table 11.1. (Continued)**

| *Code* | *Type* | *Description* |
|---|---|---|
| Scientist authenticity: Independence | Emergent | In the real world of research, scientists work independently (don't all do the same experiment, don't follow recipes, hours are not predetermined). |
| Scientist authenticity: Encountering problems | Emergent | In the real world of research, things don't always work. Problems and obstacles arise. |
| Suspense: Uncertainty of outcome | Theoretical | Not knowing if the experiment is going to work. |
| Suspense: Challenge | Emergent | Feeling that one is faced with obstacles or problems that are hard to solve. |
| Suspense: Novelty | Emergent | Doing something new, different than past classes have done. |
| Suspense: Complexity | Emergent | There is novelty, excitement in doing something complicated. |
| Suspense: Performance anxiety | Emergent | Worrying about whether or not one will succeed. |
| Suspense: Encountering problems | Emergent | Encountering difficulties, problems, obstacles. |
| Suspense: Danger | Emergent | Wanting or getting to see exciting and dangerous events. |
| Suspense: Successful results | Emergent | Wanting the experiment to work, wanting to achieve good results. |
| Other: Group work | Emergent | Getting to work together with other students. Having others to discuss with, ask and answer questions with, divide up the work with. |
| Other: Late night labs | Emergent | Having to take and be in lab in the evening when one is tired. Not getting to eat dinner until after lab (late at night). |
| Other: Lenient grading | Emergent | The perception that this lab would be graded leniently. Knowing one is not tested on this experiment. Not having follow-up or analysis questions associated with this experiment. |

### *Intrinsic Motivation*

Each student was assigned a code or label corresponding to his motivational level (degree of intrinsic motivation) with three categories of motivational levels created high, low, and mixed/unclear.

A student was assigned a motivational level of high if data from multiple sources consistently showed both positive affect as well as a high degree of persistence. For example, if a student repeatedly stated that he enjoyed the curriculum, gave detailed descriptions of the curricular features that he saw as positive, and spent two full-length laboratory periods responding to obstacles by modifying, repeating, and tweaking his experiment, this student would be coded as having a high level of motivation.

On the other hand, if a student consistently showed a lack of positive affect as well as a low degree of persistence, he was assigned a code of low for his level of motivation. A lack of positive affect could have been concluded either because a student spoke negatively about the curriculum's features or if he had absolutely nothing positive to say about the curriculum. A lack of persistence would have been concluded if, when facing obstacles and challenges, a student left the laboratory early without finishing his work, did not return the second week to complete his experiment, or upon obtaining poor results, decided to run his reaction only once.

In some cases, a student's motivational level was ambiguous or unclear either because the student's affective response was fairly neutral or because his oral explanations were hard to follow. In other cases, students' motivational levels were hard to pin down because the students demonstrated contradictory responses. For example, there were students whose affective responses were positive, but whose degrees of persistence were low. This may have been due to a contradictory or conflicted response on the part of a student or a desire to exaggerate and speak more positively to me, the researcher. In either case, when a student demonstrated an ambiguous or contradictory response, that student was coded as "mixed/unclear" because it was hard to tell what his true motivational response was.

### *Perception of Curricular Goals*

Codes also were created for any and all curricular features that students described as motivating or a-motivating. These curricular features were coded as either being about autonomy, authenticity, suspense, or other. The curricular feature of authenticity was subdivided into two codes, one for student authenticity (authentic and relevant to the experiences of students) and one for scientist authenticity (authentic to the practices of scientists). Subheadings (subcodes) were used to categorize the types of autonomy, student authenticity, scientist authenticity, and suspense that students described. For example, a few of the student authenticity subheadings that were created were a reaction of personal relevance and the student scenario (see Table 11.1).

In the course of coding student responses to the curricular features, I discovered that students described many different types of autonomy, authenticity, and suspense, the majority of which had not been theoretically predicted. For example, I had planned the curriculum to include autonomy of choice as a potentially motivating feature. Students, however, also spoke of a number of other types of autonomy that were motivating to them (e.g., independence, ownership, responsibility, etc.). Because curricular features that had not been theoretically predicted emerged from the data, I further categorized each curricular feature as being either theoretical or emergent. (See Table 11.1 for the listing and description of all the curricular codes created.)

In addition to the types of autonomy, authenticity, and suspense that were unanticipated and emerged from the data, there were three additional, unrelated curricular features that emerged that students spoke of as either motivating or a-motivating. These were: (1) the ability to work in groups, (2) late night labs, and (3) leniency in grading.

### *Achievement Goal Orientations*

Each student was assigned a series of achievement goal orientation labels or codes, one for each academic subject or course that the student mentioned. These labels were extrinsic, mastery, and mixed/unclear. For example, a student could be coded as mastery in biology, as mixed/unclear in history, and as extrinsic in English composition and Hebrew language.

Codes were determined as follows: If a student described himself as looking for an easy instructor or as just looking to get an easy A in a particular subject area, this subject area was labeled as extrinsic. If a student spoke of taking a particular course or subject area because of interest or a desire to learn, this course or subject was labeled as mastery. If a student described contradictory goals, if his description was unclear, or if it did not appear to fit a mastery or extrinsic category, then this subject area was coded as mixed/unclear.

### *Inter-Rater Reliability*

In order to ensure the reliability of the coding process, a second, independent coder was assigned the task of coding the raw data (laboratory reports, interview transcripts, and field notes) of a randomly selected subset of students (six students, 19% of the sample). The coder was a social scientist not affiliated with the Institute of Judaism. The coder coded students for their motivational levels (high, low, mixed/unclear), curricular features that were motivating or a-motivating to them, and their achievement goal orientations. The overall degree of inter-rater reliability obtained in this process was 86%. Coding of curricular features had an inter-rater reliability level of 84% (agreement in 124 out of 148 codes). Coding of achievement goal orientations had 93% agreement (*Kappa* = 0.870, *SE* = 0.068) and coding of motivational levels had 100% agreement (*Kappa* = 1.000, *SE* = 0.000).

## FINDINGS

### Student Perceptions of Curricular Goals and Features

#### *Autonomy of Choice*

Almost without exception, students spoke positively about the fact that the curriculum had provided them with a higher level of autonomy than a traditional experiment. Specifically, with regards to getting to choose their own experiment, 74% of students spoke favorably about this aspect of the curriculum.

The finding that such a large percent of students spoke positively about getting to choose their own experiment was somewhat surprising, given that 58% of students were also critical of, and somewhat frustrated by, the choosing process itself. These students were unanimous in their opinion that the choosing process did not go well. Many found themselves overwhelmed by the large number of reactions to choose from and by the large number of unfamiliar reactions.

Further evidence of student dissatisfaction with the choosing process was the fact that during their interviews, more than one third of students made suggestions regarding how to improve the process. The main suggestion was to narrow the range of choices to a small list of about 20 experiments. Ten students made this specific suggestion. Four students also suggested delaying the choosing process until later in the semester, so that students would have a stronger foundation in organic chemistry when the time came to choose their experiments.

The students' feedback was interesting because it not only clearly conveyed the message that the number of choices of experiments overwhelmed students, but also revealed that students were more focused on having the ability to choose, rather than on having a wide range of topics from which to choose. In effect, by asking to be given a short list of experiments from which to choose, students were stating that the act of getting to choose (autonomy of choice) was more important to them than having a wealth of experiments from which they could choose a topic of interest and personal relevance (student authenticity).

Participant observation data regarding the choosing process was in agreement with what was reported by students in their interviews. Based on the field notes and journal entries, further evidence showed that the choosing process did not go well and was not properly scaffolded. Students seemed to not have sufficient guidance in how to go about searching and finding experiments. Initially, when the journal notes were compiled, I was not yet aware of the frustration students were experiencing, but simply noted that it seemed that scientifically authentic but practical issues such as the cost of chemicals or the hazards of certain experimental conditions got in the way of students' autonomy of choice. At the time, it appeared, perhaps, that these scientifically authentic limitations (e.g., having to worry about safety, cost, or the availability of equipment) might have engaged or motivated students. Later evidence from the interviews, however, indicated that the authenticity of real-world obstacles did not motivate students but rather caused them to be frustrated.

### *Other Forms of Autonomy*

Given that over 70% of students spoke of appreciating the opportunity to choose their own experiment, it would appear that autonomy was an important motivating factor. Further evidence of how important and motivating autonomy was for students was that 84% of students spoke positively about other types of autonomy they felt they had experienced within the project-based curriculum (see Table 11.2). Sixty-eight percent of students spoke of feelings of independence, ownership, or responsibility, and 32% spoke of the ability to make procedural decisions or procedural modifications. Fifty-five percent spoke of the opportunity to think for themselves, to apply prior knowledge, to make decisions, or to solve problems.

**Table 11.2. Summary of Student Responses to the Autonomous Aspects of the Curriculum**

| *Types of autonomy* | *% of students who viewed this as positive* |
|---|---|
| Autonomy overall | 90% |
| Choice | 74% |
| Independence, ownership, responsibility | 68% |
| Thinking, making decisions | 55% |
| Making procedural modifications | 32% |

In discussing the forms of autonomy they had experienced as motivating, students spoke of types of autonomy that I had not anticipated. While I expected students would have to modify their procedures, I did not expect they would perceive of this as an opportunity to act more autonomously or independently. Students spoke of getting to make decisions and modifications as something motivating and positive, and clearly voiced the feeling that this freedom was generally not available to them in their other lab courses.

A further surprise was that having only three students work on each particular experiment resulted in students experiencing feelings of independence, ownership, and responsibility. Students spoke of feeling that the experiment was up to them, that they couldn't ask neighbors for help, and that even the instructor was unfamiliar with the experiment. These aspects of the curriculum gave students feelings of independence, ownership, and responsibility.

### Student Authenticity

In this study, two curricular elements were deliberately incorporated in order that students experience personal relevance and authenticity. These features were the student scenario (that students' experiments might be adopted by the chemistry department and used by future IJ students) and that students could choose experiments of personal interest to them. In spite of these curricular goals, however, interview data demonstrated that personal authenticity was rarely experienced by students.

In the course of interviews, only four students mentioned the scenario. Three mentioned it only in passing, while one spoke about it in more detail. Somehow, either students did not buy into the idea of the scenario, or they were indifferent to and unengaged by the idea of it.

Similarly, few students actually chose a reaction of personal interest or relevance to them. As noted above, because of problems with the choosing process, many students chose their experiments by default. Only five students spoke of choosing an experiment of personal interest. Of these five students, only one demonstrated a strong interest in his particular reaction. Two students did not discuss their specific reactions at all. Two others reported only vague memories of how they had become interested in and had chosen their particular reactions.

But personal relevance was important to students. Forty-five percent of students spoke of wishing that lab experiments in general, and the project-based curriculum in particular, were more relevant to their needs and interests. Specifically, students wished that experiments were more relevant to the organic lecture course and to biological topics. Nine students spoke about the need for the organic laboratory to cover topics that were important to the organic lecture course. To students, doing well in the lecture course was extremely important. Students made it clear therefore that experiments that were relevant to and reinforcing the material of the organic lecture course were very important to them.

As to the project-based curriculum, students were mixed in their beliefs about whether or not it was relevant to and reinforcing of the ideas of the organic lecture course. A few felt that it connected well to the lecture, but others disagreed. One student questioned whether or not the curriculum, where students were all involved with different chemical reactions, took time away from explicitly learning about specific reactions that were covered in the lecture course.

Students also spoke about their interest in and desire for experiments and topics of biological relevance. Six students gave examples of experiments they would have liked to do in the organic lab, or would like to have chosen when conducting the project-based curriculum. One student spoke of wishing that he could have synthesized ATP, a molecule whose function and universality really intrigued him. Another spoke of fruitlessly searching the *Organic Syntheses* reference text looking for some type of DNA or protein synthesis, to no avail. A third student spoke of wishing to research the active ingredient of poison ivy, a molecule he unfortunately had significant personal experience with.

Although, in general, students did not experience the curriculum as personally relevant, a few students described ways in which the curriculum had personal relevance for them. The aspects of the curriculum students found personally relevant, however, were not those theoretically predicted. For example, two students spoke about the autonomous aspects of the curriculum as being highly relevant to them (having to face problems, make choices, and make decisions). They saw the autonomous features of the curriculum as skills highly necessary and crucial to themselves and their classmates, who intended to become future physicians. A third student spoke of other autonomous aspects of the curriculum he felt were highly relevant. To him, what was relevant, important, and motivating was that students got to choose their own experiments. He described his college experience as one in which he was constantly following the rules, instructions, and decisions set for him by others. He, therefore, strongly appreciated the opportunity to choose his own experiment, which the project-based curriculum allowed

### *Scientist Authenticity*

The project-based curriculum was designed so students would conduct tasks authentic to the practices of synthetic organic chemists (e.g., scaling down, modifying, and adapting a literature procedure), but also so students would experience an authentic research situation of trying to get a reaction to work, without knowing in advance whether or not it would be successful. The extent to which students actually perceived themselves to be experiencing authentic scientific tasks and situations was highly varied.

To begin with, only three of the students interviewed had conducted prior chemistry research. Others who had conducted research (19 in total) had conducted biological or medical research. And, while all students were taking Organic Chemistry Laboratory II when this study was conducted, it was not apparent from the interviews that students even recognized that the tasks they were doing were common, typical tasks of synthetic organic chemists.

Because students were most familiar with medical and biological research, when asked how the project-based curriculum was similar to or different than scientific research, their answers varied based on their perceptions of what medical or biological research was like. To further complicate matters, some students with prior research experience interpreted and understood research to be the tasks they carried out when conducting research. These tasks sometimes involved a high degree of challenge, responsibility, and thinking, but often were simply tedious, repetitive, and mindless. Other students defined research as the tasks they observed others (principal investigators, graduate students, postdoctoral students) to be conducting. And students who lacked research experience frequently described and defined research based on their projections or imaginings of what medical or biological research was like.

The most common ways in which students compared the project-based curriculum to scientific research were the extent to which the curriculum provided them with autonomy (to make decisions, have choices, or work independently) or with suspense (problems arose, unexpected events happened). The comparisons made by students varied. For example, four students felt that the suspense and uncertainty of the curriculum mirrored that of scientific research, whereas seven students felt that its autonomous aspects mirrored research. On the other hand, eight students felt that the curriculum was not at all like scientific research and eight other students were mixed in their opinions. Regardless of these differences, no student spoke of, or appeared to recognize, that the tasks he had conducted (modifying and testing out a literature procedure) were authentic, commonplace activities of practicing organic chemists.

Even among students who saw the autonomous and suspenseful aspects of the curriculum as reminiscent of authentic scientific practice, not all of these students saw these curricular aspects as positive and motivating. While six of the seven students described above saw the autonomous aspects of the curriculum as motivating, only one of the four described above saw the suspenseful aspects of the curriculum as positive.

### *Suspense*

Suspense was another curricular feature deliberately included in order to enhance student motivation. Of the students interviewed, 48% spoke of suspense as being a motivating factor. The suspenseful aspects of the curriculum most commonly seen as motivating by students were uncertainty of outcome, novelty, and challenge.

Some students, however, did not view suspense as something positive and motivating. Five students had mixed feelings about the suspense of the curriculum, and three found it to be a negative and a-motivating factor. To the latter students, challenge and uncertainty of outcome resulted in feelings of frustration, discouragement, and disappointment, whereas novelty and surprise only led to increased hazards and risks, resulting in feelings of nervousness and fear.

The majority of students perceived of the suspenseful aspects of the curriculum as either something negative or something positive. Two students, however, expressed the opinion that the curriculum did not provide them with enough suspense, particularly in terms of its degree of novelty and complexity. Six remaining students did not appear to experience suspense at all. These students did not find their reactions challenging and/or they were very confident that their experiments would work and give good results.

### Group Work

When the project-based curriculum was designed, it was decided that, for practical reasons, group work would be a necessary component of the curriculum. Group work was not an intended curricular goal, nor was it included for the purpose of trying to enhance student motivation. In their interviews, however, numerous students spoke about group work and gave their reactions to this aspect of the curriculum.

Student responses reflected a wide range of sentiments. Seven students spoke of group work as something enjoyable, either because of its general social nature, or because of the specific opportunity it provided to work with friends. In contrast, three students spoke of disliking their groups. They felt that some group members were out of sync with others, and were either not as hard working as others or not as engaged as others. Four additional students spoke of their general preference for working alone, or of their overall ambivalence towards group work, because of a tendency of some students to take advantage and not do their fair share.

A larger group of 13 students had a more middle-of-the-road reaction to group work. They did not find group work motivating or a-motivating, but rather saw it as necessary, stating that having others to strategize with, and rely on, was needed with this more independently oriented curriculum. These students felt that working alone would have been difficult or scary.

When speaking about the group work, students focused on the teamwork they experienced when carrying out their reactions. But this teamwork was not a required part of the curriculum. Students were required only to work together when choosing their experiments, but were free to work independently when actually running their reactions. Almost all students took it upon themselves to work together when conducting their reactions. With one exception, each group worked together to set up its reactions and to measure out and dispense its chemicals. Some of this collaboration was, perhaps, necessary because each group had only one set of chemicals. Almost all students took it upon themselves, however, to brainstorm and problem solve together, rather than individually. This behavior is consistent with the explanation given by students above, that teamwork was necessary in this curriculum in order to obtain guidance and support and in order to tackle and solve problems.

### *Reduced Work Load/Lenient Grading*

A few students spoke about other unanticipated curricular features. Three students voiced the opinion that the project-based curriculum required less effort of them as compared to more traditional experiments. Two of these students saw this reduction in workload as a positive feature, while one saw it as a negative feature.

Two students expressed the idea that they felt that the project-based curriculum was going to be graded more leniently than other experiments. One of these students indicated that this caused him to be less invested in the curriculum because he was less concerned about his grade. The second student, however, felt that his perception of more lenient grading caused him to be more invested, because he was more relaxed and therefore more able to work independently.

## Intrinsic Motivation

In addition to investigating students' responses to the curricular elements of autonomy, authenticity, and suspense, students' motivational levels when experiencing the curriculum were examined and explored. Using the motivational narrative descriptions described in the data analysis section above, 16 students were categorized as high in intrinsic motivation, 3 were categorized as low, and 12 were categorized as mixed/unclear (see Table 11.3.) In the following paragraphs, three examples are presented: one of a student high in motivation, one of a student low in motivation, and one of a student mixed/unclear in motivation. Only pseudonyms are used.

Judah was a student who demonstrated high intrinsic motivation. When he ran his reaction, he obtained only a very small amount of product, and he obtained a product that was highly contaminated. He returned to the laboratory the next week to continue to work on the experiment. First, he attempted to purify his product using recrystallization. This was quite difficult because he had very little material and because his material was so impure. Then, later on during that laboratory period, I discovered that Judah and his group-mates had added the wrong quantity of one crucial reagent when they had conducted their reactions the previous week. After the mistake was discovered, Judah immediately reran his reaction from scratch in an attempt to improve upon his prior results.

**Table 11.3. Students' Motivational Levels**

| *Motivational Level* | *Number of Students* |
|---|---|
| High | 16 |
| Low | 3 |
| Mixed/Unclear | 12 |

Judah's behavior demonstrated a high level of motivation in that he continued to work hard and put in effort despite facing obstacles and setbacks. Judah's affective response during the interview also was consistent with a high level of motivation. Judah stated:

> I liked it. I didn't hate anything. I liked it a lot.... It kind of allowed us to take what we've learned and actually apply it.... Those are the type of things I love doing. I love application of knowledge.... So that's what I liked about Orgsyn. It's that it allowed me to really take, it kind of allowed me to develop, to be an independent developer of my own project and to take pride in something. This was mine. Not anyone else's. It was mine and I had to do all the work for it, and I had to figure it out. And if it didn't work, I couldn't go to my friend next door to me and say, hey what did you do for step 4? ... so it was kind of like a complete independent experiment, which I really enjoyed.

In this quotation, Judah repeatedly used a number of positive sounding verbs to describe his affective response to the curriculum. He used the word "like" a total of four times, the word "love" a total of two times, and the word "enjoy" once.

In contrast to Judah, another student named Benjamin demonstrated a low level of intrinsic motivation, both in his behavior as well as in his affective response. Benjamin attended the laboratory during both periods and conducted his reaction twice, but each week he left the laboratory in the middle of the period for approximately one hour. During this hour and during other times within the laboratory period, Benjamin took a back seat and allowed his group-mates to take on some of his share of the set-up and clean-up work. At the end of the second laboratory period, Benjamin's group-mates decided to recrystalize their products because they were fairly impure. Benjamin, however, chose not to recrystalize his product, despite the fact that his product was extremely impure and melted about 20 degrees lower than it should have.

Benjamin's affective response to the curriculum also demonstrated low intrinsic motivation. As Benjamin put it:

> I didn't like the fact that specifically in my experiment that we had trouble And we didn't actually end up getting what we wanted...it was a little bit of annoying.... Cause it kind of puts a damper on the whole procedure when you can't get it right. And you don't feel like it was worth it.... And we were there every night so late. And then the second time we were also there so late. And we couldn't get anything done. It was just frustrating.

Benjamin used words and phrases such as "didn't like," "annoying," "puts a damper," and "frustrating" to describe his feelings about the curriculum.

A third student, Jonathan, exhibited a mixed/unclear motivational response. Jonathan came to the laboratory during both laboratory periods and ran his reaction a total of two times. When he ran the reaction the second week, however, he chose not to distill the crude project that he obtained, which appeared to be highly contaminated with starting material and in need of purification. Instead, he concluded that his reaction had not worked, and that his crude product consisted solely of starting material. In his laboratory report, he fictitiously reported a low group yield of crude product, perhaps to give the impression that he and his group-mates did not have sufficient material to conduct a distillation.

In contrast, however, Jonathan put in a good deal of effort in between the first and second laboratory periods trying to figure out how to improve his results. After his first run was unsuccessful during the first period, Jonathan contacted me and arranged to meet with me twice to discuss his procedure. During the meetings, we discussed what might have gone wrong, what changes he might make for the second attempt, and how he could improve his results.

In the face of challenge and poor results, Jonathan put in a mixed effort. Not surprisingly, his affective response to the curriculum was also mixed. Overall, he liked the idea of the curriculum and the opportunities it provided, stating:

> I think Orgsyn was a great idea. I think that it encourages students to be more involved...in the organic lab experience. I think because you have to pick your own project, you feel a little attached to the results.... I think a lot of us are not, have not been trained to think outside of the box...and haven't been taught how to design our own experiments.... So in that respect I think it's good.... It starts getting people to think. Making them feel like they are kind of in charge of and responsible for carrying out their own experiments.

But he found his actual experience to be disappointing and discouraging:

> By the very nature that you're more interested in the project, you want to see results. So our project, we didn't get any results.... When we see that it [the distillation temperature] didn't go above 80 [degrees] and we were getting tar, that's when I think we kind of just threw in the towel. And I think we were not motivated to come back a second time.... I understand that not every experiment is going to work. But I really felt ours was a dud.

## Intrinsic Motivation & Student Perceptions of the Curricular Goals

Students' motivational responses to the curriculum varied, with a majority experiencing either a high level of intrinsic motivation or a mixed/unclear level of intrinsic motivation. Nonetheless, students' motivational responses clearly showed a connection to their perceptions of the curriculum's goals of enhanced autonomy, authenticity, and suspense (see Table 11.4).

**Table 11.4. Relationship of Motivational Level to Perceptions of Curriculum**

| | | *Motivational* | *Level* |
|---|---|---|---|
| *Perceptions of Curriculum* | *High* | *Mixed/Unclear* | *Low* |
| Positive regarding all three curricular features (autonomy, authenticity, and suspense) | 69% | 17% | 0% |
| Positive regarding two curricular features (autonomy and either authenticity or suspense) | 25% | 17% | 0% |
| Positive regarding autonomy only | 6% | 50% | 67% |

There was a vast difference between how the highly motivated students perceived the curriculum and how the least motivated students perceived the curriculum. Over two-thirds (69%) of the highly motivated students experienced all three curricular features (autonomy, authenticity, and suspense) as positive. An additional 25% of these highly motivated students experienced two of the three curricular features as positive (either autonomy and suspense, or autonomy and authenticity.) In contrast, none of the students who were coded as low in motivation perceived of authenticity and suspense as something positive, although two-thirds of them did experience autonomy as something positive.

Students who were mixed/unclear in the motivational levels were more varied in their responses to the curriculum. Over 80% of the mixed/unclear students viewed autonomy as a positive feature of the curriculum. However, 50% of them had mixed or negative reactions to the suspenseful aspects of the curriculum.

## Achievement Goal Orientations

Because it was hypothesized that students' achievement goal orientations might moderate their responses to the curriculum, student goal orientations were examined. Findings regarding student goal orientations are presented briefly here, but are explored in detail elsewhere (Horowitz, 2010).

As was described in the data analysis section above, initially students were assigned codes (mastery, extrinsic, or mixed/unclear) for each subject area of which they spoke. Following this initial coding, however, the narrative summaries and codes for each student were examined and reviewed in detail to see what trends they revealed. Three general categories of achievement goal orientations were found to emerge from the data. These goal orientations were a mastery goal orientation, an extrinsic goal orientation, and a generalized extrinsic with selected mastery goal orientation (see Table 11.5). About one-third of students (10 in total) were found to possess an overall mastery goal orientation, while only about 10% (four students) were found to possess an overall extrinsic goal orientation. About 50% of students (15 in total) were found to possess what was termed a generalized extrinsic with selected mastery orientation, namely that they were mainly extrinsic goal oriented, except in subject areas that were of interest to them. Two students did not fit any of the above categories and were left unassigned.

The following student comments illustrate how students described and characterized their respective achievement goal orientations. As one mastery-oriented student put it:

> I do like to get good grades.... But the reason why I'm particularly dissatisfied when I do [not get good grades], is...because...my goal is to effectively learn the material. .. The reason why I am in the honors program, and the reason I have not shied away from difficult courses ... [is] I'm here ... to learn.... By choosing certain courses, and certain majors, I could have been out of here very fast with a 4.0 GPA. That wasn't the point.... The reason why I'm spending four years ... is specifically because I ... want to learn.

**Table 11.5. Students' Achievement Goal Orientations**

| *Goal Orientation* | *Number of Students* |
|---|---|
| Mastery | 10 |
| Extrinsic | 4 |
| Generalized extrinsic + selected mastery | 15 |
| Unassigned | 2 |

The comments of this mastery-oriented student stand in sharp contrast to those of an extrinsically oriented classmate, who stated:

> Obviously you want a good grade. You don't want to take a hard course.... So I took sociology...because...with sociology you don't have so much work.... A guy that I took [for Hebrew].... I didn't do well on the tests.... But I got an A. Because I knew if he liked you, that you would do well.... That's why I took my Comp [English Composition professor] twice.... She loved me.... She had the highest opinion of me. So you know, you take her.... But also in your major.... I have looked for easier teachers in Bio[logy].... Because it's your major. And actually med schools look at your major more.

Generalized extrinsic with selected mastery students were more mixed in their descriptions, as can be seen from the comments of the following two students:

> When it's just to fulfill core requirements, we will choose easy courses.... For sciences, I took classes that I was more interested in...even though I knew my grade might suffer. Biochemistry. I took that, even though I knew it would be hard course.... Alright, it might hurt my GPA. But in the long run I think it's worth it.

> English literature, I'm not a big fan of. It's my weakest subject. So I'll try and take the easiest guy. But psychology, I find kind of interesting. And, if this class is not so easy, but an interesting topic, then I will take it.

## Moderating Effect of Achievement Goal Orientation

Students' achievement goal orientations moderated their motivational responses to the project-based curriculum (see Table 11.6). When compared to the overall sample, mastery-oriented students were the most likely to demonstrate high intrinsic motivation. Of the 10 mastery-oriented students, 7 were found to demonstrate high intrinsic motivation. In contrast, only two of the four extrinsically oriented students demonstrated high intrinsic motivation. And less than one third of the students categorized as generalized extrinsic with selected mastery demonstrated high intrinsic motivation. The largest group of students were those who had a generalized extrinsic with selected mastery goal orientation and who demonstrated a mixed/unclear motivational response.

**Table 11.6. Relationship of Students' Motivational Levels to their Achievement Goal Orientations**

| *Goal Orientation* | *Motivational Level* | | |
|---|---|---|---|
| | *High* | *Mixed/Unclear* | *Low* |
| Mastery | 7 | 3 | 0 |
| Generalized extrinsic + selected mastery | 4 | 9 | 2 |
| Extrinsic | 2 | 1 | 1 |
| Unassigned | 2 | 0 | 0 |

## Moderating Effects of Academic and Professional Interests

Because an overwhelming percent of participants were pre-med (87%), a lack of variety in career goals made it impossible to assess the moderating affect of career goals upon student motivation. Additionally, almost 70% of students were biology majors, with the next most popular majors (chemistry and psychology) containing only four and three students, respectively. With so many students majoring in biology, it was hard to detect a moderating effect due to a particular academic major. And, while one might have anticipated that majoring in chemistry might moderate students' motivational responses, this was not discernable in this study.

## Other Moderating Factors: Degree of Challenge

One moderating factor that had not been theoretically predicted but emerged from the data was the degree of challenge experienced by the student. It was found that in a number of instances, students who felt moderately challenged were more likely to demonstrate high intrinsic motivation, whereas students whose projects turned out to be highly challenging or extremely easy were more likely to demonstrate a mixed/unclear motivational level.

Both Jonathan (described above) and his group-mate David, for example, exhibited a mixed/unclear motivational level, in which they spoke positively about many aspects of the curriculum, but at the same time felt discouraged and frustrated by its high level of difficulty and their inability to get their reaction to work. On the other hand, two highly motivated students, Shmuel and Yitzchak, whose experiments were much more successful and less challenging than those of Jonathan and David, spoke very positively about their experiments and described facing and overcoming challenges as being energizing, stimulating, and motivating.

A third case was a student named Jordan who exhibited a mixed/unclear motivational level despite repeatedly describing himself as a student who loved organic chemistry and who loved the natural sciences. Jordan felt under-challenged and not motivated by his experiment. He described feeling disappointed that his group-mates had picked such a simple experiment, and an experiment whose chemistry was simplistic and uncomplicated.

A further finding was that it was not simply a matter of the degree of difficulty of a particular experiment that moderated the response of students. Rather, it was students' perceptions of the degree of difficulty that moderated their responses. Two group-mates, Chaim and Avram, for example, who were both high mastery-oriented students, both carried out the same simple, straightforward experiment. Avram was a highly experienced student who had had multiple research experiences. He exhibited a mixed/unclear motivational response, perhaps due to finding his project too simplistic or not challenging enough. Chaim, an insecure student who lacked research experience, experienced the same project as challenging and highly motivating. In this instance, it was not the actual level of experimental difficulty that moderated the response of the students, but rather it was their perceptions of its degree of difficulty.

This perception of challenge or of degree of difficulty appeared to moderate the response of a number of students. In some cases, the obstacles faced by students loomed very large in their eyes (e.g., Jonathan and David), yet, in other cases, these obstacles were taken in stride (e.g., Yitzchak). It is likely that students' perceptions of the degree of challenge varied partly due to their level of skill and confidence, as appears to be the case in the example of Chaim and Avram given above, rather than just simply due to variability in the actual experimental difficulty.

This result, that student perceptions of degree of challenge were linked to their level of skill and confidence, is consistent with flow theory (Csikszentmihalyi, 1997). Flow theory is a theory of intrinsic motivation that argues it is only when degree of challenge and level of skill are appropriately matched that a person will experience high intrinsic motivation or a state of flow. In contrast, if the ratio of skill to challenge is too great, boredom is experienced. If the ratio of skill to challenge is too low, anxiety is experienced (Csikszentmihalyi, 1997). Thus, the results of this study are consistent with flow theory and indicate that skill-to challenge ratio is an additional moderating factor in this project-based curriculum.

## DISCUSSION OF RESULTS

### Student Perceptions of Curricular Goals and Features

The response of students to the intended goals of the project-based curriculum was striking. While the vast majority of students responded positively to the autonomous aspects of the curriculum, the student response to the suspenseful aspects of the curriculum was mixed. Additionally, students did not find the curriculum to be authentic or relevant to their personal experiences, nor did they recognize that the tasks they were engaged were scientifically authentic.

#### *Autonomy*

Although the students of this study were a somewhat unique population, their response to this curriculum may be of interest to other science curriculum developers. The students in this study were clearly motivated by and desiring of more autonomy, so much so that, despite a frustrating and overwhelming choosing process, they were in favor of being able to choose their own experiments. In addition to freedom of choice, students also experienced other forms of autonomy as positive, such as feelings of ownership, independence, and self-reliance.

Those of us involved with the development of college-level science laboratories need to take note of these findings. We need to think of practical ways in which student autonomy can be enhanced within the context of the college science laboratory. In particular, ways need to be found to provide students with adequate numbers of choices so that they do not feel controlled or constrained by too few choices, but do not feel discouraged and overwhelmed by too many (Patall, Cooper, & Robinson, 2008).

#### *Suspense, Challenge, and Flow*

Unlike the positive response of students to the autonomy of the curriculum, the student response to its suspense was mixed, with some students frustrated and overwhelmed by experiments that did not work. It could be argued that the level of challenge of each experiment was not matched appropriately to the degree of skill of each student (Csikszentmihalyi, 1997), and that, if an appropriate degree of challenge had been provided, then students would have been motivated by the curriculum's suspense.

Providing a good match between student skill and experimental challenge within the context of the synthetic organic chemistry is, unfortunately, extremely difficult. This occurs because it is hard to rank the level of difficulty of most syntheses, and because it is often impossible to predict the obstacles and challenges that will arise within any one particular synthesis.

Even if it were possible to predict or control the degree of difficulty of different organic syntheses, ascertaining and then matching experiments to the degree of skill of students would be another next to impossible challenge. It would be very difficult to measure students' skills and their self-confidence in their own skills, so as to determine and be able to match the experimental challenges to students' skills.

It is not just in the organic chemistry laboratory that appropriate levels of challenge are needed in order to motivate, and not discourage or overwhelm students. Advocates of project-based and other non-traditional approaches to science instruction have pointed out that often inquiry-based science curricula require much more of students, in terms of cognitive engagement, sustained effort, and intrinsic motivation (Blumenfeld et al., 2006; Edelson et al., 1999). Students, like anyone else, need to experience success. When course work is too open-ended, students may feel lost or overwhelmed. When students only experience failure or a lack of results, they may feel discouraged or disappointed. Unlike more experienced adults, students often lack the skills and knowledge necessary to learn and move forward from failures and challenges. And, while a complete lack of suspense and surprise does not provide the ideal motivational environment for students, challenges must be appropriate in their level of difficulty if they are going to stimulate and motivate students (Deci, 1992; Malone & Lepper, 1987).

Flow theory argues that there is an optimal ratio of skill to challenge at which flow, or high levels of intrinsic motivation, can be achieved (Csikszentmihalyi, 1997). But often, such flow is achieved by persons already highly interested in and knowledgeable about particular disciplines or tasks (e.g., mountain climbing, chess playing) (Csikszentmihalyi, 1975). These persons are experienced enough to create their own challenges and to create challenges at an appropriate level of difficulty. Perhaps suspense, or the challenge that goes along with it, is not an appropriate method of motivating novices or persons who do not have a developed personal interest (Hidi & Renninger, 2006) in a particular subject.

### *Student Authenticity*

Like suspense, authenticity was another feature of the curriculum that fell short of its intended goal of enhancing student motivation. While the curriculum included two elements intended to provide student authenticity (choice of a chemical reaction of personal interest and a student scenario), students did not find these elements personally relevant. The curriculum fell short of providing students with other types of personal relevance they were seeking, specifically relevance to the organic chemistry lecture course and relevance to biological topics. With regards to relevance to the organic lecture course, this curriculum sought to expand students' exposure to organic reactions beyond the range of reactions that they were exposed to in the organic lecture course. Clearly, most students did not appreciate this exposure and did not get stimulated or excited by it. Biological relevance also was not achieved with this curriculum. The timeframe of the curriculum allowed students to conduct only one or two synthetic steps, and therefore did not expose students to the more complicated and lengthy process of synthesizing biological molecules. Even if the timeframe of the curriculum had been expanded, the complexity of most biological molecules would have made it very difficult to provide students with options of biologically relevant syntheses.

One way in which this curriculum could be improved for the future and be made more relevant to students might be to provide students with more scaffolding and guidance during the selection process when students were choosing their experiments. With more guidance regarding how to go about choosing an experiment, students might be less overwhelmed by the choosing process and better able to choose experiments that were engaging, interesting, and more personally relevant to them.

### *Scientist Authenticity*

One quite revealing finding of this study was that students were seemingly completely unaware of the tasks they were conducting as authentic scientific tasks. The majority of students were biology majors so this may, perhaps, explain this lack of student awareness and lack of motivational response. This finding may have implications for other project-based science curricula. While some curricula (e.g., Means, 1998; Songer, 1998) have students collaborate with scientists, one must question whether it is indeed the case with most scientifically authentic curricula that students, especially younger students, can truly appreciate and even recognize that they are participating in and emulating the practices of research scientists.

When this curriculum was implemented, students were told that by adapting and testing their chemical reactions, they were mirroring the practices of the IJ chemistry department. A future addition for this curriculum that could be tested is to provide students with more explicit information regarding how their practices actually mirrored those of practicing synthetic chemists. It is not clear, however, if the IJ organic chemistry student population would have found this scientist authenticity to be an engaging factor, as most of them seemed to view biological and medical research as more important and interesting than chemical research.

Although the students of this study did not recognize the tasks they conducted as scientifically authentic, some perceived of themselves as experiencing scientist authenticity in the form of increased levels of autonomy and/or suspense, as was described above, while suspense might be an authentic feature of scientific research, it was not necessarily a feature desired by students.

#### *Authenticity Trumps Autonomy*

One interesting finding of this study was that authenticity "trumped" or overrode student autonomy. Namely, authentic constraints such as time, money, safety, or a lack of equipment prevented students from having true autonomy in their choice of reactions or even in the particular procedural steps they conducted. Similar types of practical constraints also limit the work of research scientists. Therefore, the constraints experienced by students were truly scientifically authentic ones. In this study, not only did students not recognize these constraints as scientifically authentic, they experienced them as negative or even a-motivating. As far back as Dewey (1910) and Schwab (1964), theorists and curriculum developers have argued against rote, cookbook, mindless, or inauthentic science curricula. But, in some ways the findings of this study are a reminder as to why cookbook science curricula exist. Practical, yet authentic, constraints make it difficult to implement truly realistic and naturalistic curricula. Science curriculum developers must remain cognizant of these types of practical constraints as they try to implement better and more engaging curricula.

### Moderating Effects of Students' Goal Orientations

It was not surprising to observe that the achievement goal orientations of students moderated their motivational responses to the curriculum. That mastery-oriented students were more motivated and engaged by the curriculum as compared to more extrinsically oriented students was not an unexpected finding. It was most interesting to observe that, across the board, almost all students were motivated by the autonomy of the curriculum. This finding might help science educators find better ways to motivate all kinds of students, but especially those who are least interested in learning, such as extrinsically oriented students.

## GENERALIZABILITY AND LIMITATIONS

The students in this study were primarily pre-med students. It is likely that they were typical of pre-med students at other United States institutions and that their motivational responses were similar to what would be observed among other pre-med students. However, the students of this study were also somewhat unique. Because they were all male, were of the Jewish faith, and attended a religious institution, they may have been somewhat different from pre-med students at other universities. Therefore, the findings of this study are not highly generalizable to pre-med students at other institutions.

## CONCLUSIONS

The students of this study were emphatic in the characterization of their science courses as being highly controlled and lacking in choice and decision making. As a result, almost all students responded positively to the autonomous aspects of this study's curriculum. Other elements of the curriculum, specifically suspense, authenticity, and the degree of challenge, had a more mixed response from students. Additional research is needed regarding how to design curricula that enhance student autonomy but also provide students with an optimal degree of challenge, such that they are not overwhelmed or understimulated by project-based curricula. Further work also is needed on how to most effectively design curricula in which students can actively choose topics of personal interest. Curricula need to consider how to better inform students of and interest them in the research practices of scientists.

The study found that it was easiest to motivate students who were already interested in learning. It is not sufficient, however, to design curricula tailored towards mastery-oriented students. The challenge is to find ways to motivate those of our students who are not focused on learning. Further research is needed on how to best design curricula so as to motivate other kinds of students, such as the more extrinsically oriented students in this study.

# APPENDIX A

## Handouts of Student Instructions

*Student Instructions: Week 1*

## Design Your Own Experiment: The Organic Syntheses Project

### Introduction:

This semester, you will have the opportunity to select, design and test out a new organic experiment. If your testing goes well, your experiment may end up being adopted by the chemistry department for use in future semesters.

### Overview:

During week 1 of the semester (while in lab), you and two teammates will choose an experiment from the reference text *Organic Syntheses* (available in lab and online at www.orgsyn.org). After obtaining instructor approval, it will be your job to scale down and modify this experiment to make it suitable for our college setting.

For your lab report (due week 2), your team will come up with a procedural plan for the experiment and a list of needed chemicals.

During weeks 11 & 12, you will test out your experiment, after the necessary chemicals have been ordered and received.

### *Organic Syntheses* Reference Text:

Choosing an experiment appropriate for use in the undergraduate organic laboratory is not easy. Many procedures found in the primary literature (i.e., journals) are unreliable, can be dangerous, and may require specialized apparatus. *Organic Syntheses*, on the other hand, is a fruitful source of organic experiments. Its procedures are checked and are therefore highly reliable. Additionally, *Organic Syntheses* frequently warns the reader about potential hazards involved in its procedures. Nonetheless, many reactions found in *Organic Syntheses* are not practical for the organic laboratory. In fact, almost all need to be scaled down (from an industrial scale) to a scale appropriate to a university setting.

### Guidelines for Choosing an Experiment:

The following is a list of guidelines that you should follow when selecting an experiment from the *Organic Syntheses* reference text. These guidelines are similar to those followed by the chemistry department when selecting, designing and adapting experiments for your use.

- Does the experiment have educational benefits?

  Does it illustrate useful chemical principles?
  Do you find it interesting?
  Do you think future students will find it interesting?

- Is this experiment safe?

  Does it involve the generation or use of dangerous chemicals (i.e., toxic gases, strong acids, carcinogens)?
  Are high temperatures (> 200°C) required?
- Is this experiment practical?

  Are all reagents available commercially, and available at a reasonable price (i.e., < $5.00/gram)?
  Can the experiment be conducted within a four-hour laboratory period?
  Does it require specialized equipment?
  Are anhydrous conditions required?

## Laboratory Report (due week 2):

Each team will submit ONE joint laboratory report.
Your laboratory report should contain the following sections:

- Purpose
- Safety Warnings
- Chemical Reaction
- Scaled Down Procedure

Read Section 7.6 in *Techniques in Organic Chemistry*, and then modify the *Organic Syntheses* procedure to a scale of 0.5 –1 grams.

- Purchase Order Form

You are required to submit both an electronic and hard copy of this form, detailing all the chemicals and solvents you will require for your experiment. A sample copy of this form can be found on the next page of this packet. An electronic copy is available via Angel.

When purchasing chemicals, first check to see if they are available from Acros at www.acros.com. If not, check Sigma-Aldrich at www.sigmaaldrich.com. Plan to only order enough of each reagent so that your team together can conduct a total of six repetitions of your scaled-down procedure. Sample Purchase Order Form.

| Student 1 | Student 2 | Day/ Instructor | Org Syn Vol | Org Syn Page | Org Syn Year |
|---|---|---|---|---|---|
| | | | | | |

| | Manufacturer | Chemical Name | Registry # (CAS #) | Catalog # | Unit Cost (include units) | Desired Quantity (include units) | Total Cost |
|---|---|---|---|---|---|---|---|
| Example | Acros | Benzophe none | 119-61-9 | 219680500 | $37.80/ 50 grams | 100 grams | $75.60 |
| Reagent 1 | | | | | | | |
| Reagent 2 | | | | | | | |
| Reagent 3 | | | | | | | |
| Reagent 4 | | | | | | | |
| Reagent 5 | | | | | | | |
| Solvent 1 | | | | | | | |
| Solvent 2 | | | | | | | |
| Solvent 3 | | | | | | | |
| Solvent 4 | | | | | | | |

*Student Instructions: Weeks 11 & 12*
***Organic Syntheses* Project**

## Overview:

You have two weeks to complete this project. Your goal is to successfully scale down the *Organic Syntheses* procedure and come up with an optimized procedure that can be utilized by future classes of organic laboratory students.

A successful procedure should achieve both high purity and high yield. Purity must be demonstrated via IR, as well as melting point or refractive index. Yield and percent yield must also be ascertained.

## Progress Report (due week 12):

Each individual student must submit his own report.
The report should contain the following sections.

- Chemical Reaction
- Procedure Followed
- Preliminary Results
- Discussion

  Discuss what worked, what didn't work, what challenges arose.

  Describe what, if any, procedural modification were made.

  Describe any unexpected circumstances or findings.

- Procedural Plan for Week 12

  Describe what revisions or modifications will be implemented during week 12.

## Final Report (due week 14):

Each individual student must submit his own report.
The report should contain the following sections:

- Chemical Reaction
- Optimized Procedure
- Final Results

actual yield, theoretical yield, percent yield

experimental melting point or refractive index, literature melting point or refractive index

IR peaks of significance

- Discussion

  Compare your results (and procedure) to that of your teammates.

  Discuss what worked, what didn't work, what challenges arose.

  Describe what, if any, procedural modification were made.

  Describe any unexpected circumstances or findings.

- Conclusions

  Was the experiment successful? Why or why not?

  Should the chemistry department adopt the experiment? What or why not?

- *Optional*

  Do you think the *Organic Syntheses* project should be continued in the future? Why or why not?

- Appendix: Labeled IR spectrum

## APPENDIX B: INTERVIEW PROTOCOL

1a. I am interested in learning (more) about your experiences with the Orgsyn project. How did your project go?

- In what ways was it successful or unsuccessful?
- What challenges did you face?

1b. How did you feel about the project?

- How confident were you that your project would work?
- What aspects of it did you enjoy? What did you find frustrating?
- How did the *Orgsyn* project compare to your other organic lab experiences this year?
- Do you think we should do this project again next year? Why or why not?

2a. How did you feel about working as part of a group?

- What were the ways in which it was beneficial/not beneficial?
- What were the ways in which your group worked well/didn't work well together?

2b. How did you and your group-mates go about choosing your project?

- Did you do any searching (online) in advance?
- Were you looking for any particular type of reaction?

3. I see from the survey that you have had/have not had prior experience doing research. How did the *Orgsyn* project compare to your ideas about/experience with real research?

   - What kinds of research experiences have you had?
   - Do you think college labs should be more open-ended/research-like? Why or why not?

4. I am interested in how mapping out how you go about choosing your courses in different subject areas. Can you tell me in which subject areas do you choose courses just to get an easy A, and in which subject areas do you choose more challenging courses because of an interest or a desire to learn?

   - I'm thinking in terms of premed requirements, your major, general IJ requirements, your Jewish studies program.
   - Which Jewish studies program are you in?
   - How many years are you planning to spend on campus?
   - I see that you've taken xxx honors courses. Can you tell me which courses and tell me about them?

5. Is there anything that hasn't come up that you would like to share with me?

## REFERENCES


Anderson, G. L., Herr, K., & Hinlen, A. S. (1994). *Studying your own school: An educator's guide to qualitative practitioner research.* Thousand Oaks, CA: Corwin Press.

Barab, S. A., Hay, K. E., Barnett, M., & Keating, J. F. (2000). Virtual solar system project: Building understanding through model building. *Journal of Research in Science Teaching, 37*(7), 719–756.

Barron, B. J. S., Schwartz, D. L., Vye, N. J., Moore, A., Petrosino, A., Zech, L., & Bransford, J. D. (1998). Doing with understanding: Lessons from research on problem- and project-based learning. *The Journal of the Learning Sciences, 7*(3 & 4), 271–311.

Blumenfeld, P. C. (1992). Classroom learning and motivation: Clarifying and expanding goal theory. *Journal of Educational Psychology, 84*(3), 272–281.

Blumenfeld, P. C., Kempler, T. M., & Krajcik, J. (2006). Motivation and cognitive engagement in learning environments. In R. K. Sawyer (Ed.), *The Cambridge handbook of the learning sciences* (pp. 475–488). New York, NY: Cambridge University Press.

Blumenfeld, P. C., Marx, R. W., Patrick, H., Krajcik, J., & Soloway, E. (1997). Teaching for understanding. In B. J. Biddle, T. L. Good & I. F. Goodson (Eds.), *International handbook of teachers and teaching* (Vol. 2, pp. 819–878). Dordrecht, The Netherlands: Kluwer Academic Publishers.

Blumenfeld, P. C., Soloway, E., Marx, R. W., Krajcik, J., Guzdial, M., & Palincsar, A. S. (1991). Motivating project-based learning: Sustaining the doing, supporting the learning. *Educational Psychologist, 26*(3 & 4), 369–398.

Carr, W. (2006). Philosophy, methodology and action research. *Journal of Philosophy of Education, 40*(4), 421–435.

Chisholm, M. G. (1975). Organic chemistry laboratory: A realistic view of problem-solving. *Journal of Chemical Education, 52*(11), 739–741.

Cochran-Smith, M., & Lytle, S. L. (1992). *Inside/Outside: Teacher research and knowledge.* New York, NY: Teachers College Press.

Cooper, M. M., & Kerns, T. S. (2006). Changing the laboratory: Effects of a laboratory course on students' attitudes and perceptions. *Journal of Chemical Education, 83*(9), 1356–1361.

Cordova, D. I., & Lepper, M. R. (1996). Intrinsic motivation and the process of learning: Beneficial effects of contextualization, personalization, and choice. *Journal of Educational Psychology, 88*(4), 715–730.

Csikszentmihalyi, M. (1975). *Beyond boredom and anxiety: The experience of play in work and games.* San Francisco, CA: Jossey-Bass.

Csikszentmihalyi, M. (1997). *Finding flow: The psychology of engagement with everyday life.* New York, NY: Perseus Books, L.L.C.

Davis, D. S., Hargrove, R. J., & Hugdahl, J. D. (1999). A research-based sophomore organic chemistry laboratory. *Journal of Chemical Education, 76*(8), 1127–1130.

Deci, E. L. (1992). The relationship of interest to the motivation of behavior: A self-determination theory perspective. In K. A. Renninger, S. Hidi, & A. Krapp (Eds.), *The role of interest in learning and development* (pp. 43–70). Mahwah, NJ: Lawrence Erlbaum.

Deci, E. L., & Flaste, R. (1995). *Why we do what we do: Understanding self-motivation.* New York, NY: Penguin Books.

Dewey, J. (1910). Science as subject matter and as method. *Science, 31*(787), 121–127.

Dobrev, A. (1996). "Course thesis" as end of the laboratory experimental program. *Journal of Chemical Education, 73*(9), 856.

Doyle, M. P. (1973). A new approach to organic laboratory project. *Journal of Chemical Education, 50*(5), 358.

Driver, R., H. (1994). Constructing scientific knowledge in the classroom. *Educational Researcher, 23*(7), 5–12.

Edelson, D. C. (2001). Learning-for-use: A framework for the design of technology-supported inquiry activities. *Journal of Research in Science Teaching, 38*(3), 355–385.

Edelson, D. C., Gordin, D. N., & Pea, R. D. (1999). Addressing the challenges of inquiry-based learning through technology and curriculum design. *The Journal of the Learning Sciences, 8*(3–4), 391–450.

Gallet, C. (1998). Problem-solving teaching in the chemistry laboratory: Leaving the cooks... *Journal of Chemical Education, 75*(1), 72–77.

Gay, A. (2004). *"No one does this for fun": Contextualization and process writing in an organic chemistry laboratory course*. Unpublished dissertation, Teachers College, Columbia University, New York.

Glagovich, N. M., & Swierczynski, A. M. (2004). Teaching failure in the laboratory. *Journal of College Science Teaching 34*(6), 45–47.

Hennessy, S. (1993). Situated cognition and cognitive apprenticeship: Implications for classroom learning. *Studies in Science Education, 22*(1), 1–41.

Herman, P., & Gomez, L. M. (2006, June). *Motivation in project-based classrooms: New measures better coupled to students' experiences.* Paper presented at the 7th International Conference of the Learning Sciences, Indiana University, Bloomington, IN.

Hickey, D. T. (1997). Motivation and contemporary socio-constructivist instructional perspectives. *Educational Psychologist, 32*(3), 175–193.

Hidi, S., & Renninger, K. A. (2006). The four-phase model of interest development. *Educational Psychologist, 41*(2), 111–127.

Hiegel, G., & Belloli, R. (1971). Independent synthesis project in the organic chemistry laboratory. *Journal of Chemical Education, 48*(12), 825–826.

Higginbotham, T., Anderson, J., Rosca, C., Jencunas, D., Copman, S., & Zinkowski, J. (2006, June–July). *The story of one urban high school's efforts to improve student attitudes, motivation, self-efficacy and perceptions of self, school, and science through project-based science instruction.* Paper presented at the 7th International Conference of the Learning Sciences, Indiana University, Bloomington, IN.

Horowitz, G. (2003). A discovery approach to three organic laboratory techniques: Extraction, recrystallization, and distillation. *Journal of Chemical Education, 80*(9), 1039–1041.

Horowitz, G. (2007). The state of organic teaching laboratories. *Journal of Chemical Education, 84*(2), 346–353.

Horowitz, G. (2010). It's not always just about the grade: Exploring the achievement goal orientations of *pre-med* students. *Journal of Experimental Education. 78*(2), 215–245

Kanter, D. E., & Reiser, B. J. (2007, April). *Designing project-based inquiry science curricula for meaningful understanding of content.* Paper presented at the Annual Conference of the American Educational Research Association. Chicago, IL.

Kanter, D., & Schreck, M. (2006). Learning content using complex data in project-based science: An example from high school biology in urban classrooms. *New Directions for Teaching and Learning, 108,* 77–91.

Katchadourian, H. A., & Boli, J. (1985). *Careerism and intellectualism among college students: Patterns of academic and career choice in the undergraduate years.* San Francisco, CA: Jossey-Bass.

Kempler, T. M. (2006). *Optimizing students' motivation in inquiry-based learning environments: The role of instructional practices.* Unpublished doctoral dissertation, University of Michigan, Ann Arbor.

Kharas, G. B. (1997). A new investigative sophomore organic laboratory involving individual research project. *Journal of Chemical Education, 74*(7), 829–831.

Krajcik, J., Blumenfeld, P. C., Marx, R. W., Bass, K. M., Fredricks, J., & Soloway, E. (1998). Inquiry in project-based science classrooms: Initial attempts by middle school students. *The Journal of the Learning Sciences,* 7(3&4), 313–350.

Kovacic, P. (1972). Special project in the undergraduate organic laboratory. *Journal of Chemical Education, 49*(12), 846.

Krapp, A., Hidi, S., & Renninger, K. A. (1992). Interest, learning and development. In K. A. Renninger, S. Hidi, & A. Krapp (Eds.), *The role of interest in learning and development* (pp. 3–19). Mahwah, NJ: Lawrence Erlbaum.

Kroll, L. (1985). Teaching the research process via organic chemistry lab project. *Journal of Chemical Education, 62*(5), 516–518.

Lepper, M. R. (1988). Motivational considerations in the study of instruction. *Cognition and Instruction, 5*(4), 289–309.

Lepper, M. R., & Henderlong, J. (2000). "Extrinsic" versus "intrinsic" motivation reconsidered. In C. Sansone & J. M. Harackiewicz (Eds.), *Intrinsic and extrinsic motivation: The Search for optimal motivation and performance* (pp. 257–307). New York, NY: Academic Press.

Linn, M., Clark, D., & Slotta, J. (2003). WISE design for knowledge integration. *Science Education, 87*(4), 517–538.

Lipstein, R. L., & Renninger, K. A. (2007). "Putting things into words": The development of 12-15-year-old students' interest for writing. In S. Hidi & P. Boscolo (Eds.), *Writing and motivation* (pp. 113–140). Boston, MA: Elsevier.

Malone, T. W. (1981). Toward a theory of intrinsically motivating instruction. *Cognitive Science, 5*(4), 333–369.

Malone, T. W., & Lepper, M. R. (1987). Making learning fun: A taxonomy of intrinsic motivation for learning. In R. E. Snow & M. J. Farr (Eds.), *Aptitude, learning and instruction* (Vol. 3, pp. 223–253). Hillsdale, NJ: Lawrence Erlbaum.

Maltbia, T. (2001). *The journey of becoming a diversity practitioner: The connection between experience, learning and competence*. Unpublished doctoral dissertation, Teachers College, Columbia University, New York.

Marx, R. W., Blumenfeld, P. C., Krajcik, J., Fishman, B. J., Soloway, E., Geier, R., & Tal, R. T. (2004). Inquiry-based science in the middle grades: Assessment of learning in urban systematic reform. *Journal of Research in Science Teaching, 41*(10), 1063–1080.

Marx, R. W., Blumenfeld, P. C., Krajcik, J., & Soloway, E. (1997). Enacting project based science. *The Elementary School Journal, 97*(4), 341–358.

Means, B. (1998). Melding authentic science, technology, and inquiry-based teaching: Experiences of the GLOBE program. *Journal of Science Education and Technology, 7*(1), 97–105.

Meece, J. L., Blumenfeld, P. C., & Hoyle, R. H. (1988). Students' goal orientation and cognitive engagement in classroom activities. *Journal of Educational Psychology, 80*(4), 514–523.

Mistler-Jackson, M., & Songer, N. B. (2000). Student motivation and internet technology: Are students empowered to learn science? *Journal of Research in Science Teaching, 37*(5), 459–479.

Mohrig, J. R. (2004). The problem with organic chemistry labs. *Journal of Chemical Education, 81*(8), 1083–1085.

Newton, T., Tracy, H. J., & Prudente, C. (2006). A research-based laboratory course in organic chemistry. *Journal of Chemical Education, 83*(12), 1844–1849.

Nolen, S. B. (1988). Reasons for studying: Motivational orientations and study strategies. *Cognition and Instruction, 5*(4), 269–287.

Nolen, S. B. (1996). Why study? How reasons for learning influence strategy selection. *Educational Psychology Review, 8*(4), 335–351.

Nolen, S. B. (2001). Constructing literacy in the kindergarten: Task structure, collaboration, and motivation. *Cognition and Instruction, 19*(1), 95–142.

Organic syntheses. (1921–present). *Organic Syntheses Collective Volumes.* New York, NY: John Wiley and Sons.

Paris, S. G., & Turner, J. C. (1994). Situated motivation. In P. R. Pintrich, D. R. Brown & C. E. Weinstein (Eds.), *Student motivation, cognition, and learning: Essays in honor of Wilbert J. McKeachie* (pp. 213–237). Hillsdale, NJ: Lawrence Erlbaum Associates.

Patall, E. A., Cooper, H., & Robinson, J. C. (2008). The effects of choice on intrinsic motivation and related outcomes: A meta-analysis of research findings. *Psychological Bulletin, 134*(2), 270–300.

Pickering, M. (1988). A physical chemist looks at organic chemistry lab. *Journal of Chemical Education, 65*(2), 143–144.

Pienta, N. J., Regitz, C., Richards, J., & Sorrell, T. N. (1993). Curriculum and experiments in an organic honors laboratory: A model for parallel honors and regular sections. *Journal of Chemical Education, 70*(10), 841–843.

Pintrich, P. R. (1991) Editor's comment. *Educational Psychologist, 26*(3 & 4), 199–205.

Pintrich, P. R., & Schunk, D. H. (2002). *Motivation in education: Theory, research and applications.* Upper Saddle River, NJ: Pearson Education.

Pitts, V. M. (2006). *Do students buy in? A study of goal and role adoption by students in project-based curricula.* Unpublished doctoral dissertation, Northwestern University, Evanston.

Pitts, V. M., & Edelson, D. C. (2004, June). *Role, goal, and activity: A framework for characterizing participation and engagement in project-based learning environments.* Paper presented at the 6th International Conference of the Learning Sciences, University of California, Los Angeles.

Polman, J. L. (2000). *Designing project-based science: Connecting learners through guided inquiry.* New York, NY: Teachers College Press.

Poon, T., McIntyre, J. P., Eller C. F., Eller, L. R., Jones, K. M., Massello, W., ... Wilkes, E. E. (1999). The preparation of a UV-light-absorbing polymer: A project-oriented laboratory experiment for the introductory organic chemistry curriculum. *Journal of Chemical Education, 76*(11), 1523–1524.

Renninger, K. A., Bachrach, J. E., & Posey, S. K. E. (2008). Learner interest and achievement motivation. In M. L. Maehr, S. A. Karabenick, & T. C. Urdan (Eds.), *Advances in motivation and achievement: Social psychological perspectives* (Vol. 15, pp. 461–491). Bradford, UK: Emerald Group.

Rivet, A. E., & Krajcik, J. S. (2004). Achieving standards in urban systemic reform: An example of a sixth grade project-based science curriculum. *Journal of Research in Science Teaching, 41*(7), 669–692.

Ruttledge, T. R. (1998). Organic chemistry lab as a research experience. *Journal of Chemical Education, 75*(12), 1575–1577.

Ryan, R. M., & Deci, E. L. (2000). Self-determination theory and the facilitation of intrinsic motivation, social development, and well-being. *American Psychologist, 55*(1), 68–78.

Schraw, G., Flowerday, T., & Lehman, S. (2001). Increasing situational interest in the classroom. *Educational Psychology Review, 13*(3), 211–224.

Schwab, J. J. (1964). The teaching of science as enquiry. In J. J. Schwab (Ed.), *The teaching of science* (pp. 2–103). Cambridge, MA: Harvard University Press.

Shaffer, D. W., & Resnick, M. (1999). "Thick" authenticity: New media and authentic learning. *Journal of Interactive Learning Research, 10*(2), 195–215.

Sherin, B., Kanter, D., Schwarz, J., Stieff, M., Herman, P., & MacKenzie, S. (2002, October). *Conceptual dynamics in project-based science.* Paper presented at the 5th International Conference of the Learning Sciences, Seattle, Washington.

Sivan, E. (1986). Motivation in social constructivist theory. *Educational Psychologist,* 21(3), 209–233.

Songer, N. B. (1996). Exploring learning opportunities in coordinated network-enhanced classrooms: A case of kids as global scientists. *The Journal of the Learning Sciences, 5*(4), 297–327.

Songer, N. B. (1998). Can technology bring students closer to science? In B. J. Fraser & K. G. Tobin (Eds.), *International handbook of science education* (Vol. I, pp. 333–347). Dordrecht, The Netherlands: Kluwer Academic.

Tobias, S. (1990). *They're not dumb, they're different: Stalking the second tier.* Tucson, AZ: Research Corporation.

Turner, J. C. (1995). The influence of classroom contexts on young children's motivation for literacy. *Reading Research Quarterly, 30*(3), 410–441.

Vygotsky, L. S. (1978). *Mind in society: The development of higher psychological processes.* Cambridge, MA: Harvard University Press.

Wade, L. G. (1979). Chemistry without a cookbook: An effective alternative. *Journal of Chemical Education, 56*(12), 825–826.

Williams, M., & Linn, M. (2002). WISE inquiry in fifth grade biology. *Research in Science Education, 32*(4), 415–436.

CHAPTER 12

# PROBLEM SOLVING BY DEVELOPING GUIDED RESEARCH IN INTRODUCTORY UNIVERSITY PHYSICS COURSES

**Jenaro Guisasola, Mikel Ceberio, and Kristina Zuza**


## ABSTRACT

Compared to traditional instruction on solving problems, this chapter proposes a methodology known as "problem solving by developing guided research" that is framed within the wide range of problem-solving proposals in inquiry-based problem solving. This chapter firstly aims to investigate how implementing a problem solving strategy based on developing guided research can help students to improve their skills in scientific methodology procedures to solve problems. Secondly, it aims to provide evidence of improvements in problem solving within experimental groups compared to groups of students following traditional teaching, and thirdly, to explore the effect of the problem-solving proposal on students' attitudes. Evidence is provided from the results of implementing this methodology among first-year university engineering students.


*Research Based Undergraduate Science Teaching*, pp. 379–402

Physics education usually states that solving problems leads to understanding physics and is a reliable strategy to prove the learning achieved by students (Maloney, 2011). Research, however, has repeatedly shown that many students in university general physics courses who use traditional teaching formats learn to solve the type of quantitative problems found at the end of each textbook chapter. These students, however, usually are incapable of explaining the meaning of their own numerical solutions to the problems. In this respect, McDermott (1991) says that successfully calculating numerical solutions does not imply a corresponding level of conceptual understanding. Traditional instruction in physics courses focused on problem solving ignores important procedure-based goals and usually looks at mathematical operations (Dufresne & Gerace, 2004; Guisasola, Ceberio, & Almudí, & Zubimendi, 2006). This traditional approach usually leads to poor achievement in terms of developing skills and procedures associated with problem solving and the meaning of physics (Dancy & Henderson, 2007; Guisasola, Ceberio, & Zubimendi, 2006; Huffman, 1997; Van Heuvelen, 1991). As Etkina, Karelina, and Ruibal-Villasenor (2008) propose, "in science and engineering education, students have to acquire not only conceptual and qualitative understanding of physics principles but also the science process abilities that are needed to construct explanatory models" (p. 1). It will be necessary to design learning sequences that do not just explain the problem's numerical resolution but also utilize a teaching strategy giving students the opportunity to use scientific procedures such as putting forward a hypothesis about the relevance of a model for the problem proposed, suggesting a number of different strategies to explain it, and justifying the answer based on evidence.

The aims of this chapter are, firstly, to investigate how implementing the problem-solving strategy based on developing guided research (Guisasola, Furió, & Ceberio, 2008) can help students to learn scientific methodology procedures to solve problems. Secondly, it aims to provide data to determine if there have been any improvements in problem solving among experimental groups compared to groups of students following traditional first-year engineering teaching, and thirdly, it will explore how the problem solving by developing guided research proposal affects students' attitudes.

The acquisition development process for the elements of scientific methodologies and for understanding concepts is explored within a theoretical framework based on a problem-based learning in small groups focus (PBL) (Duch, 1996; Gil & Martínez-Torregrosa, 1983; Hsu, Brewe, Foster, & Harper, 2004; Raine & Collett, 2003; Van Kampen, Banahan, Kelly, McLoughlin & O'Leary, 2004) that places importance on prior knowledge and is developed within the context of situations close to reality (Savery & Duffy, 1995).

## THEORETICAL FRAMEWORK

Researchers have demonstrated the restrictions of traditional teaching on problem solving among students. University students do not usually find it difficult to use mathematics and formulas, not surprising considering that their instruction focused on these components (Kim & Pak, 2002). After solving a large array of problems, however, the most common alternative conceptions detected by the authors still remain intact. Concern for university students' poor learning in terms of problem solving, plus their weak understanding of concepts and laws of physics, have led to developing different teaching strategies for problem solving (Jonassen, 2007; Maloney, 2011; Taconis, Ferguson-Hessler, & Broekkamp, 2001).

It has been frequently mentioned that teaching paper-and-pencil problem solving does not usually look at what a problem is, and so might limit the strategies used. The research community has tackled this issue, and, in accordance with Garret (1987), the concept of problem is defined generically as an uncertain situation that leads anyone in this situation to behave in a way (problem solving) that attempts to discover a solution (result) and so reduce the tension inherent in this uncertainty.

A set of research projects on problem solving by experts and novices used qualitative analysis identifying successive representations of the problem to establish the characteristics of experts' problem solving (Larkin, 1981; Larkin & Reif, 1979). This approach assumes that expertise implies optimum use of cognitive resources and methodological skills available in the discipline (Ferguson-Hesler & De Jong, 1990; Hegarty, 1991). Within this line of research, Van Heuvelen (1991) suggests explicit teaching of problem solving, including qualitative analysis and multiple representation, in the method known as Overview, Case Study (OCS), used in introductory physics courses at the university level. Reif (1995) suggests that the explicit method to be taught for problem solving should be divided into three main phases: (1) initial analysis of the problem, including multiple representations of the situation; (2) construction of a solution making use, if necessary, of dividing the problem up into sub-problems; and (3) checking the result. In accordance with Maloney (2011), the main contribution from this line of research is the quantity of information on how people solve problems, and its main limitation is that much of the research in this line examining experts and novices has not been carried out in the classroom context.

Other types of teaching strategies have been developed that consider that comprehensive learning is achieved by interaction with the environment and that knowledge evolves through social interaction and evaluation of the feasibility of individual knowledge (Savery & Duffy, 1995). From this perspective, Qin, Johnson, and Johnson (1995) suggested using the problems to structure learning, using the teacher as a meta-cognitive trainer, and working together in groups. The structure of problems solved with this approach present the following characteristics: (1) the initial situation approached does not provide all the information required to come to a solution; (2) there is no one path to solve the task proposed; (3) as new information comes in, the problem definition changes; and (4) the students are never completely sure that they have made the right selection from the possible solution options (Arambula-Greenfield, 1996; Gallagher, Stepien, Sher, & Workman, 1995).

In a review of the research on problem solving, Taconis et al. (2001) show that the effective treatments are based on guiding students to acquire appropriate strategies and criteria that can be used to evaluate their own solving process and its result. Effective treatments also provide students with feedback that will allow them to continue acquiring procedures to solve problems. The results from the aforementioned research have contributed towards the design of our focus for problem solving as developing guided research (Boilevin, 2005; Gil & Martínez-Torregrosa, 1983; Guisasola, Ceberio, & Zubimendi, 2006, 2006b). Below, we will briefly describe the structure of the teaching strategy and its implementation.

## STRUCTURE OF THE TEACHING FOR PROBLEM SOLVING BASED ON DEVELOPING GUIDED RESEARCH

Problem solving as developing guided research firstly looks at what the problem is understood to be. Then it concludes that a situation can only be conceived as a problem insofar as the result is unknown and that, a priori, we do not have a solution: a situation for which there are no obvious solutions (Hayes, 1981).

Table 12.1 briefly highlights some of the basic suppositions of this didactic proposal. The characteristics presented should not be taken as heuristic but as methodological guidelines to solve the problem. The strategy used aims to summarize and move forward in the search for consistency between different contributions from educational research on problem solving (Maloney, 2011). It is used both for improving problem-solving skills and to improve understanding of the concepts and laws, strengthening the qualitative aspects of the problem in preparation for algorithmic resolution.

Implementation is developed using the following spiral model: (1) Researchers plan the transposition from the model's generic characteristics to specifying them into classroom activities; in other words, they design a first proposal for a statement and possible solution adapted to the model, as presented in the example from the Appendix; (2) researchers train teachers on the proposed methodology in seminars and teachers work together to solve the problems, taking the opportunity of interacting with each other and with the researchers to analyze the teaching methodology features and define the reference version for each problem to be used in the classroom; and (3) two of the researchers, and the teachers, implement them in the classroom.

**Table 12.1. General Characteristics of Scientific Treatment That Can Be Transferred to the Field of Teaching**

<table>
<tr><th colspan="2">Problem Situation for Which There Are No Obvious Solutions<br>Including aspects from Science, Technology and Society (STS) in the problem statements: Improving the attitude and increasing the interest towards science and its learning</th></tr>
<tr><th>Methodological Guidelines for Solving Problem</th><th>Possible Actions to Carry Out</th></tr>
<tr><td>From an ambiguous problematic situation to a dimensioned problem: Qualitative analysis</td><td>- Clarify the goal in terms of what you're looking for<br>- Size up the situation to model it and simplify it<br>- Outline a theoretical reference framework<br>- Identify variables, search for data<br>- Identify parts of the problem</td></tr>
<tr><td>The role of the hypotheses in problem solving</td><td>- Predict how the system might change<br>- Establish dependency relationships between variables<br>- Analyze borderline cases with special physical relevance</td></tr>
<tr><td>Searching of strategies as attempts for resolution</td><td>- Present a sequential description of actions<br>- Identify the laws and fundamental principles<br>- Assess possible alternative lines</td></tr>
<tr><td>Regarding consistency in the solving process: Analysis of results</td><td>- Analyze the plausibility of the value of the answer<br>- Analyze the dimensional consistency of the answer<br>- Analyze the dependency of the answer with the problem parameters according to the hypotheses<br>- Analyze whether the same answer is obtained with another solving method</td></tr>
<tr><td>From solving a problem to approaching others: New perspectives</td><td>- Tackle at another level of depth<br>- Approach other models<br>- Set new problems</td></tr>
</table>

The proposal for solving problems by developing guided research has to be applied in the classroom by structuring it into small work teams of three or four students (Heller, Keith, & Anderson, 1992; Qin et al., 1995) who interact together, with the teacher, and with other sources of information. Implementing this scientific strategy to tackle a problem in the classroom context can be considered, in terms of time, as divided into four phases. In the first phase, the teacher sets the students an open problem situation that might arouse their interest (Park & Lee, 2004). In their respective work teams, these students attempt to transform the situation set into a problem they can tackle. The teacher guides the discussion between the teams to reach a consensus.

In the second phase, each work team tackles the problem, analyzing it. They start to come up with a first hypotheses and assess any possible solving strategies. They make a first attempt during which interactions occur between the different students in the group and between them and the different sources of information such as textbooks, the Internet, and so on. Information is shared in class, with the teacher highlighting the conceptual and methodological difficulties in solving the problem, guiding the students to keep them on track, and helping them to overcome possible obstacles.

In a third phase, a few days after the classroom information sharing, and after teamwork outside the classroom, the students present and defend their solutions to the problem in front of the rest of the class. The teacher acts as a guide and moderator for the possible discussions that emerge between the students using questions to guide problem solving and to bring out the knowledge and scientific skills involved.

Finally, in the fourth phase, individual written reports are drawn up summarizing the problem solving and verbalizing the main arguments, simplifications, hypotheses, and justifications used for the scientific work.

During the whole process, the teacher presents the same problem to the class and the different student teams work simultaneously on each phase of the process.

## DESCRIPTION OF THE SAMPLE AND THE STRATEGIES

The research reported here was done with first-year engineering students at the University of the Basque Country. All first-year engineering students had previously completed two years of physics in high school. The physics syllabus at a Spanish high school, for students aiming to subsequently take science and engineering degrees, covers mechanics and electromagnetism subjects over a two-year period and with a level similar to A-level courses in the United Kingdom. In engineering degrees in Spain, the number of male students exceeds the number of female students by far. The percentage of men and women is similar in all groups: 80% men and 20% women. The average student age is nineteen. The University of the Basque Country is a public Spanish university, meaning that the government subsidizes 90% of the cost of the studies and the vast majority of students come from middle- and upper middle-class families. Regarding student performance for first year engineering, about 70% of students pass the exams with good marks. The exam is based on standard textbook problems similar to those set at the end of each chapter in the textbooks (e.g., Tipler & Mosca, 2004). The students were randomly divided into five groups. As stated by Ferguson and Takane (1989), random distribution of students who have had the same secondary education is sufficient to ensure the same level of knowledge with the experimental groups. To ascertain the students' initial knowledge of physics, however, we gave sample students from the control and experimental groups questionnaire entitled *Force Concept Inventory* (Hestenes, Wells, & Swackhamer , 1992) at the beginning of the first semester and the questionnaire, *Conceptual Survey of Electricity and Magnetism* (CSEM), developed by Maloney, O'Kuma, Hieggelke, and Van Heuvelen (2001), at the beginning of the second semester. Both questionnaires have been shown to be a reliable assessment tool (Lasry, Rosenfield, Dedic, Dahan, & Reshef, 2011; Planinic, 2006). The results we obtained showed that students' knowledge of physics can be described as memory-based learning of concepts, laws, rules, and procedures, which can be useful for solving standard problems and examination exercises, but does not give them sufficient comprehension to apply these concepts to different contexts and phenomena. There were no meaningful differences in correct answers between the three sample groups. We can, therefore, conclude that all the groups had approximately the same level of academic competence.

The first-year physics students received two hours of lectures, 1.5 hour of problems class, and two hours of lab work per week, for 15 weeks in the area of mechanics (first semester) and another 15 weeks in the area of electromagnetism (second semester). The classes were taught by teachers from the physics department with broad teaching and research experience who, in all cases, had passed at least one public selection exam to become members of the staff at the University of the Basque Country. Both the students who received traditional classes and the experimental students used the same textbook (Tipler & Mosca, 2004) for the lectures and for solving problems at the end of the chapter, with the same program for all of them.

The students in the experimental groups tackled a total of 12 open problems in accordance with the methodology proposed, making use of the hours in problem class but without receiving extra classes. The difference with the control group lies in how their time was divided up between the different activities, prioritizing open problem solving over blackboard solving of standard textbook problems, plus the inclusion of activities consisting of solving problem situations in lectures. This procedure means that the experimental students also used the given methodology to learn the new curriculum concepts. This meant that they could apply some aspects of the methodology in approximately half the sessions of the lectures, thereby making it easier to assume the methodological proposal more quickly.

The study described here had a total sample of 177 first-year engineering students split into five experimental classes with between 30 and 40 students per class (EG groups in the tables). Another 97 students took part, split into three classes that received traditional teaching (CG groups in the tables).

The data were compiled in the third year of implementing this methodology with first year engineering students, from September to June (two semesters), meaning that the teaching staff accumulated significant experience in implanting the methodology. The main instrument used as a data source in this study involved solving written physics problems (Rivard, 1994). We understand that, as this is a context with less pressure, it is more appropriate than an oral session as students can reasonably explain the steps given in the problem solving. Data, however, also have been used from recordings of interviews held with three or four experimental students who voluntarily solved problems out loud. This has given us a further look into the reasoning and procedures students used to solve first-year engineering physics problems. In another work that is being developed, these interviews will be analyzed in greater depth. This chapter will not transcribe parts of the interviews, although it must be understood that the ideas expressed in them coincide with the examples that we are including in the next section, extracted from the written resolutions.

## RESULTS

### Analysis of the Reasoning and Procedures Used by the Experimental Students when Solving Problems

To investigate how the students in the experimental group reasoned when solving physics problems and how they used the concepts and laws involved, we asked them to solve the problems featured in Figure 12.1 in the different examinations for the engineering physics fundaments subject. The exams generally featured statements focused on the theoretical part of physics with a more standard structure as shown in Figure 12.1. We understand that the experimental students, after appropriate instruction, should be capable of applying the methodology to tackle any type of statement, both open and usual.

| |
|---|
| **Mass-rod collision:** We have a rod that can turn freely on a vertical plane around a fixed axis that goes through one of its ends. It is moved from its balance position and it is released. If, when it passes through the vertical, the free end of the rod collides with a small ball of plasticine that is at rest on a horizontal table, leaving them stuck together, determine the speed of the collision output. |
| **Parallel wires:** We have two very long parallel wires with an equal charge although different signs uniformly distributed along their length. Obtain the expression for the electric field at a point situated between the two wires and contained in the plane defined by them. |
| **Moving coil:** A rectangular conducting coil moves horizontally to the right passing from one region where there is a uniform magnetic field $B_1$, perpendicular to the coil and entering another region where there is no magnetic field. What force would have to be applied on the coil to maintain its constant speed? |

**Figure 12.1.** Proposed problems.

The analysis involved a semi-quantitative evaluation based on protocols designed for analyzing written answers in concordance with the scientific capabilities appearing in Table 12.1 (Cook & Reichardt, 1986). In this analysis, we have attempted to divide the answers into partial sections that can be categorized as an "explanation" and that have been grouped into categories (see Table 12.2) to determine the level developed by the students in problem solving (Cortazzi, 1993). The criteria for applying this categorization are distinguished for each problem aiming to compile the main procedure aspects involved in line with correct solving and the characteristics already shown in Table 12.1. One member of the research team first read the students' answers and then derived from them a draft set of description categories for each problem in relation to characteristics of Table 12.1. Three weeks later, the same researcher again read the students' answers and tentatively allocated each answer to one of the draft categories. The intra-rater reliability Kappa coefficient (Cohen, Manion, & Morrison, 2007) was calculated for this period of time, obtaining a mean value of 0.82 for all the problems, which is satisfactory for a confidence level of 95%. After discussing the tentative categories with two other members of the research team, we held a training session in which we examined 10% of the sample. This small sample was selected by the researcher who established the first tentative classification, according to the relevance of some answers for the categories defined. Next, the members of the research team went on to analyze all questionnaires independently achieving the mean Kappa Cohen reliability coefficient of 0.81, which reveals very good consistency in the judges' criteria for setting the categories described. Finally, a meeting was held during which the answers were classified and answer allocations were compared. Any disagreements about category description or answer allocations were solved by referring to the answers, as the only evidence of students' understanding. The focus was on the students' skills taking the students' answer as a whole, rather than on the occurrence of particular statements corresponding to a specific category description. An iterative process was used to produce final category descriptions that reflected similar understanding among answers allocated to each category and the differences between the categories.

The development levels for the scientific skills used to solve problems have been organized cumulatively in order to search for overall resolution consistency, without assessing the cases of right answers that are unconnected and disassociated from a general approach. We do not, for example, consider a student analyzing a result without having made reasoned hypotheses beforehand to be progress. Therefore, when we include a student in a category, he or she has had to satisfactorily meet the requirements of the former category. So, a student from category 4 who "analyzes the result" meets the strategic requirements of category 3, comes up with a hypothesis from category 2, and fits the qualitative analysis from category 1.

After implementing the alternative proposal, a minority of students (see Table 12.2) persisted with resolution habits based on the mathematical equations without describing the situation and without beginning with qualitative analysis.

Level 1 includes the answers that, while carrying out a qualitative analysis of the situation, do not complete a sufficient hypothesis proposal. For example, from among the students who were classified in Group 1 at this level 1, in the analysis of the mass–rod collision problem we find approaches such as:

> It can be seen that we have an elastic collision as the plasticine ball turns with the bar. Before this collision, a circular movement accelerated the bar, as the weight accelerates it. We can suppose that we release the rod from an angle $\theta$ or a height h.

This student produces a correct graphic diagram of the situation where he also indicates the variables relating to the ball mass, the rod mass, the length of the rod, and initial angle of displacement for the rod against the vertical. However, the variables hypothesis that he comes up with does not meet the requirements explicitly given in Table 12.1 relating to making hypotheses, and so he does not make it to level 2.

**Table 12.2. Levels of Using Scientific Skills When Solving Problems (data are given according to the problem that was answered in the exam by each of the 5 EG experimental groups)**

| | | % of Answers | | | | | | | | | | |
|---|---|---|---|---|---|---|---|---|---|---|---|---|
| | | *Mass-rod Collision Problem* | | | *Parallel Wires Problem* | | | | *Moving Coil Problem* | | | |
| *Level* | *Answer category* | *EG1 N=30* | *EG2 N=35* | *EG3 N=38* | *EG2 N=35* | *EG3 N=38* | *EG4 N=40* | *EG5 N=34* | *EG2 N=35* | *EG3 N=38* | *EG4 N=40* | *EG5 N=34* |
| 0 | No qualitative analysis | 13.5 | 15.0 | 12.5 | 10.0 | 8.0 | 12.5 | 20.5 | 5.0 | 4.5 | 5.5 | 9.5 |
| 1 | Qualitative analysis performed | 23.0 | 24.5 | 22.0 | 24.5 | 21.5 | 22.5 | 26.5 | 10.5 | 8.0 | 14.5 | 12.5 |
| 2 | Make hypothesis | 30.0 | 28.5 | 27.0 | 18.0 | 18.0 | 20.0 | 6.0 | 31.5 | 34.0 | 31.5 | 28.0 |
| 3 | Draw up solution strategies | 13.5 | 10.0 | 15.5 | 17.5 | 24.0 | 22.5 | 17.5 | 33.0 | 32.5 | 23.0 | 37.5 |
| 4 | Analyse results | 20.0 | 22.0 | 23.0 | 30.0 | 28.5 | 22.5 | 29.5 | 20.0 | 21.0 | 25.5 | 12.5 |

In the next answer category level (level 2), we have included the students who, having completed a qualitative analysis of the situation, also were able to make reasoned hypotheses on the problem solution. One example of satisfactory hypotheses made by the students when solving problem 1 would be:

> I believe that the result will appear depending on the following magnitudes: The larger the ball mass, the lower the speed after the collision; the larger the rod mass, the more force it will take in the collision; regarding the angle or the initial height, the more we displace it the more speed it will gain before the collision; the length of the rod will affect its inertia momentum.

Once the open problem situation has been analyzed from the qualitative point of view, and after making reasoned hypotheses, the students included in level 3 drew up their justified resolution strategies, considering alternatives. These students wrote assessments such as:

> On the other hand, we should suppose that the rotation of the rod in principle can be solved in different ways. The first, using the momentum of the forces to obtain the value of the angular acceleration. If this is constant, meaning if it remains according to constant magnitudes, it is easy to obtain the speed by kinematics. Another possibility is to apply energy conservation as the only force that the work does, the weight, is conservative (the reactions along the axis are not displaced).

and "I would tackle the second part of the problem using conservation of the angular momentum, because linear momentum is not constant because the bar is held from above and there are strong reactions in the collision."

The highest level of using scientific procedures (level 4) supposes that the qualitative analysis, making hypotheses, and drawing up strategies phases have been completed successfully and finally, once the result has been achieved, coherence is analyzed. An example of the final results analysis phase properly carried out would be for the mass–rod collision problem:

> If we look at the result, we can see that the two masses, the rod length and the angle all appear. The acceleration of gravity also appears that I did not consider before (in the hypotheses) but it seems normal to me because it is constant. It can be seen that if we increase the ball mass, it will be harder for it to move after the collision. If we increase the angle, it is seen that the speed is higher as we had said before (in the hypotheses). It can be seen that the rod mass and length appear in the numerator and the denominator. I believe that this is why it influences both ways, through inertia and through the force that they produce in the collision.

As the categorization levels for the use of the procedures during solving have been organized cumulatively, we would obtain the number of students who performed correct qualitative analysis by adding together the students included in levels 1, 2, 3, and 4. Those who have made hypotheses that are consistent with our criteria will be those obtained by adding the levels 2, 3 and 4 and so on, with the other categories.

In general, the data indicate that, when the solving methodology is developed throughout the course, the use of scientific skills tends to be consolidated. The moving coil problem was set for students in the final exam of the course. Over three quarters of the students made at least one qualitative approach and around half went through the hypothesis procedures and drew up strategies for levels 2 and 3.

## DO EXPERIMENTAL STUDENTS SOLVE PHYSICS PROBLEMS BETTER THAN THE CONTROL STUDENTS?

It could be argued that the improvement in the use of scientific procedures does not have to have a favorable effect on obtaining correct results. In this respect, the analysis presented beforehand does not assure that the students tackling the problems with this methodological focus know how to correctly solve standard problems such as those appearing in textbooks. For this reason, an analysis was run on solving three standard problems (Figure 12.2) set in an exam situation evaluating whether, from the academic point of view and independently of whether scientific procedures were used, the result obtained was correct.

The results compiled in Table 12.3 correspond to solutions from 177 students from the experimental groups (EG) and 97 students from the control groups (CG). The degree of significance to compare whether there were significant differences in the percentage data between the experimental and control groups was devised by calculating the statistic $t$ for the usual level of confidence of 5% or less, having obtained significant statistical differences in all cases. We should not underestimate, however, the poor results in some problems or sections of them such as the problem of the spherical-shell or the second section of the wire-circuit problem.

## STUDENTS' ATTITUDES TOWARD TEACHING PROBLEM SOLVING AS DEVELOPING GUIDED RESEARCH

In order to make a stable methodological change, it is fundamental that there is a change in student attitudes in parallel, not only to obtain better results in problem solving (although this will clearly help the attitude change) but also when evaluating the new methodology more positively (Koballa & Glynn, 2008; Lawrenz & Cohen, 1985). We cannot forget that the student has to *want* to learn. We asked the experimental students about the content and teaching strategies worked on in problem classes, about the way in which the open problems have been worked on in class, and finally, about their overall satisfaction with the solution method. In Table 12.4, we present the main categories identified in the analysis of the students' answers, the percentage corresponding to each of them, and sample answers.

**Spring-masses.** A block with mass $m_1$ on a rough table, is connected to a spring with constant k. The block $m_1$ is attached to another block $m_2$ by a massless string running over a pulley as shown in the figure. If the system is released from rest when the spring is not stretched, what is the speed of the system when the spring has stretched a distance x?

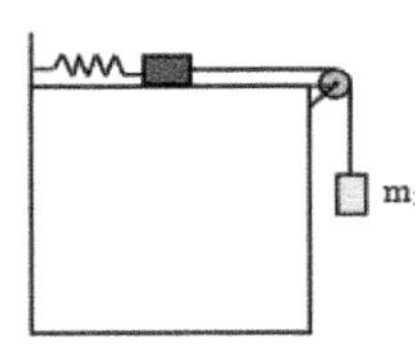


**Spherical-shell.** An initially discharged thick spherical conducting shell, has an internal radius **a** and external radius **b**; in the center of the spherical cavity there is a point charge **+q**.

a) justify how the charge will be distributed on the thick spherical-shell once equilibrium has been achieved.

b) b) Obtain the electric field on points in the regions outside **b**, between **a** and **b** and inside **a**.

**Wire-circuit.** A long straight wire is arranged around the axis carrying a current $I_1$. A rectangular circuit, located as indicated in the diagram, transports a current $I_2$. Determine the magnetic force on each of the segments of current (1) and (2) of this circuit, due only to the effect of the wire.

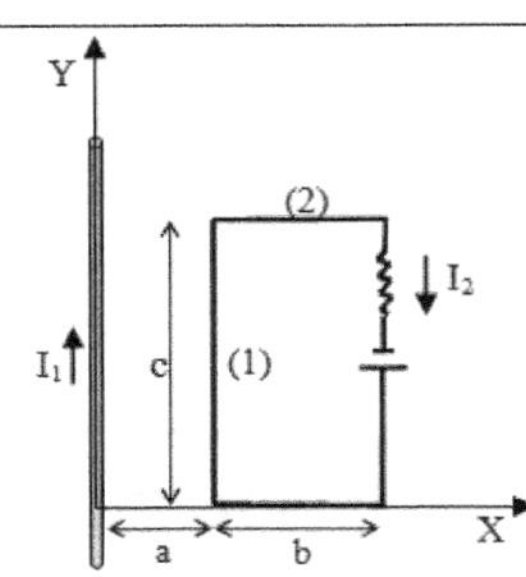


**Figure 12.2.** Proposed standard problems.

**Table 12.3. Comparative Results for the Correct Result Obtained by the Experimental Groups (EG) and Control Groups (CG)**

| *Problem* | | *% Correct Result* | | |
|---|---|---|---|---|
| | | *EG (N = 147)* | *CG (N = 97)* | *t* |
| Spring-masses | | 36.5 | 4.0 | << 0.05 |
| Spherical-shell | | 56.5 | 35.5 | << 0.05 |
| Wire-circuit | (1) | 81.0 | 47.5 | << 0.05 |
| | (2) | 33.5 | 7.5 | << 0.05 |

**Table 12.4. Categorization of Students' Opinions on the Proposal**

| *Category* | *Percentage (N=104)* | *Answer Samples* |
|---|---|---|
| The procedures that have been worked on in open problems (qualitative analysis, making hypotheses, solving strategies, analysis of results) have helped me to think more deeply and get a better understanding of the concepts, laws and principles of physics. | 78% | *"The open problems have helped me to look in greater depth at the laws and formulas ..."*<br>*"It was useful for me to solve the problem qualitatively ... firstly knowing what the problem is like and then seeing which concepts and laws come into play ..."* |
| The way of working in class and the teacher's questions have helped me to solve the problems. | 72% | *"Personally for me, this type of problem has helped me a lot and I've learnt more working in a group, discussing the concepts, etc. The difficulty was pretty high, but this made me think more. In general, I believe in doing this type of problems."*<br>*"Discussing each problem as a group is a good way of thinking about the laws you have to apply"* |
| I've learned nothing new. It's best to do the problems as usual. | 13% | *"You don't have to follow so many steps to solve a problem; it takes a lot of time and I have other subjects to study."* |
| They do not answer | 5% | |

The results provide evidence that the majority of students think that the teaching received helped them to learn and practice some characteristics of the scientific methodology, giving them a better understanding of the concepts and laws of physics and, at the same time, helping them to solve problems better, including standard problems.

## CONCLUSIONS

This study presents evidence that teaching problem solving as developing guided research allowed the majority of students in this sample to develop skills related to scientific methodology and interpret meaningfully the concepts and laws involved in physics problems. When drawing conclusions and implications for teaching, however, it is necessary to bear in mind that the experiments were carried out with a small number of students at a single university. Thus, we cannot produce evidence nor generalize to more general contexts. Our study has not been designed to present conclusive evidence on all university students' difficulties with solving physics problems. There may be difficulties due to other factors not explicitly taken into account in this study. We have checked that the results obtained in this study match results found in other studies using teaching strategies for supporting development of students' scientific skills in problem solving (Arambula-Greenfield, 1996; Gallagher et al., 1995; Qin et al., 1995).

The effect of teaching problem solving as developing guided research into the best use of students' scientific skills has been explored by means of analyzing the answers to open problems. The results from Table 12.2 tell us that a significant percentage of the experimental students used expert type scientific procedures such as qualitative analysis of the problem (Leonard, Dufresne, & Mestre, 1996), making hypotheses (Seroussi, 1995; Wenham, 1993), or analysis of the results (Reif, 1995). In addition, from analyzing their answers, it can be deduced that a majority of these students reasoned and justified, based on evidence, their decisions during resolution.

Eventually, this way of focusing problem solving as a scientific activity could replace the students' usual strategy based on searching data for an ad hoc formula. We have also detected, however, that the exclusively mathematical solving habits using direct application of equations are difficult to change in this sample. A significant proportion of students continued to solve problems as a mechanical application of equations to problem stereotypes. This is perhaps further evidence that, for students to develop the skills required for the scientific field, it is necessary to practice them repeatedly, and to do this, learning environments have to be created where the student is asked to analyze tests or data, compare solutions given by different groups, and justify the strategies employed. Our data suggest that it may be necessary to change teaching strategies over the whole stage of instruction and not only during one academic year.

The implementation of the problem-solving strategy improved these students' results regarding obtaining a correct solution from the academic point of view, in standard textbook problems. The scientifically justified choice of decisions to solve the problem means that the number of correct solutions among experimental students was significantly higher than the control students. We suggest that applying reasoning from the scientific activity improves the understanding of the laws and concepts to be applied in the context of the problem.

The results obtained when implementing this way of solving problems with our students allow us to state that the majority took on a favorable attitude to learning problem solving, demonstrating their abilities to be proactive, participative, and with the will to overcome any learning difficulties. This is important as implementing new problem-solving strategies requires students to actively participate, work together, and have a positive attitude.

Finally, it is necessary to indicate that it is difficult to introduce the methodology at the start of the course. It requires a greater investment of time than traditional teaching. Students must perceive that the alternative methodology is a true reflection of the evaluation tests. For this reason, once the first four weeks of the academic year have passed, an exam will be given that tackles key procedure aspects of the methodology, its implementation steps up the pace, and students become familiar with the new way of working. This implies using the same basic curriculum agreed on by our university department for all introductory physics courses in engineering, although it is true that experimental students dedicate less time to repetitive exercises.

Although this study has not analyzed the implications of extending this form of problem solving to the new teaching staff in university teaching, we think that it could be positive for their professional development. It is necessary to highlight that its application must be accompanied by prior training and mentoring. In general, prior work is required between the group of teachers who discuss and take into account the different possibilities that can arise in the open problem. Training also is required in designing questions and the way of approaching them with students. The way of asking should represent, on the one hand, a challenge to stimulate reflection and search for solutions and, on the other, not stretch beyond the students' cognitive possibilities. This necessary balance represents a challenge for the group of teachers involved that requires taking into account results from research into teaching physics and applying them within the context of problem solving.

# APPENDIX

## Example of an Open Problem

In the last few years interest has increased for jumping off bridges, towers, or helicopters held only by an elastic rope around your ankles. This is known as bungee jumping. Design a safety device for this extreme "sport."

## Teacher's Guide

### *From an Ambiguous Problematic Situation to a Dimensioned Problem: Qualitative Analysis*

The fact of eliminating the statement's data means that decisions have to be made and the situation has to be modeled. Qualitative analysis involves making an effort to seek which variables influence and to select the unknown. The teacher guides the initial discussion by coming up with a question such as the following:

*What would make bungee jumping a safe activity?* (see Figure 12.3)

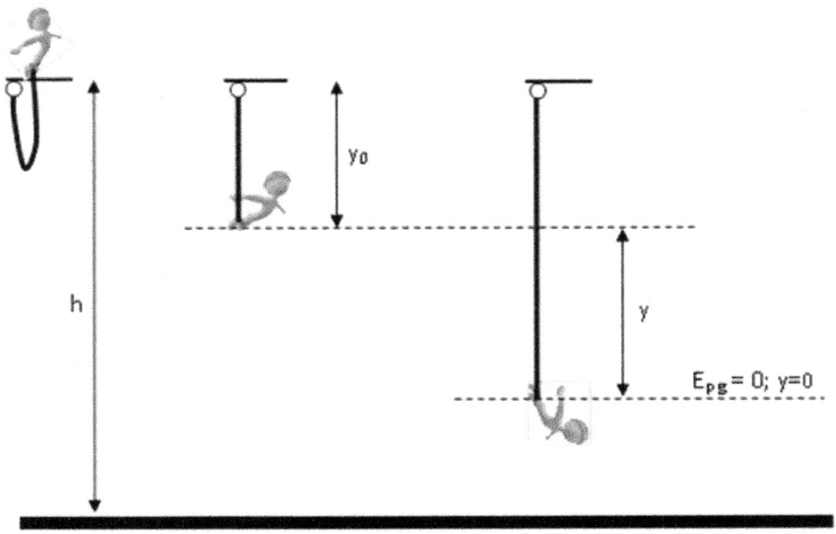


**Figure 12.3.** What would make bungee jumping a safe activity?

The students, working with other students in their team and later as a consequence of interactions with the teacher and students from other teams, contribute arguments along the following line:

a. For the jump to be safe, the sum of the height, $y_0$, that is dropped until the elastic starts to stretch (coinciding with the natural length of the rope) plus the distance $y$ dropped during the stretching of the rubber, has to be less than the initial height h of the jumper over the ground.
b. One way of making the problem operational would be to obtain the total height ($y_t = y_0 + y$) depending on the suspended mass and the elasticity of the rope, and compare it with h, with a wide safety margin, for example, 30% of $h$.
c. Another option would be to assess the elasticity that the chosen rope would have to have for ($y_0 + y$) to be less than ($h - 0.3h$), with the same aforementioned safety margin.
d. In order to simplify the calculations, we consider that the rope responds to Hooke's law ($F = -kx$) and that its mass is negligible. In addition, we consider that the jumper with mass m has a height that is much less than the height $h$ of the jump, so we can treat it as a point particle. We can also say that air resistance is negligible.
e. In accordance with our modeling, at any point during the fall while the elastic rope is being stretched, the interactions on the jumper will be the gravitational force that the earth exerts on him and the elastic force with which the rope (spring) pulls on him from above.

### *The Role of the Hypotheses in Problem Solving*

The hypotheses are what focus and guide resolution, indicating the parameters to take into account. It is the hypotheses and the whole body of knowledge that they are based on that will allow us to analyse the result and the whole process.

It is also true that sometimes, even very often, students introduce "incorrect" ideas when they formulate hypotheses, but this, far from being negative, maybe constitutes the most efficient way of bringing to light alternative conceptions and reasoning.

When analyzing the bungee jump, it is the actual students who, encouraged by the teacher (*On which factors will the height dropped depend?* they might ask for example), speculate with hypothetical reasoning such as the following:

f. Logically, the longer the rope, the greater the distance $y_0$ that the jumper free-falls and the greater $y_t$ will be.

g. The greater the mass m of the jumper, the more weight will be exerted on the rope (the spring in our model) and the greater the distance y that this will stretch, and so $y_t$ will be greater.
h. On the other hand, it is important to assess the elasticity of the rope that, in our simplification, we associate with a spring. The greater the value for $k$, the less the stretch y of the spring will be for the same force exerted on it and so $y_t$ will be less.

### In Search of Strategies as Attempts for Resolution

The solving strategies are not automatically derived from the theoretical principles but are also tentative constructions, working from the qualitative approach made, from the hypotheses formulated and the knowledge that is possessed, but they require imagination and testing. This is the time, therefore, to try and obtain the distance dropped depending on the variables that condition it. For this reason, we set the questions:

*How can we determine this distance?*

The student teams should assess the following options:

i. Alternative 1: If we approach Newton's Second Law, as the elastic force is variable with the position y, the resulting acceleration will also vary with the position. Therefore, the kinematics to be used to obtain the elongation does not correspond to that of a one-dimensional movement with constant acceleration. Here we have a Simple Harmonic Motion whose characteristics will be tackled later in the program of the course.

$$ky - mg =; a = \frac{k}{m} y - g$$

j. Alternative 2: The most appropriate line of solution in this case will be the energy-based strategy. The only forces that do any work during the fall (we have said that air resistance is negligible) are gravitational force and elastic force, both conservative, so that the mechanical energy is conserved during the process. The potential gravitational energy at the moment of the jump (considering the jump from rest) will be transformed into kinetic energy and elastic potential energy during the drop. At the point of maximum elongation of the spring, it will only have potential elastic energy.
k. If we take as level zero for potential gravitational energy the lowest position that the jumper will occupy, we can express:

$$mg(y + y_0) = \frac{1}{2} ky^2 - mgy - mgy_0 = 0$$

The solution for this equation:

$$y = \frac{mg \pm \sqrt{mg^2 + 2kmgy_0}}{k} = \frac{mg \pm \sqrt{mg(mg + 2ky_0)}}{k} \quad (1)$$

***On the Consistency of the Solving Process: Analysis of Results***

The analysis of results from a problem situation is inherent to the scientific activity. The teacher could guide this analysis by setting the students a series of questions that should give way to reasoning of type l, m, and n:

*Is it reasonable that there are two acceptable values for the distance y as indicated by equation (1)?*

l. As, $\sqrt{mg^2 + 2kmgy_0} > mg$ the negative solution is lacking physical sense and so the altitude dropped would be given by:

$$y = \frac{mg + \sqrt{mg(mg + 2ky_0)}}{k} \quad (2)$$

*Are our hypotheses consistent?*

m. In the expression obtained we can appreciate that, as we envisaged, y increases when the suspended mass m increases and when the length of the rope $y_0$ increases. The distance dropped, however, decreases when the spring's elastic constant is increased,

*In which conditions can the jump be considered to be safe?*

n. If, as we envisaged in the initial analysis, we wish to have a safety margin of 30% of height h from which the jump is made, it must be observed that the total height ($y_t = y_0 + y$) is less than or equal to $h - 0.3h$:

$$y_0 + \frac{mg + \sqrt{mg(mg + 2ky_0)}}{k} \leq 0{,}7h$$

***From Solving a Problem to Approaching Others: New Perspectives***

As in any research, the results can bring up new problems. Student and teacher creativity must once again come into play to consider other open perspectives after the solution.

At this point we can ask our students to judge whether the jump is safe in a possible real situation—for example, if the mass of the jumper is 75kg, the rope used has a length of 15m and an elastic constant (considered as a spring) of 50N/m, and the bridge from which he jumps stands 80m over the river.

On the other hand, it is important to highlight that a jump that is safe for one person might not be for another person with greater mass. In this case, it would be possible to take the jump using ropes with other elastic characteristics.

## REFERENCES


Arambula-Greenfield, T. (1996). Implementing problem-based learning in a college science class. *Research in Science Education, 25*(4), 383–393.

Boilevin, J. M. (2005). Enseigner la physique par la situation problème ou par problème ouvert. *Aster, 40*, 13–37.

Cohen, L., Manion, L., & Morrison, K. (2007). *Research methods in education* (6th ed.). London, UK: Routledge/Taylor & Francis Group.

Cook, T. D. & Reichardt, Ch. S. (1986). *Qualitative and quantitative methods in evaluation research*. London, UK: Sage.

Cortazzi, M. (1993). *Narrative analysis*. London, UK: Palmer Press.

Dancy, M., & Henderson, C. (2007). Framework for articulating instructional practices and conceptions. *Physical Review Special Topics-Physics Education Research, 3*(1), 010103.

Duch, B. J. (1996). Problem-based learning in physics: The power of students teaching students. *Science Teaching, 26*, 529–541.

Dufresne, R. J., & Gerace, W. J. (2004). Assessing-to-learn: Formative assessment in physics instruction. *The Physics Teacher, 42*, 428–433.

Etkina E., Karelina A., & Ruibal-Villasenor M. (2008). How long does it take? A study of student acquisition of scientific abilities. *Physical Review Special Topics: Physics Education Research, 4*, 020108.

Ferguson, G. A., & Takane, Y. (1989). *Statistical analysis in psychology and education*. New York, NY: McGraw Hill International Editions/Psychology Series.

Ferguson-Hessler, M., & De Jong, T. (1990). Studying physics texts: Differences in study processes between good and poor performers. *Cognition and Instruction*, 7, 41–54.

Gallagher S. A., Stepien W. J., Sher, B. T., & Workman, D. (1995). Implementing problem-based learning in science classroom. *School Science and Mathematics, 95*(3), 136–146.

Garret, R. M. (1987). Issues in science education: problem solving, creativity and originality. *International Journal of Science Education, 9*, (2), 125–137.

Gil, D., & Martínez-Torregrosa, J. (1983). A model for problem solving in accordance with scientific methodology. *European Journal of Science Education, 5*(4), 447–455.

Guisasola, J, Ceberio, M., & Zubimendi J. L. (2006). University students' strategies for constructing hypothesis when tackling paper-and-pencil tasks in physics. *Research in Science Education, 36*(3), 163–186.

Guisasola J., Ceberic M., Almudi J. M., & Zubimendi J. L. (2006). *La enseñanza de problemas de física en primer curso de universidad* [Physics problem solving teaching at first year of university]. Barcelona, Spain: Editorial Octaedro.

Guisasola, J., Furió, C., & Ceberio, M. (2008). Science education based on developing guided research. In M. V. Thomase (Ed.), *Science education in focus* (pp. 173–201). New York, NY: Nova Science.

Hayes, J. R. (1981). *The complete problem solver*. Philadelphia, PA: Franklin Institute Press.

Hegarty, M. (1991). Knowledge and processes in mechanical problem solving. In R. J. Sternberg & P. A. Frensch (Ed.), *Complex problem solving: Principles and mechanisms* (pp. 253–285). Hillsdale, NJ: Lawrence Erlbaum.

Hestenes, D., Wells, M., & Swackhamer, G. (1992). Force concept inventory. *Physics Teacher, 30*, 141–158.

Heller, P., Keith, R., & Anderson, S. (1992). Teaching problem solving through cooperative grouping. Part 1: Group versus individual problem solving. *American Journal of Physics, 60*(7), 627–636.

Hsu, L., Brewe, E., Foster, T. M., & Harper, K. A. (2004). Resource letter RPS-1: Research in problem solving. *American Journal of Physics, 72* (9), 1147–1156.

Huffman, D. (1997). Effect of explicit problem solving instruction on high school students` Problem-solving performance and conceptual understanding of physics. *Journal of Research in Science Teaching, 34*(6), 551–570.

Jonassen, D. H. (2007). What makes scientific problems difficult? In D. H. Jonassen (Ed.), *Learning to solve complex scientific problems* (Part I, pp. 23–39). Mahwah, NJ: Lawrence Erlbaum.

Kim, E., & Pak, S.-J. (2002). Students do not overcome conceptual difficulties after solving 1000 traditional problems. *American Journal of Physics, 70*(7), 759–765.

Koballa, T. R., & Glynn, S. M. (2008). Attitudinal and motivational constructs in science learning. In S. K. Abell & N. G. Lederman (Ed.), *Handbook of Research on science Education* (pp. 75–102). New York, NY: Routledge.

Larkin, J. H. (1981). Enriching formal knowledge: A model for learning to solve textbook physics problems. In J. R. Anderson (Ed.), *Cognitive skills and their acquisition* (pp. 311–334). Hillsdale, NJ: Erlbaum.

Larkin, J. H., & Reif, F. (1979). Understanding and teaching problem solving in physics. *European Journal of Science Education, 1*(2), 191–203.

Lasry, N., Rosenfield, S., Dedic, H., Dahan, A., & Reshef, O. (2011). The puzzling reliability of the force concept inventory. *American Journal of Physics, 79*(9), 909–912.

Lawrenz, F., & Cohen, H. (1985). The effect of methods classes and practice teaching on students attitudes toward science and knowledge of science processes, *Science Education 69*(1), 105–113.

Leonard, W. J., Dufresne, R. J., & Mestre, J. P. (1996). Using qualitative problem-solving strategies to highlight the role of conceptual knowledge in solving problems. *American Journal of Physics, 64*, 1495–1503.

Maloney, D. (2011). An overview of physics education research on problem solving. In Getting Started in PER, series Reviews in PER 2 (1). Retrieved from http://www.per-central.org/items/detail.cfm?ID=11457

Maloney, D. P., O'Kuma, T. L., Hieggelke, C. J., & Van Heuvelen, A. (2001). Surveying students' conceptual knowledge of electricity and magnetism. *Physics Educational Reearch, American Journal of Physics Supplement, 69*(7), 12–23.

McDermott, L. (1991). Millikan lecture 1990: What we teach and what is learned—closing the gap. *American Journal of Physics, 59*, 301–315.

Park, F., & Lee, L. (2004). Analysing cognitive or non-cognitive factors involved in the process of physics problem-solving in an everyday context. *International Journal of Science Education, 26*(13), 1577–1595.

Planinic, M. (2006). Assessment of difficulties of some conceptual areas from electricity and magnetism using the Conceptual Survey of Electricity and Magnetism. *American Journal of Physics, 74*(12), 1143–1148

Qin, Z., Johnson, D. W., & Johnson R. T. (1995). Cooperative versus competitive efforts and problem solving. *Review of Educational Research, 65*(2), 129–143.

Raine, D. J., & Collett, J. (2003). Problem-based learning in astrophysics. *European Journal of Physics, 24*, S41–S46.

Reif, F. (1995). Millikan lecture 1994: Understanding and teaching important scientific thought processes. *American Journal of Physics, 63*(1), 17–32.

Rivard, L. P. (1994). A review of writing to learn science: implications for practice and research. *Journal of Research in Science Teaching, 31*, 969–983.

Savery, J. R., & Duffy, T. M. (1995). Problem based learning: An instructional model and its constructivist framework. *Educational technology, 35*, 31–36.

Seroussi, D. E. (1995). Heuristic hypotheses in problem solving: An example of conceptual issues about scientific procedures. *Science Education, 79*(6), 595–609.

Taconis, R., Ferguson-Hessler, M. G. M., & Broekkamp, H. (2001). Teaching science problem solving: An overview of experimental work. *Journal of Research in Science Teaching, 38*(4), 442–468.

Tipler, P. A., & Mosca, G. (2004). *Physics for scientists and engineers* (5th ed.). New York, NY: W.H. Freeman and Company.

Van Heuvelen, A. (1991). Learning to think like a physicist: A review of research based instructional strategies. Part II: Design of instructional strategies. *American Journal of Physics, 59*, 891–897.

Van Kampen, P., Banahan, C., Kelly, M., Ellish McLoughlin, E., & O'Leary, O. (2004). Teaching a single physics module through problem based learning in a lecture-based curriculum. *American Journal of Physics, 72*(6), 829–834.

Wenham, M. (1993). The nature and the role of hypotheses in school science investigations. *International Journal of Science Education, 15*(3), 231–240.

CHAPTER 13

# INTEGRATING RESEARCH INTO UNDERGRADUATE COURSES

## Current Practices and Future Directions

**Iris Alkaher and Erin L. Dolan**

**ABSTRACT**

Course-based undergraduate research experiences, or CUREs, are increasingly common because they engage undergraduates in research at schools that are lacking a research infrastructure or that cannot accommodate large undergraduate populations in internship-style research. Course-based undergraduate research experiences are lauded for their scientific and instructional authenticity as they present cognitive demands that learners would encounter in the real world and engage students in scientific practices that encourage them to view themselves as scientists and position themselves as contributors to the scientific body of knowledge. In addition, CUREs may influence students' academic and career paths more than internship-style research experiences, which typically serve to confirm students' prior academic or career choices. Here, efforts to integrate science research experiences into undergraduate courses are reviewed. Theory informing the design and implementation of CUREs is described; CURE models, mechanisms, and impacts are summarized; and an evaluation of research on CUREs is offered, including the quality of the measures and shortcomings or gaps. Results are

*Research Based Undergraduate Science Teaching*, pp. 403–434

described from our own qualitative, interpretive study of how CUREs can be tools for undergraduate self-authorship, specifically: (1) ways in which undergraduates' views about the source of scientific knowledge change as they participate in course-based research, (2) ways in which undergraduates' scientific identities change as they participate in course-based research, and (3) ways in which undergraduates' views about their relationship with science change as they participate in course-based research. The profiles of six undergraduates who represent the range of developmental transformation toward self-authorship are described. The chapter concludes with recommendations for further study and practice of CUREs.

Course-based undergraduate research experiences, or CUREs, are increasingly common because they engage undergraduates in research at schools that are lacking substantial research infrastructure or that simply cannot accommodate large undergraduate populations in internship-style research (Desai, Gaston, Stiles, Stewart, Laine, & Quick, 2008; Wood, 2003). Although they are designed according to many of the same principles as inquiry laboratory activities (Anderson, 2002), CUREs have both science learning and science research goals. Similar to research apprenticeships, CUREs aim to engage students in the processes and practices of science while contributing to the body of scientific knowledge. When implemented at the introductory level, CUREs can engage students in research who may not pursue further education or careers in science (Harrison, Dunbar, Ratmansky, Boyd, & Lopatto, 2011; Marcus, Hughes, McElroy, & Wyatt, 2010). Thus, CUREs may influence students' academic and career paths more than research apprenticeships, which typically serve to confirm students' prior intentions (Hunter, Laursen, & Seymour, 2007; Lopatto, 2004, 2007; Seymour, Hunter, Laursen, & DeAntoni, 2004). Here, we review CURE models and approaches as well as theories informing the design of CUREs and findings from the study of CUREs. We describe our work on CUREs as a mechanism for encouraging undergraduates' developmental transformation toward self-authorship. We end by making recommendations for CURE practice and further study of CUREs.

## CALLS FOR ACTION

The National Science Foundation (NSF) and Howard Hughes Medical Institute in the United States, the Higher Education Funding Council and Higher Education Academy in the United Kingdom, and other stakeholders in undergraduate science education from around the globe are calling for increased engagement of undergraduates in research (e.g., Brewer et al., 2011; Healey & Jenkins, 2009). Between 2007 and 2010, NSF funded

29 projects that involve undergraduates in conducting bench, field, or informatics research in the context of their science courses (McCook, 2011; Wei & Woodin, 2011; Woodin, Carter, & Fletcher, 2010). These calls emphasize the importance of authenticity of learning experiences: activities that present cognitive demands learners would encounter in the real world (Barab & Duffy, 2000; Lave & Wenger, 1991). Participation in authentic scientific practices can help students view themselves as scientists and position themselves as contributors to the scientific body of knowledge (Barab & Duffy, 2000; Chinn & Malhotra, 2002; Rahm, Miller, Hartley, & Moore, 2003; Roth & Barton, 2004). Senge (1994) defined learning environments that allow for the practice of real-world activities as practice fields. The design of meaningful and effective practice fields involves the following principles (summarized in Barab & Duffy, 2000):

- Learners must be actively engaged in domain-related practices.
- Learners should work in collaborative and social environments that enable them to discuss their ideas through a process of continual negotiation.
- Learners must be given and must assume ownership for solving the problem or addressing the issues presented.
- Learners should have many opportunities for reflection during the experience to think about and evaluate their moves in "real time."
- Instructors should provide scaffolding that supports students' reflection on their thinking and actions.
- Learners must be introduced to the context of the problem or issue and its relevance to the society in motivating and challenging ways.

CUREs are thought to follow many of these principles by providing a rich, authentic problem space for student learning (Turvey & Shaw, 1995), while putting students in charge of identifying, at least in part, the focus and purpose of their investigations (Rahm et al., 2003; Roth, van Eijk, Reis, & Hsu, 2008).

## CURE APPROACHES AND OUTCOMES

The scientific problems that CUREs address have several common characteristics. Most importantly, CUREs are structured to fit the constraints of undergraduate classrooms: the scientific methods are scalable in terms of space and expense and streamlined to enable participation by learners with a range of expertise via one-to-many instead of one-to-one mentorship. Many CUREs make use of freely available data and software, such as projects in which students assemble and annotate genome sequences (e.g., Genomics Education Partnership; http://gep.wustl.edu/). Others involve massively parallel observation, such as sequencing and analysis of whole genomes (e.g., GCAT-SEEKquence; http://lycofs01.lycoming.edu/~gcat-seek/) or collection and analysis of ecological or behavioral datasets (e.g., Online Research in Biology; http://www.birds.cornell.edu/orb/investigations). Thus, CUREs offer opportunities of scale, in which scientific questions can be addressed that would otherwise be untenable for a single research group or even through collaboration across research groups (Goodner, Wheeler, Hall, & Slater, 2003; Hatfull et al., 2006; Marris, 2010). In other cases, students do experiments or think about research in ways that experts do not, heading the research in entirely different and potentially fruitful directions (e.g., Owens, Alerding, Crosby, Bandara, Westwood, & Winkel, 2008). Table 13.1 includes a number of examples of CUREs, and more detailed catalogs are available elsewhere (e.g., Wei & Woodin, 2011). In 2011 the NSF has awarded a grant for the establishment of a network of people and programs integrating research experiences into undergraduate courses (http://www.nsf.gov/awardsearch/showAward.do?AwardNumber=1061874). One objective of this project is to generate a web-accessible clearinghouse of CUREs, which is available at: https://curenet.franklin.uga.edu.

The most widely utilized instrument for documenting the outcomes of CUREs is the *Classroom Undergraduate Research Experiences* (CURE) *Survey* (Lopatto, 2010). The CURE survey comprises three elements: an instructor report of the extent to which the learning experience resembles the practice of science research (i.e., the outcomes of the research are unknown, students have some input into the focus or design of the research), a student report of learning gains, and a student report of attitudes toward science. The student portions are series of Likert-type items about students' attitudes toward science and their educational and career interests, as well as students' perceptions of the nature of the learning experience, the nature of science, their learning styles, and the science-related skills they developed as a result of participating in a CURE.

Studies making use of the CURE survey, other Likert-type instruments, or student grade point averages or assignment or exam grades (e.g., Caruso, Sandoz, & Kelsey, 2009; Drew & Triplett, 2008; Goodner et al., 2003; Harrison et al., 2011; Lopatto et al., 2008, Lopatto, 2010; Shaffer et al., 2010) indicate that students who participate in CUREs make many of the same gains as students who participate in more intensive but less scalable research apprenticeships (Hunter et al., 2007; Junge, Quinones, Kakietek, Teodorescu, & Marsteller, 2010; Lopatto, 2004, 2007, 2010; Seymour et al., 2004). Specifically, CURE participants report increased interest in science, preparation for further education or careers in science, abilities to "think and work like a scientist," and understanding of the

**Table 13.1. Examples of Course-Based Undergraduate Research Experiences (see Wei & Woodin (2011) for a detailed list of biology CUREs)**

| *Discipline* | *Project* | *Website* | *References* |
|---|---|---|---|
| Biology | | | |
| Bioinformatics, Genomics, and Synthetic Biology | Bring Your Own Cassava | None available | Siritunga, Montero-Rojas, Carrero, Toro, Velez et al. (in press) |
| | Genome Consortium for Active Teaching | http://www.bio.davidson.edu/projects/gcat/gcat.html | Buonaccorsi, Boyle, Grove, Praul, Sakk, et al. (2011); Campbell, Eckdahl, Fowlks, Heyer, Mays Hoopes, et al. (2006) |
| | Genomics Education Partnership | | Lopatto et al. (2008), Shaffer et al. (2010) |
| | PhageHunters / Science Education Alliance | http://www.hhmi.org/grants/sea | Harrison et al. (2011) |
| | UCLA Undergraduate Consortium for Functional Genomics | http://www.bruinfly.ucla.edu/ | Chen, Call, Beyer, Bui, Cespedes, et al. (2005) |
| | Undergraduate Research in Microbial Genome Analysis | http://www.jgi.doe.gov/education/genomeannotation.html | Ditty, Kvaal, Goodner, Freyermuth, Bailey, et al., (2010); Kerfeld & Simons (2007) |
| | Upper Green River Barcode of Life Project | http://bioweb.wku.edu/faculty/marcus/Barcode.html | Marcus et al. (2010) |
| Ecology | Online Research in Biology | http://www.birds.cornell.edu/orb/ | Wilson, Trautmann, MaKinster, & Barker (2010) |
| Plant Science | PB-CURE [pseudonym] | [removed for blind review] | Authors (2011) |
| Chemistry | | | |
| General and organic chemistry | Center for Authentic Science Practice in Education | http://www.purdue.edu/discoverypark/caspie/ | Russell, Bentley, Wink, & Weaver (2009) |

*(Table continues on next page)*

**Table 13.1. (Continued)**

| *Discipline* | *Project* | *Website* | *References* |
|---|---|---|---|
| Environmental Science | | | |
| Water quality | Community-Based Participatory Research in Indian Country | None available | Cummins, Doyle, Kindness, Lefthand, Bear Don't Walk, et al. (20 |
| Geoscience | | | |
| Igneous petrology | Fort Lewis College Geology 390 | None available | Gonzales & Semken (2006) |
| Geology | Lake Superior State University curriculum | http://www.lssu.edu/geology/newcurriculum | Kelso & Brown (2008, 2009) |

nature and processes of science. A few studies utilized a quasi-experimental design in which assignment or exam grades were compared between students enrolled in CURE versus non-CURE versions of the same course, with CURE participants achieving at a higher rate than nonparticipants (e.g., Marcus et al., 2010; Wink & Weaver, 2008). Few, if any of these studies explore the variables other than CURE participation that could be influencing the outcomes being studied. Little information is available about the instruments used in these studies, including their dimensionality, construct validity, or reliability. We were not able to identify any studies of CUREs using quantitative measures for which validity and reliability information is available.

## THEORETICAL FOUNDATIONS FOR CURES

For the most part, the connection between the scientific focus of a particular CURE and the ongoing work of the scientific community is clearly articulated. The extent to which CUREs are grounded in learning theory is less clear. Sociocultural and situated theories of learning (e.g., Bandura, 1977; Lave & Wenger, 1991; Rogoff, 1990; Vygotsky, 1978) are useful for explaining processes and outcomes of undergraduate involvement in science research, which typically takes the form of an apprenticeship. These theories and their applicability to CUREs and our own research are discussed here.

Sociocultural learning theories emphasize the importance of social interactions to learning, often between a learner and a more knowledgeable individual. The more knowledgeable individual models behaviors for the learner to observe and imitate (Bandura, 1977) and offers feedback and support as the learner "internalizes" these behaviors (Vygotsky, 1978). In the context of science research, target behaviors typically include those common to scientific practice across disciplines, such as designing, conducting, interpreting, and communicating about research, as well as those unique to particular disciplines or sub-disciplines, such as making meaning of e-values when comparing gene sequences. Research groups typically include multiple knowledgeable individuals, such as research staff, graduate students, postdoctoral associates, and the lead investigator (i.e., faculty member), who can model behaviors that span the range of expertise. In CUREs, instructors and classmates offer the most physically immediate models, which may limit the potential for learning. Instructors, for example, may not have specialized or up-to-date expertise in the scientific domain of the CURE they are implementing. Also, classmates may model inappropriate behaviors, or undergraduates may not view the expert behaviors of classmates as such. Bandura (1977) posited that the effectiveness of modeling depends on the learner's attention, motivation, and memory. The authentic nature of research apprenticeships has been shown to enhance students' motivation and improve their attitudes toward science (Hunter et al., 2007; Junge et al., 2010; Lopatto, 2004, 2007), but the evidence from CUREs is more limited and equivocal (e.g., Harrison et al., 2011; Lopatto, 2010; Shaffer et al., 2010).

The community of practice theory presented by Lave and Wenger (1991) is useful for explaining learning through research apprenticeships. Specifically, the faculty head of a research group, the graduate and postdoctoral students, and the undergraduates fit Lave and Wenger's characterization of "triadic group relations" among "masters" (or "old-timers"), "young masters" o(r "journeymen"), and "apprentices" (or "newcomers"), respectively. As individuals develop expertise, they move along the continuum from apprentice to master, engaging in practices more central to the work of the community, specifically, the design, conduct, and communication of research. The group recruits newcomers who learn from its old-timers, and socially reproduces the unit (Brown & Duguid, 1991). Community of practice theory may have more limited applicability to CUREs. Wenger, McDermott, and Snyder (2002) describe levels of participation within a community of practice, from central to peripheral, as coordinator, core, active, and peripheral. In a typical academic research apprenticeship structure, the faculty head plays the role of coordinator as the focal point for the group and organizer of its overarching research agenda. Research technicians, graduate students, and postdoctoral associates constitute the core of the group through their active participation in its work and their role as auxiliaries to the faculty. These layers of community participation are most likely absent in the CURE structure. In apprenticeships, undergraduates become "deeply engaged" (Wenger et al., 2002, p. 56) by choosing to participate and dedicating many hours per week in their apprenticeship work. In contrast, undergraduates enrolled in CUREs may not be aware of the community of practice they are joining and may be enrolling for purposes other than an interest in the community's practice (e.g., the course is required for their major). A disconnect between the focus and practice of a CURE and participating students' interests presents greater concern when the construct of intersubjectivity is considered (Rogoff, 1990). Intersubjectivity is the sharing of focus and purpose between the learner and the more experienced individual. Intersubjectivity is assumed to be inherent to science research groups (i.e., apprenticeship-style research experiences) because members elect to join the group, which is investigating a common, overarching research problem. The motives of undergraduates enrolled in CUREs are likely to be more diverse. Undergraduates who seek out research apprenticeships are likely to value the outcomes of these experiences. Undergraduates who enroll in courses that involve research experiences (i.e., CUREs) may not be interested in learning scientific practices, in the scientific focus of the CURE, or in the research direction that would be most sensible to an expert in the CURE's scientific domain.

Course-based undergraduate research experiences are complex, social, and situated environments, and thus can also be understood from a situated cognition perspective (Greeno, 1997). These experiences aim to engage students in the "ordinary practices of the culture" (Brown, Collins, & Duguid, 1989, p. 34) by using scientific information and materials that are being generated and studied actively by the scientific community and by engaging in discourse with members of that community (Collins, 2006; Lave & Wenger, 1991). Situated learning envisions learning as a "process of enculturation or individual participation in socially organized practices, through which specialized local knowledge, rituals, practices, and vocabulary are developed" (Hennessy, 1993, p. 2). Thus, through CUREs, it is argued that students have opportunities to learn both explicit and tacit knowledge and skills through practice guided by more expert others (Collins, Brown, & Newman, 1989).

In our own research, described below, we begin to explore how a course-based research experience facilitates students' self-authorship (Baxter Magolda, 1992, 2003), including how students define themselves in relation to science. Self-authorship is defined by Baxter Magolda (1992) as a transformation from accepting knowledge as authority to constructing knowledge oneself, and from viewing knowledge as absolute to seeing knowledge as evolving and learning as a lifelong process. Self-authorship also involves a transition from defining self based on others' perceptions to defining self based on internally constructed values. In the case of science class, this transition could be operationalized as a move away from seeing science grades as an indicator of one's potential to learn and practice science, or viewing the body of science as limited to the content of science class. Baxter Magolda (2004) argues that learning should be seen as a "transformation rather than exclusively skill acquisition" (p. 2), and this kind of shift may be necessary for undergraduates to integrate the processes and knowledge of science productively into their professional, personal, and civic lives. Our work illustrates how some undergraduates used a course-based research experience to begin to compose a new role for themselves with respect to science, including their views of personal abilities to construct scientific knowledge and use science in their daily lives even if they did not plan to pursue further education or careers in science.

## CURES AS CONTEXTS FOR SELF-AUTHORSHIP

To begin to understand how CUREs can be tools for undergraduate self-authorship (Baxter Magolda, 1999), we conducted a qualitative, interpretive study of how a diverse group of undergraduates used a course-based research experience to reposition themselves in relation to science. We asked: (1) In what ways do undergraduates' views about the source of scientific knowledge change as they participate in course-based research? (2) In what ways do undergraduates' views of their own scientific identities change as they participate in course-based research? and (3) In what ways do undergraduates' views about their relationship with science change as they participate in course-based research? We intensively describe the profiles of six undergraduates to illustrate the range of transformations we observed.

## CONTEXT AND PARTICIPANTS

The context for this study is a plant biology CURE (PB-CURE; the names of the CURE project and the participants involved in this study are pseudonyms). This CURE was initiated in 2007 at a large public research university in the Mid-Atlantic region of the United States to engage undergraduates in research in the context of their science courses. The goals of the PB-CURE are to: (1) involve undergraduates and their instructors in inquiry that has both science learning and science research objectives; (2) promote student understanding of the relationships among genes, biochemicals, organisms, and ecosystems; and (3) examine the impacts of engaging undergraduates in course-based research. Through PB-CURE, students address the unanswered question of how disabling certain genes in *Arabidopsis* plants, which are widely studied in plant biology, influences the plants' interactions with root and leaf herbivores. With guidance from their instructors, students design and conduct their own investigations to identify differences in herbivory of wild-type plants, which includes all functional genes, versus mutant plants, which include one altered gene. Students report their findings to scientists interested in the altered genes. Thus, through PB-CURE, students have the opportunity to make discoveries that relate to their scientists partners' research and be a part of a broader scientific community, while learning concepts in plant biology, genetics, ecology, and the processes and nature of science (Dolan & Grady, 2010; Dolan, Lally, Brooks, & Tax, 2008; Lally, Brooks, Tax, & Dolan, 2007).

The PB-CURE project was implemented in undergraduate courses in three different Mid-Atlantic institutions during the 2008–2009 academic year. All of the 83 students enrolled in these courses were invited to participate in the study, and almost all agreed to do so. In order to create a broadly representative sample, the first author then selected 34 students (18 science majors and 16 non-majors) who differed in their self-reported science backgrounds, abilities, and interests. The courses and participants are described in more detail below.

### Introductory Biology Laboratory Course at a Two-Year Junior College

This course was required for all students, enrolling 15 students in each of three sections per semester, or 45 students per semester, most of whom were in their first year of college. Twenty students participated in this study, with 11 indicating that they intended to major in science and 9 indicating they intended to pursue a non-science major. The students attended one three-hour session per week in a typical laboratory classroom, and were not required to enroll in a complementary theory or lecture course. PB-CURE was the primary course curriculum, although the course included other lessons on anatomical, physiological, genetic, and ecological aspects of the plants and herbivores used in PB-CURE. Although the students varied in their academic achievement and science background knowledge, most were lower or moderately achieving and had not completed other undergraduate science coursework. Many were the first in their families to attend college or received need-based scholarships to support their undergraduate education.

### Introductory Biology Lecture and Laboratory Course at a Four-Year Liberal Arts College

Sixteen students were enrolled in this course, three of whom were science majors and thirteen of whom were non-science majors. Six students participated in this study: one junior science major and five non-majors (two juniors and three sophomores). Students met twice a week for three hours each time for a combination of lecture and laboratory instruction. The course content comprised plant and animal biology, with CURE as the major focus of laboratory learning and the only lab experience that spanned multiple class sessions. The course included a few related and unrelated demonstrations. Students in this course varied in their academic achievement.

## Upper Level Plant Biology Course at a Public Research University

This course enrolled 22 life science majors who were taking it as either a requirement or an elective, depending on the major (e.g., horticulture, biology, environmental sciences). Eight students participated in this study (four juniors and four seniors). Students met twice a week for 75 minutes each time. The course curriculum included organism and ecosystem level concepts, as in the introductory courses, but also included molecular and cellular concepts such as the functions of enzymes and other biochemicals in plants. The PB-CURE project was the only laboratory learning experience in this course, which was otherwise taught in a lecture format. In comparison with students in the other two colleges, these students were higher achieving and had completed more science coursework. Some had completed science research internships. The students in this institution were studying plants being actively investigated by plant scientists at the same institution, who visited the class and discussed the students' experiments with them. This type of interaction did not occur at the other institutions in this study.

The instructors learned about PB-CURE from the director of the program (second author) and believed that engaging their students in course-based research could improve their students' understanding of the processes of science, prepare their students to pursue research internships in the future, and enrich their own teaching. The instructors embedded PB-CURE in their courses for approximately 10 weeks, during which they served mainly as facilitators. In the first three weeks, the instructors taught concepts related to PB-CURE, such as characteristics of *Arabidopsis* and the herbivores and the meaning of genes and mutations, using lectures and class discussions. In the next seven weeks, the students worked in groups of two to four to design and conduct their experiments, including: (1) planting their *Arabidopsis*; (2) brainstorming and selecting research questions and generating hypotheses; (3) developing experimental plans, including selecting variables to measure, planning ways to observe and collect their data, and drafting research proposals for feedback and revision; (4) infesting plants with herbivores according to their revised plans; (5) making observations, recording data, and taking care of their plants; (6) analyzing their data, interpreting their results, developing conclusions, and considering the implications of their results; and (7) presenting their final research in class and/or writing a final report or paper.

## DATA COLLECTION AND ANALYSIS

In order to identify how undergraduates used their course-based research experiences to reposition themselves in relation to science, the first author conducted pre- and post-semi-structured interviews with each student (Merriam, 1998; Patton, 2002). Interviews were conducted prior to each student's involvement in PB-CURE and were used to document his or her scientific background, career plans, conceptions of science and research, attitudes toward science, and expectations about being involved in a CURE. Post-interviews were conducted after students completed all PB-CURE-related work. This interview was used to characterize students' engagement in CURE activities and explore how students repositioned themselves with respect to science, including their conceptions of and attitudes toward science. All interviews were audiotaped and transcribed verbatim for analysis. Full protocols of the interviews are included in the Appendix. All PB-CURE activities were videotaped in order to capture students' actions and discussions throughout the experience. Video recordings were used to develop more complete pictures of students' experiences with PB-CURE and confirm the statements students made about their involvement in PB-CURE.

Interview data were the primary focus of analysis, which was conducted through a process of inductive content analysis (Patton, 2002) by iterative coding and data checking to ensure the credibility of the findings (Denzin & Lincoln, 2005). First, the first author reviewed the interview transcripts for repeating ideas and common themes and created an initial set of codes for analysis. Then, two other researchers (second author and a postdoctoral researcher studying a similar course-based research program) used the initial set of codes to analyze the transcripts independently, generating new codes as necessary. The three coders then discussed their coding, affirming the initial set of codes, agreeing on the establishment of new codes or the combining of existing codes, or removing codes not supported by the data. Full agreement was obtained among the three coders with regards to the codes and their definitions. After this first iteration, the first author generated tentative themes regarding students' conceptions of science (Patton, 2002). The two authors re-reviewed the interview data set independently to confirm that the themes were representative of the full data set.

In order to address our research questions, we selected a subset of students whose experiences represented diverse outcomes related to self-authorship. We profiled each student as an instance of transformation or lack thereof. In describing each student's profile, we highlight changes, or lack of change, in their views about sources of scientific knowledge (research question 1), their own scientific identities (research question 2), and their relationships to science (research question 3).

## RESULTS

Here, we present profiles of six students whose experiences represent the broadest range of transformation, or lack thereof. Each profile illustrates how the student views sources of scientific knowledge, how the student's scientific identity manifests, and how the student sees himself or herself in relation to science. Although each profile emphasizes a specific transformational theme, these themes were apparent to greater or lesser extents in all of the students participating in the study. Students' pseudonyms were selected to indicate the extent of their transformation. Names that start with "D" represent students who are Developing self-authorship, learners whom Baxter Magolda (1992) characterized as independent or contextual knowers. In this study, developing students transitioned to more contextualized views of scientific knowledge and came to see a role for themselves in the construction of this knowledge. Names that start with "A" represent what Baxter Magolda (1992) characterized as Absolute knowers, or learners who view knowledge as certain and derived from external authorities. In this study, absolute knowers held onto their views that scientific knowledge was to be acquired from the instructor and that they had no role to play in the construction of scientific knowledge. Finally, names that start with "T" represent students who Baxter Magolda (1992) characterized as Transitional knowers. In this study, transitional students are those who see a more active role for themselves as learners, but still see some aspects of scientific knowledge as certain and rely on instructors communicate particularities of what is to be learned.

### David, an Introductory Student: Everyone Can Do Science, Including Me!

David was a 19-year-old African American sophomore at the two-year college. At the time of the study, he had not declared a major but was confident that he would not major in science. Although he liked learning about organisms in school, he found scientific topics and principles difficult to understand, especially "when numbers and calculations are involved." He reported that he did not do well in high school science courses. Prior to his involvement in PB-CURE, David defined science as learning "how nature works" and viewed engagement in research as "exploring something that you did not know before." David expected that being involved in PB-CURE would be interesting but challenging because he had not excelled in science in the past and was not confident about his scientific background.

During PB-CURE, David preferred to work with two students who he believed were more accomplished in science and had stronger scientific backgrounds than his own. He reported that, regardless of the good teamwork and help he received from his group members, he felt behind, especially in early phases of PB-CURE, because of differences in the background knowledge and skills of his group members. As he progressed in PB-CURE, David became more engaged in PB-CURE tasks and the nature of his engagement changed. At first, David mainly asked questions aimed at learning basic scientific facts and concepts (e.g., definitions of hypothesis, control, dependent and independent variables, or the characteristics of *Arabidopsis* plants and spider mites). As the research continued, he was more enthusiastic as he observed the development of the plants and collected data: "It was amazing because I have never known a plant to be able to live such a short life. That was pretty cool. I was excited every day to see my plants." As his group analyzed their data, David was more talkative, making suggestions about how to explain results, improve the experiment in the future, and build on what the group had found. In his post-interview, David emphasized that participating in PB-CURE changed his perception of his ability to be involved in science:

> [PB-CURE] helped me to lower the pedestal that I have for a lot of scientific methods. Experimentation isn't just for professors and high-priced scientist. What I learned was that anybody can do an experiment, it's just a matter of if you want to do it, what you want to do and how badly you want to pursue it.... I probably don't try things just because I'm scared of failure and I'm scared of somebody doing it better than me.... [I learned that] I can actually do something that I've never done before and I even enjoy it.

David's comments illustrate that he came to see himself as a generator of scientific knowledge, and could see himself as a scientist.

David also indicated that he was more capable of recognizing and diagnosing phenomena he learned about during PB-CURE in nonscientific contexts in his life:

> I work at a grocery store in the produce department. Spider mites could pose eggs, a large problem for agricultural companies. And, who knows? Maybe if I go to work one day and some of our supply comes off the truck damaged, I would say, "Maybe there's spider mite infestation."

David's profile illustrates the development he made in relation to science. He transitioned to seeing himself as a producer of scientific knowledge and user of this knowledge in his daily life.

### Doug, an Introductory Student: Science Is Useful in My Daily Life

Doug was an 18-year-old Caucasian student in his first year at the two-year college. He grew up in a rural town in the Mid-Atlantic region of the U.S. and came from a family of farmers. He loves living in the country and planned to become a "scout camp ranger." He originally intended to major in forestry studies. When he realized that camp rangers did not need to major in science, he decided to pursue training as an electrician. Doug reported that he liked studying plants and was a solid science student in high school, but that most science instruction included too much complexity that was unrelated to his daily life and career interests. Doug said, for example, "Learning science is much more than learning about how plants live, it is understanding complicated processes related to natural phenomena.... It is too complicated and I don't need it."

As he started his work with PB-CURE, Doug was anxious to find a research question that would fit his interests and his family's needs, and would be connected to his daily life. He was interested in testing the effects of what he called "natural insecticides" on herbivory of *Arabidopsis* plants, because his own family had faced issues with insect damage of their personal garden and their agricultural crops. Doug explained, "I was curious to see whether or not the cedar would cause the spider mite to say, 'Hey, I really do not want to be here,' and go to another place." His family had discussed adding cedar chips to their soil as a natural insecticide, and he wanted to investigate whether adding cedar chips to the soil of wild-type and mutant *Arabidopsis* plants would affect the attraction of spider mites to the plants. As he discussed potential research questions with his group members, he suggested this idea and emphasized how the results would be relevant to farmers, including his own family. The group decided to pursue his idea, and Doug became the de facto leader of their research. Doug supplied the cedar chips for his group's investigation and was eager to observe the plants and to collect, analyze, and discuss the group's data. He volunteered to speak on behalf of his group as the class presented their research to one another. In his post-interview, Doug described how he could use what he learned in PB-CURE in his everyday life:

> Through our experiment I learned that the plants were not going to be able to grow very effectively in that soil [that contains cedar].... What we do [in our farm] is add other things to the soil just to see what way it would grow better.... My father and I would talk about adding cedar.... I think we might not try that now, because I know that there is higher possibility that cedar will negatively affect the plants.... It was this topic that got me totally involved in the experiment.

Like David, Doug came to see himself as a producer of scientific knowledge. Prior to his involvement in PB-CURE, Doug saw science as something that was too complex for his needs. By pursuing a line of inquiry that was of personal interest, Doug came to see learning science as worthwhile for solving a problem in his daily life. Doug also became more motivated to learn more about plant–insect relationships and suggested the need for further research "to find how you can limit the negative effects of the cedar on the plants themselves but increase the possible effects of its natural deterrence for insects," which was his original interest. Through his PB-CURE work, Doug came to see himself as producer of scientific knowledge and as a user of this knowledge in his daily life to solve real-world problems.

## Dorothy, an Upper-Level Science Major: I Can Engage in Scientific Discourse with Real Scientists

Dorothy, a Caucasian in her mid-fifties, was a junior science major at the research university. After 20 years of working as an X-ray technician in the medical center of the university, Dorothy decided to return to school because she was eager to learn, although she did not plan to switch careers. Dorothy always liked science, including plant biology, but had not been involved in research previously. Dorothy chose to work with two high-achieving students who were involved in a CURE in another course because she believed their experience would help the group be successful in PB-CURE. Dorothy was highly engaged in the research and very involved in whole-class and small-group discussions, asking the instructors and her group members many questions that reflected curiosity and eagerness to learn about the plants and herbivores and about the processes of science. She was very enthusiastic and motivated and was especially excited about observing the plants and finding interesting results. She explains, for example, "I was excited to see the plants and how they were doing, and when we got purple knots [i.e., herbivore clusters that developed pigment], everybody was thrilled."

Dorothy was especially interested in discussing her results with two plant scientists who visited the class to talk with students about their PB-CURE work. These scientists, who worked at the university where Dorothy was enrolled, were experts on the genes the students were studying as part of PB-CURE. Dorothy explained how this discussion made her feel involved in something authentic and helpful: "She [one of the scientists] was so enthusiastic about research, just so supportive, and that helped to make you feel important, like you were contributing to something [real]." Dorothy's response suggests that interaction with experts in the scientific domain of the CURE may be especially impactful for participating students, legitimizing the authenticity of their coursework, as Dorothy explained:

> I was involved in a real scientific research most certainly and they (i.e., the scientists) attempted to bring me into it as best they could with the trust that they have because they had to give the students a little bit of trust [with respect to the creditability and validity of the students' findings].

Dorothy came to see herself as capable of contributing to "real" scientists' work, including engaging in meaningful dialogue with practicing scientists.

## Dian, an Introductory Student: Science Is a Discourse and I'm Learning the Language

Dian, an African American in her late-twenties, was a junior majoring in criminal justice at the four-year college. She started her undergraduate degree after she got married and had a family. She did not like science in high school but was more positive after enjoying several science courses in college. She, however, still viewed science as a difficult discipline. Dian expected PB-CURE to resemble the social science research she had conducted as a requirement of her major.

}Dian's most substantive transformation was in her developing awareness that science involves a unique language and that this language is negotiated among collaborators and their professional communities to support co-construction of knowledge. She explained that she came to understand the importance of creating "common language" among individuals to facilitate consistency, reliability, and accuracy in recording and interpreting data. As Dian addressed measuring changes in the *Arabidopsis* leaves as part of collecting data, she explained:

> What is green to me, or light green, is not going to be the same to another person.... We have to figure out a way for everybody to speak the same language, which is the problem in science because everybody is studying science all over the world. There has got to be some kind of common language or form. That way you know what everybody means.

Dian advocated for her group to create a color grid in order to record information about their plants' color in a way that would be more meaningful. When her group members suggested recording leaves as "damaged" and "not damaged," she asked them to be more accurate, explaining, "We could easily just write 'not damaged' and 'damaged'... but when we do our [color] grid, we can see that they (i.e., all of the damaged leaves) don't look the same."

Dian further explained how she used her newly acquired language to engage in scientific discourse with people outside of the class: "I have spoken to people who are horticulture majors and they started talking about the different structures of the plants. And I felt, 'Oh! I can actually understand a little bit.' I could really talk to them." Dian's response indicates how learning scientific discourse through engagement in PB-CURE empowered her to engage in scientific discourse outside of the PB-CURE classroom.

Another shift in Dian's conception of science related to her understanding of the meaning and role of testing and "proving" hypotheses in science. Prior to participating in PB-CURE, Dian viewed engagement in research as "learning how to conduct experiments and test a hypothesis through action…and being able to prove the hypothesis is important." At the end of PB-CURE, as she realized her results were different than what she expected, she reported that she learned:

> Even if my hypothesis is not proven, it is still doing science and thinking like real scientists.... It does not even matter if it is proven right or if it is not That is the glory of doing an experiment and testing the hypothesis. There is no wrong answer.

Dian's response suggests that, through her involvement in PB-CURE, she transitioned from seeing scientific knowledge as absolute, either "right" or "wrong," to constructed through the testing and discussion of ideas by individuals within a community.

## Allyson, an Introductory Student: I Don't Like Science and I Don't Understand It!

Allyson, a Caucasian 20-year-old, was a junior majoring in social work at the four-year college. Allyson viewed science as difficult to understand and irrelevant to her major. Her achievement in science courses in high school was always moderate to low. She saw little connection between the scientific focus of PB-CURE and her educational and career interests, and she maintained this stance throughout her involvement in PB-CURE:

> I do not like this class so much. I do not really understand it. Probably, what is hard for me is remembering all the different [information] like plant life cycles and reproduction. I cannot even remember what they are called, and I am just having a hard time with it because we talk more about plants and not so much about the human body and I am not interested in plants. My major is mainly studying people and how they work together, and I do not see how plants are related to that.

Allyson also reported that her experience with PB-CURE confirmed her suspicions that doing science was beyond her capacity and that she was unable to make sense of science without intensive guidance:

> It was kind of hard being put on your own and not really having too much guidance from the professor. I felt lost.... I expected him to make sure that I am putting down the right data.... He could be more specific, letting us know he wants everybody to measure color, height, and root system, instead of asking us to collect data on any of three variables according to our choice.

Allyson also reported she did not significantly improve her scientific understanding and believed she had learned nothing because she could not identify differences between wild-type and mutant plants grown in treatment versus control conditions. She explained, "I probably did not learn much about the research because I did not really see a significant difference between [mutant and wild-type] plants. They were looking pretty much the same." Her response indicates that she held a novice conception of science: that science is about finding differences between experimental conditions.

Allyson was particularly frustrated by the scientific terminology and complexity of PB-CURE, which exacerbated her conception of scientific knowledge as something she was supposed to acquire from an external authority and which only had value as measured by course assessments. She explained:

> What was challenging was understanding the information that was being given to us.... Things like gene sequence, what a second metabolite is, what are first level consumers, second level consumers, and what are the differences, I guess you would say jargon of biology.... I felt low (in terms of achievement) when compared to [my classmates] because I felt like they have more experience.... They would say that they were as lost as I am, but I guess I didn't really believe that because I would look at their quiz grades and I was like, "Well, you understood it enough."

Although Allyson was largely negative about her PB-CURE experience, she reported realizing several benefits. She explained how she enjoyed observing her own plants and seeing their development: "I was excited every day. I would come in early just to see my plants. It was like my little baby." She added that investigating plants "made me have more respect and understanding of different organisms." She also reported that she improved her learning habits: "This experiment improves general organization skills.... I had a habit of writing shorthand and I realized that I really needed to take actual thorough notes and numbers. It has helped me organize my notes and my thoughts." For the most part, Allyson's comments indicated a more instrumental view of the experience: learning is about skill acquisition. At least one comment she made, however, hinted that she was beginning to develop in her intellectual flexibility, which is a hallmark of self-authorship. Specifically, she reported that the collaborative nature of PB-CURE "required communication, you have to share, you have to listen to opinions, listen to other people."

Allyson's profile demonstrates that not all the students involved in the PB-CURE repositioned themselves in relation to science. Some of the non-major students, who were low science achievers like Allyson, did not see the relevance of science to their career or future plans, and did not appear to progress toward self-authorship in the domain of science. Although Allyson identified some positive outcomes of participating in PB-CURE, she did not change her initial conception of science as certain and as a pursuit that was beyond her capabilities.

### Thomas, an Upper-Level Student: Science Is Frustrating, but I Know That's How Science Works

Thomas, a 21-year-old Caucasian, was a senior majoring in biology at the research university. Thomas was involved in research during high school, completed several laboratory courses as an undergraduate, and had participated in undergraduate research apprenticeships. Thomas enjoyed science, and prior to starting his PB-CURE work, he saw PB-CURE as an opportunity to do "real science" in class and as useful preparation for pursuing a career in science. After completing his PB-CURE work, Thomas saw PB-CURE as both an opportunity and a challenge. He acknowledged that challenges and frustrations were integral elements of doing science. Thomas, for example, reported that his involvement in PB-CURE confirmed his conceptions of certain aspects of science, but that he found this frustrating:

> There were so many lurking variables that could have affected that [the number of spider mites on the plants], it was hard to make conclusions. And, while you could say, "Oh, these are our conclusions," somebody else came back and said, "Well, you didn't measure this variable, it (i.e., the outcome we saw) could be due to this." What are you supposed to respond to that? "Yes, you're right, I can't control that." So, that was a little bit frustrating but that's how science works.

Thomas' frustration resulted, in part, from his expectation that coursework entails more instructional guidance. He acknowledged that problems are inherent to the process of science, but viewed the instructors, rather than himself, as the problem solvers. He described his frustration in choosing the most appropriate variables to measure:

> When we were planning the experiment, we planned to measure juvenile versus adult number of eggs.... We weren't told that that's basically impossible to do.... We had to go back when we were halfway through the experiment, saying, "Oh, we can't do any of this, what are we going to collect now?" That kind of scrambling was hard and a little more guidance would eliminate that issue.

Thomas's profile illustrates that he is transitional with respect to scientific self-authorship. He sees himself as capable of producing science knowledge, but has yet to see himself capable of the more complex work of science, such as trouble-shooting and problem solving. Thomas was capable of high achievement in didactic course settings, and most likely expected the instructor in PB-CURE to communicate the knowledge to be learned and acted upon, as in his other courses. He believed that his instructor was withholding knowledge (rather than this knowledge being unknown), forcing him to "reinvent the wheel" during his investigation.

## CROSS-PROFILE ANALYSIS AND DISCUSSION

The results of this study demonstrate how participating in a CURE provided opportunities for undergraduates at both introductory and upper levels to make progress toward self-authorship in multiple ways. First, developing students, who were represented here in the profiles of David, Doug, Dorothy, and Dian, came to view themselves, rather than external authorities, as sources of scientific knowledge. Second, developing students assumed scientific identities that included abilities to produce scientific knowledge (e.g., Doug and Dorothy), make use of scientific knowledge (e.g., David and Doug), and engage in scientific discourse (e.g., Dorothy and Dian). Third, developing students repositioned themselves in relation to science: coming to see scientific knowledge as relevant to their daily lives (e.g., David and Doug) and the practice of science as within their capabilities (e.g., David, Doug, Dorothy, and Dian). Not all undergraduates in this study made progress toward self-authorship. Some students remained absolute knowers, as illustrated in Allyson's profile. They held on to their conceptions of scientific knowledge as something to be acquired from an external authority and irrelevant to their professional interests and daily lives. Other students were transitional knowers, as illustrated in Thomas's profile. They saw a role for themselves in constructing scientific knowledge and participating in the practice of science, but expected instructors to guide these processes.

Although this study was not designed to determine which aspects of PB-CURE were most critical for self-authorship, the students' comments are consistent with the interpretation that opportunities for ownership and for discourse with peers and experts were important for fostering these students' transitions toward self-authorship. Regarding ownership, students had the flexibility to choose their particular research questions and designs. If students capitalized on the flexibility, as Doug did by investigating the impact of cedar chip application on herbivory of the plants, then the CURE was useful for supporting transformation toward self-authorship. If students did not capitalize on the flexibility, or saw the scientific focus of the CURE as too distant from their interests (e.g., Allyson and Thomas), then development toward self-authorship was less evident. Both Dorothy and Dian emphasized the importance of discourse during PB-CURE. For Dorothy, talking with scientists actively investigating the genes she was studying in her PB-CURE investigation was integral to her belief that her findings had meaning beyond the classroom. For Dian, talking with her group members and needing to communicate their results to others outside of the group helped her recognize that the practice of science involved the use of a common language. By learning this language, she felt better prepared to engage in scientific conversations that she believed were previously inaccessible to her.

The developing students in this study saw a greater relevance for science in their daily lives, even if they did not plan to pursue further education or careers in science. This finding is particularly notable as many studies of students' views of science curricula conclude that students see science as impersonal, socially sterile, and irrelevant (Cleaves, 2005; Osborne & Collins, 2001). Thus, CUREs may be especially powerful tools for non-science majors to use in defining the relevance of science to their daily lives. As was apparent in Allyson's profile, CUREs may also present special risks for non-science majors. Rather than taking ownership of her PB-CURE investigation in a way that connected to her interests, Allyson confirmed her beliefs that science was irrelevant to her and beyond her capacity.

## FURTHER STUDY AND PRACTICE OF CURES

Although our results demonstrate the potential of CUREs as tools for undergraduate self-authorship, they also raise a number of questions about how CUREs should be practiced. Doug's profile illustrates the power of designing CUREs with sufficient flexibility to allow for student ownership, yet some might argue that the design of PB-CURE was too flexible. Doug transformed his research into an investigation that was only tangentially related to the overarching research question addressed by the design

PB-CURE: What are the functions of genes in *Arabidopsis*? This instance should challenge the undergraduate science education community to think carefully how CUREs should be designed and implemented to best to balance ownership from the student perspective and connectedness with or contribution to the scientific domain of the CURE. It is also important to consider whether this balance should differ in courses that enroll majors versus non-majors. Analysis of these issues should explore ways of establishing shared focus and purpose (i.e., intersubjectivity; Rogoff, 1990) between learners, who may enroll in courses for a variety of reasons other than interest in science research, and CURE designers and instructors.

Dorothy's profile illustrates the importance of affording students opportunities to discuss their work with experts in the scientific domain of the CURE. Indeed, undergraduates should have conduits to share their CURE data and to frame it within a larger body of knowledge in order to maximize the authenticity and legitimacy of their work (Lave & Wenger, 1991; Roth et al., 2008). How can this be accomplished? Should undergraduates share their results and interpretations through mechanisms that are dialogic in nature, such as in-person conversations or web-mediated discussions, either synchronously (e.g., video conference or text-based chatting) or asynchronously (e.g., discussion board posts)? Should these discussions include particular individuals, such as the CURE designer or instructor or other scientists with a particular interest in the CURE domain? Or, can interactions with the broader scientific communities, such as through posts in publicly available scientific databases like Genbank (http://www.ncbi.nlm.nih.gov/genbank/) or the *Arabidopsis* Information Resource (http://www.*Arabidopsis*.org/), yield meaningful student outcomes? To what extent are these databases both intellectually and technically accessible to undergraduates participating in CUREs, who are not likely to be experts in the scientific domain of their CURE? To what extent should user interfaces be developed to encourage involvement of non-experts, as has been done with citizen scientist projects such as the Cornell Lab of Ornithology (http://www.birds.cornell.edu)? Research on interactions between undergraduates involved in CUREs and the broader scientific community is needed to understand the impacts of interactions with experts on student learning in CURE contexts and to identify CURE designs and practices that support effective student–expert interactions. Although it may offer fewer opportunities for expert modeling, the social structure of CUREs (i.e., one instructor and many students) presents more opportunities for peer-to-peer interaction than typical research apprenticeships. Further study of student–student interactions in the context of CUREs also is needed to understand the unique contributions of these interactions to student learning and to identify CURE designs and practices that support effective student–student interactions.

Dian's profile illustrates how she came to recognize science as a discourse. To what extent do different CURE designs offer opportunities for students to negotiate the language of their CURE investigations? CUREs present opportunities for students to operate at the edge of what is known scientifically. In some CUREs, the research methods and data types are stipulated by the CURE design (e.g., Genomics Education Partnership); thus, student negotiation is most likely focused on interpretation of the data. In contrast, PB-CURE integrates the study of herbivory and *Arabidopsis* gene function. Because the herbivory of *Arabidopsis* has not been widely studied, appropriate terminology and methodological explanations were not always available in the research literature for students to use in describing their work. While Dian viewed this as an opportunity to author aspects of her investigation, Thomas expected the CURE designer and instructor to define and limit the methodologies students could employ. Further study of the relationship between CURE structure and student outcomes is needed to understand the importance of designing CUREs that involve student negotiation about different aspects of research.

Further research is also needed to understand the student outcomes of participating in CUREs and whether different learners realize different outcomes. For example, do science majors and non-majors differ in their learning gains? Do students who are interested in the focus of their CURE make more gains than those who are less interested, regardless of their major? Are students more likely to realize benefits of participating in a CURE if explicit connections are made between the scientific focus of the CURE and students' daily lives and educational or career interests? Are students who participate in CUREs better prepared to reap the benefits of participating in subsequent research apprenticeship? Research addressing these questions should make use of measures for which information about reliability, validity, and dimensionality is available, such as measures of students' experimental design abilities (Sirum & Humberg, 2011), or views of scientific inquiry and the nature of science (Lederman, 2007; Lederman, Abd-El-Khalick, Bell, & Schwartz, 2002; Lombrozo, Thanukos, & Weisberg, 2008; Neumann, Neumann, & Nehm, 2011; Schwartz, Lederman, & Lederman, 2008).

As more instructors teach using CUREs, research is needed to understand the extent to which different aspects of instructor expertise influences student learning. For example, do instructors need specialized expertise in the scientific domain of the CURE they are implementing, or is expertise in the design and conduct of science research sufficient to guide students in their CURE work? In addition, what pedagogical content knowledge must instructors develop to teach effectively using CUREs? Also, what skills or competencies do instructors need to assist students like Allyson and Thomas in making progress toward self-authorship? Instructors who are familiar with the challenges students with varying backgrounds and interests face in navigating the uncertainty and complexity of science research may be better prepared to support them in doing so successfully. Further research also is needed to identify and characterize instructor outcomes of teaching using CUREs. Previous research has shown that CUREs offer opportunities of scale, in which scientific questions can be addressed that would otherwise be unwieldy for single investigators or small groups of investigators (e.g., Goodner et al., 2003; Hatfull et al., 2006; Marris, 2010). It is possible that data and ideas generated by CUREs may help faculty move their research programs in unanticipated directions. CUREs may also help faculty who are not otherwise engaged in research to maintain connections with the broader research community, including keeping up to date on current research and maintaining an identity as a scientist. Finally, research is needed to identify and characterize outcomes for graduate students who serve as teaching assistants (TAs) for CUREs. For example, CURE TAs may experience many of the positive outcomes observed for graduate students who mentor undergraduates in research (Dolan & Johnson, 2009, 2010) or who serve as teaching assistants in inquiry laboratory courses (French & Russell, 2002).

## APPENDIX

### *Pre-interview*

*Students' experiences with science and expectations about being involved in course-based research*

1. Tell me about your previous experiences with science. [Probe about science class, science research experiences.]
2. Have you had previous experiences with course-based research? Please describe.
3. What do you expect to find beneficial / interesting / challenging about doing this CURE investigation?
4. Do you see this CURE as real scientific research? Please explain.

*Students' conceptions of and attitudes toward science*

5. What are your educational / career plans?
6. Do you like science? Please explain.
7. Do you see science as relevant to your career or daily life? Please explain.

8. What do you find interesting / challenging about science? [If relevant: About science research?]

*Students' views of student and instructor roles in this CURE*

9. Do you think it is important / relevant for you to participate in science research? Please explain.
10. What is your role in your CURE investigation? What is your instructor's role?

## Post-interview

*Students' experiences with PB-CURE*

11. What did you enjoy about this CURE? What aspects of the CURE were not enjoyable?
12. What aspects of your CURE experience were successful? Not successful?
13. How did you benefit from participating in this CURE? What did you find challenging?
14. What aspects of your CURE experience were interesting? Not interesting?
15. What did you learn from doing your CURE investigation? [Probe about scientific concepts, terms, skills.]
16. What do you wish you had learned from your CURE investigation? [Probe about scientific concepts, terms, skills.]
17. How is PB-CURE similar to your other science learning experiences? How is it different?

*Students' conceptions of and attitudes toward science*

18. Do you see any connections between this CURE and your daily life? Please explain
19. Do you see any connections between this CURE and your educational/career plans? Please explain.
20. How is this CURE similar to what you expect science to be like? How is it different?
21. Has your CURE experience changed your view of science? Please explain.

*Students' views of student and instructor roles in PB-CURE*

22. Do you think that you were involved in a real scientific research? Please explain.
23. What would you change about what you did during this CURE, if anything?
24. What would you change about what your instructor did during this CURE, if anything?

## REFERENCES

Anderson, R. D. (2002). Reforming science teaching: what research says about inquiry. *Journal of Science Teacher Education, 13,* 1–12.

Bandura, A. (1977). *Social learning theory.* New York, NY: General Learning Press.

Barab, S. A., & Duffy, T. (2000). From practice fields to communities of practice. In D. Jonassen & S. M. Land (Eds.), *Theoretical foundations of learning environments* (pp. 25–56). Mahwah, NJ: Lawrence Erlbaum Associates.

Baxter Magolda, M. B. (1992). *Knowing and reasoning in college: Gender-related patterns in students' intellectual development.* San Francisco, CA: Jossey-Bass.

Baxter Magolda, M. B. (1999). *Creating contexts for learning and self-authorship: Constructive developmental pedagogy.* Nashville, TN: Vanderbilt University Press.

Baxter Magolda, M. B. (2003). Identity and learning: Student affairs' role in transforming higher education. *Journal of College Student Development, 44,* 231–247.

Baxter Magolda, M. B. (2004). Self-authorship as the common goal. In M. B. Baxter Magolda & P. M. King (Eds.), *Learning partnerships: Theory and models of practice to educator for self-authorship* (pp. 1–35). Sterling, VA: Stylus.

Brown, J. S., Collins, A., & Duguid, P. (1989). Situated cognition and the culture of learning. Educational Researcher, 18, 32-42.

Brown, J. S., & Duguid, P. (1991). Organizational learning and communities-of-practice: Toward a unified view of working, learning, and innovation. *Organization Science, 2,* 40–57.

Brewer, C. A., Smith, D., Bauerle, C., DePass, A., Lynn, D., O'Connor, C., … & Wubah, D. (2011). *Vision and change in undergraduate biology education: A call to action.* Washington, DC: AAAS. Retrieved from http://visionandchange.org/finalreport

Caruso, S. M., Sandoz, J., & Kelsey, J. (2009). Non-STEM undergraduates become enthusiastic phage-hunters. *CBE – Life Sciences Education, 8,* 278–282.

Chinn, C. A., & Malhotra, B. A. (2002). Epistemologically authentic inquiry in schools: A theoretical framework for evaluating inquiry tasks. *Science Education, 86,* 175–218.

Cleaves, A. (2005). The formation of science choices in secondary school. *International Journal of Science Education, 27,* 471–486.

Collins, A. (2006). Cognitive apprenticeship. In R. K. Sawyer (Ed.), *The Cambridge handbook of the learning sciences* (pp. 47–59). New York, NY: Cambridge University Press.

Collins, A., Brown, J. S., & Newman, S. E. (1989). Cognitive apprenticeship: Teaching the crafts of reading, writing, and mathematics. In L. B. Resnick (Ed.), *Knowing, learning, and instruction: Essays in honor of Robert Glaser.* Hillsdale, NJ: Erlbaum.

Cummins, C., Doyle, J., Kindness, L., Lefthand, M. J., Bear Don't Walk, U. J., Bends, A., ... & Eggers, M. J. (2010). Community-based participatory research in Indian Country: improving health through water quality research and awareness. *Family & Community Health, 33,* 166–174.

Denzin, N. K., & Lincoln, Y. S. (Eds.). (2005). *The Sage handbook of qualitative research.* Thousand Oaks, CA: Sage.

Desai, K. V., Gatson, S. N., Stiles, T. W., Stewart, R. H., Laine, G. A., & Quick, C. M. (2008). Integrating research and education at research-extensive universities with research-intensive communities. *Advances in Physiology Education, 32,* 136–141.

Dolan, E. L., & Grady, J. (2010). Recognizing students' scientific reasoning: A tool for categorizing the complexity of reasoning during teaching by inquiry. *Journal of Science Teacher Education, 21,* 31–55.

Dolan, E., & Johnson, D. (2009). Toward a holistic view of undergraduate research experiences: An exploratory study of impact on graduate/postdoctoral mentors. *Journal of Science Education and Technology, 18,* 487–500.

Dolan, E., & Johnson, D. (2010). The undergraduate–postgraduate–faculty triad: Unique functions and tensions within a science research community of practice. *CBE—Life Sciences Education, 9,* 443–453.

Dolan, E. L., Lally, D. J., Brooks, E., & Tax, F. E. (2008). PREPping students for authentic science. *The Science Teacher 75,* 38–43.

Drew, J. C., & Triplett, E. W. (2008). Whole genome sequencing in the undergraduate classroom: Outcomes and lessons from a pilot course. *Journal of Microbiology and Biology Education, 9,* 3–11.

French, D., & Russell, C. (2002). Do graduate teaching assistants benefit from teaching inquiry-based laboratories? *Bioscience, 52,* 1036–1041.

Goodner, B. W., Wheeler, C. A., Hall, P. J., & Slater, S. C. (2003). Massively parallel undergraduates for bacterial genome finishing. *ASM News, 69,* 584–585.

Greeno, J. G. (1997) Response: On claims that answer the wrong questions. *Educational Researcher, 26,* 5–17.

Harrison, M., Dunbar, D., Ratmansky, L., Boyd, K., & Lopatto, D. (2011). Classroom-based science research at the introductory level: Changes in career choices and attitude. *CBE—Life Sciences Education, 10,* 279–286.

Hatfull, G. F., Pedulla, M. L., Jacobs-Sera, D., Cichon, P. M., Foley, A., Ford, M. E., ... & Hendrix, R. W. (2006). Exploring the mycobacteriophage metaproteome: Phage genomics as an educational platform. *PLoS Genetics, 2,* e92.

Healey, M., & Jenkins, A. (2009). *Developing undergraduate research and inquiry.* York, UK: Higher Education Academy.

Hennessy, S. (1993). Situated cognition and cognitive apprenticeship: Implications for classroom. Studies in Science Education, 22, 1–41.

Hunter, A.-B., Laursen, S. L., & Seymour, E. (2007). Becoming a scientist: The role of undergraduate research in students' cognitive, personal, and professional development. *Science Education, 91,* 36–74.

Junge, B., Quinones, C., Kakietek, J., Teodorescu, D., & Martseller, P. (2010). Promoting undergraduate interest, preparedness, and professional pursuit in the sciences: An outcomes evaluation of the SURE program at Emory University. *CBE—Life Sciences Education, 9,* 119–132.

Lally, D. J., Brooks, E., Tax, F. E., & Dolan, E. L. (2007). Sowing the seeds of dialogue: Public engagement through plant science. *Plant Cell, 19,* 2311–2319.

Lave, J., & Wenger, E. (1991). *Situated learning: Legitimate peripheral participation.* New York, NY: Cambridge University Press.

Lederman, N. G. (2007). Nature of science: Past, present, and future. *In* S. K. Abell *&* N. G. Lederman *(Eds.), Handbook of research on science education* (pp. 831–879). Mahwah, NJ: Lawrence Erlbaum Associates.

Lederman, N. G., Abd-El-Khalick, F., Bell, R. L., & Schwartz, R. S. (2002). Views of nature of science questionnaire: Toward valid and meaningful assessment of learners' conceptions of nature of science. *Journal of Research in Science Teaching, 39,* 497–521.

Lombrozo, T., Thanukos, A., & Weisberg, M. (2008). The importance of understanding the nature of science for accepting evolution. *Evolution: Education and Outreach, 1,* 290–298.

Lopatto, D. (2004). Survey of undergraduate research experiences (SURE): First findings. *Cell Biology Education, 3,* 270–277.

Lopatto, D. (2007). Undergraduate research experiences support science career decisions and active learning. *CBE—Life Sciences Education, 6,* 297–306.

Lopatto, D. (2010). *Science in solution: The impact of undergraduate research on student learning.* Tucson, AZ: Research Corporation for Science Advancement.

Lopatto, D., Alvarez, C., Barnard, D., Chandrasekaran, C., Chung, H.-M., Du, C., ... & Elgin, S. C. R. (2008). Undergraduate research: Genomics Education Partnership. *Science, 322,* 684–685.

Marcus, J. M., Hughes, T. M., McElroy, D. M., & Wyatt, R. E. (2010, Jan/Feb.). Engaging first-year undergraduates in hands-on research experiences: The Upper Green River Barcode of Life project. *Journal of College Science Teaching 39,* 39–45.

Marris, E. (2010, August 10). Birds flock online. *Nature News, 395.* Retrieved from http://www.nature.com/news/2010/100810/full/news.2010.395.html

McCook, A. (2011). Two-year colleges are jumping into the U.S. research pool. *Science, 333,* 1572–1573.

Merriam, S. (1998). *Qualitative research and case study applications in education.* San Francisco, CA: Jossey-Bass.

Neumann, I., Neumann, K., & Nehm, R. (2011). Evaluating instrument quality in science education: Rasch-based analyses of a nature of science test. *International Journal of Science Education, 33,* 1373–1405.

Osborne, J., & Collins, S. (2001). Pupils' views of the role and value of the science curriculum: A focus-group study. *International Journal of Science Education, 23,* 441–467.

Owens, D. K., Alerding, A. B., Crosby, K. C., Bandara, A. B., Westwood, J. H., & Winkel, B. S. J. (2008). Functional analysis of a predicted flavonol synthase gene family in *Arabidopsis*: *Plant Physiology, 147,* 1046–1061.

Patton, M. Q. (2002). *Qualitative research and evaluation methods.* Thousand Oaks, CA: Sage.

Rahm, J., Miller, H. C., Hartley, L., & Moore, J. C. (2003). The value of an emergent notion of authenticity: Examples from two student/teacher-scientist partnership programs. *Journal of Research in Science Teaching, 40,* 737–756.

Rogoff, B. (1990). *Apprenticeship in thinking.* New York, NY: Oxford University Press.

Roth, W. M., & Barton, A. C. (2004). *Rethinking scientific literacy.* New York, NY: Routledge.

Roth, W.-M., van Eijck, M., Reis, G., & Hsu, P.-L. (2008). *Authentic science revisited.* Rotterdam, The Netherlands: Sense.

Shaffer, C. D., Alvarez, C., Bailey, C., Barnard, D., Bhalla, S., Chandrasekaran, C., ... & Elgin, S. C. R. (2010). The Genomics Education Partnership: Successful integration of research into laboratory classes at a diverse group of undergraduate institutions. *CBE—Life Sciences Education, 9,* 55–69.

Schwartz, R., Lederman, N. G., & Lederman, J. S. (2008, April). *An instrument to assess views of scientific inquiry: The VOSI questionnaire.* Paper presented at the annual meeting of the National Association for Research in Science Teaching, Baltimore, MD.

Senge, P. (1994). *The fifth discipline fieldbook: Strategies and tools for building a learning organization.* New York, NY: Doubleday.

Seymour, E. L., Hunter, A.-B., Laursen, S., & DeAntoni, T. (2004). Establishing the benefits of research experiences for undergraduates: First findings from a three-year study. *Science Education, 88,* 493–594.

Sirum, K., & Humburg, J. (2011). The experimental design ability test (EDAT). *Bioscene, 37,* 8–16.

Turvey, M. T., & Shaw, R. E. (1995). Toward an ecological physics and a physical psychology. In R. L. Solso & D. W. Massaro (Eds.), *The science of the mind: 2001 and beyond* (pp. 144–170). New York, NY: Oxford University Press.

Vygotsky, L. S. (1978). *Mind in society.* Cambridge, MA: Harvard University Press.

Wei, C. A., & Woodin, T. (2011). Undergraduate research experiences in biology: Alternatives to the apprenticeship model. *CBE—Life Sciences Education, 10,* 123–131.

Wenger, E., McDermott, R., & Snyder, W. M. (2002). *Cultivating communities of practice: A guide to managing knowledge.* Boston, MA: Harvard Business School.

Wilson, C. R., Trautmann, N. M., MaKinster, J. G., & Barker, B. (2010). Science pipes: A world of data at your fingertips. Exploring biodiversity with online visualization and analysis tools. *The Science Teacher, 77,* 34–39.

Wink, D. J., & Weaver, G. C. (2008, October). *Evaluation of the Center for Authentic Science Practice in Education (CASPiE) model of undergraduate research.* Paper presented at the National Research Council's Workshop Linking Evidence to Promising Practices in STEM Undergraduate Education, Washington, DC.

Wood, W. B. (2003). Inquiry-based undergraduate teaching in the life sciences at large research universities: A perspective on the Boyer Commission report. *Cell Biology Education, 2,* 112–116.

Woodin, T., Carter, V. C., & Fletcher, L. (2010). Vision and change in biology undergraduate education, a call for action - initial responses. *CBE—Life Sciences Education, 9,* 71–73.

# PART IV

## PROFESSIONAL DEVELOPMENT OF LEARNING ASSISTANTS, TEACHING ASSISTANTS, AND INSTRUCTORS

CHAPTER 14

# DEVELOPMENT OF BIOLOGY GRADUATE LABORATORY ASSISTANTS' CONCEPTUALIZATIONS OF TEACHING SCIENCE AS INQUIRY

**Kristen R. Miller and J. Steve Oliver**

## ABSTRACT

This qualitative study examines how four novice graduate student instructors of undergraduate non-majors' biology laboratory sections developed their conceptualizations of teaching science as inquiry over the course of a one-semester teaching assignment. While teaching, the graduate laboratory assistants (GLAs) were involved in an instructor preparation program on teaching science as inquiry. Three of the GLAs developed pedagogically appropriate conceptualizations of teaching science as inquiry, but the fourth was only able to conceptualize inquiry in relation to his work as a scientist. These results indicate that professional development on teaching science as inquiry can shape GLAs' beliefs about what it means to teach science as inquiry, but other factors may also affect how they conceptualize inquiry teaching practices. Future research should include further analysis of the outside factors identified in this study and whether the developed conceptualizations affect the actual teaching in which instructors engage.

*Research Based Undergraduate Science Teaching*, pp. 437–461

Teaching assistantships serve multiple purposes for graduate students: They provide a paycheck, often allow for a waiver of tuition and fees, and afford valuable teaching experiences. The assignments undertaken by a graduate teaching assistant generally involve the education of undergraduate students. In the sciences, the assistants often teach laboratory sections within introductory level courses, but the assignments may vary extensively across and within disciplines (Harris & McEwen, 2009). The teaching preparation of these instructors appears to vary as much as the actual teaching assignments. If any sort of teaching preparation is provided, it most often occurs at the university level, meaning it is not program or department specific (Gray & Buerkel-Rothfuss, 1991). The research we report here is focused on graduate teaching assistants teaching introductory biology laboratories with the benefit of a departmental-specific teaching preparation workshop. The laboratory instruction in the course that they were prepared to teach was built around student inquiry, in large measure a novel conception of instruction to the teaching assistants.

It is not uncommon for university science departments to attempt to reform their instructional practices in introductory courses by incorporating more "inquiry" into the lecture and/or the laboratories. Common paths to the addition of inquiry include recognition of the *National Science Education Standards* (NSES) (National Research Council [NRC], 1996) as the mandate for the reform. For many programs, however, incorporating inquiry is a formidable challenge because higher education science instructors often have little to no teaching training (Sunal et al., 2001; Tanner & Allen, 2006). McKeachie (1990) reported that higher education instructors tend to have only slight familiarity with current teaching methodologies. Further, if one takes a look at the immense literature base on science education reform in the last 20 years, it quickly becomes apparent that the word, *inquiry*, has been encumbered with many meanings. As Windschitl (2004) described it, "ideas about inquiry are partly 'in the head' (with different people understanding different aspects), partly embodied in the practices of the classroom, and partly codified in various community-wide discourses" (p. 483). Inquiry generally appears to reflect:

> at least three distinct categories of activities—what scientists do (e.g., conducting investigations using scientific methods), how students learn (e.g., actively inquiring through thinking and doing into a phenomenon or problem, often mirroring the processes used by scientists), and a pedagogical approach that teachers employ (e.g., designing or using curricula that allow for extended investigations). (Minner, Levy, & Century, 2010, p. 3)

These meanings of inquiry may overlap, or be kept as exclusive categories, depending upon who describes them, and significant variation within each of these categories can be found within the science teacher education research literature (Anderson, 2002).

Teaching science as inquiry is what Chiappetta (1997) also refers to as scientific inquiry. Students are involved in "active student learning and the importance of understanding a scientific topic. Here the content becomes a critical aspect of the inquiry" (p. 23). In order to teach science as inquiry, activities in which students engage and the teaching methods used to demonstrate those activities should show that science is a process of inquiry similar to how scientists "do" science. Engaging students in this inquiry process can draw a parallel in application to how students might approach problems in their own lives that need to be investigated (Flick, 2003, as cited in Abrams, Southerland, & Silva, 2008, p. xvi). A difficult task in teaching science as inquiry is that it requires students to utilize metacognitive skills to actively learn how to inquire (i.e., how to ask relevant questions and justify explanations or ideas).

The faculty of the division of biological sciences at the University of Georgia (UGA) specialize in teaching and research related to biology education. They collectively have redirected the goals of all undergraduate introductory biology laboratory courses toward "doing" science. This redirection has resulted in a shift in each laboratory course toward students' conducting laboratory activities in which they are (1) asking testable questions, (2) defining contexts for these questions, (3) defining methodologies, (4) collecting data, (5) justifying conclusions made, and (6) engaging in argumentation and communication of findings (National Research Council [NRC], 1996; 2012). Fairly early in this redirection process, it became apparent that the graduate laboratory teaching assistants (GLAs) who instructed these laboratory courses were having many more instructional challenges compared to those experienced in the previous laboratory structure. In response, an ongoing teaching preparation program was developed that provided an initial introduction into the teaching of biology as inquiry as well as ongoing teaching support to the GLAs throughout the duration of their teaching assignments. Across a given teaching assignment, this support begins with a pre-semester workshop preparing GLAs for the multiple meanings of inquiry in their instruction. Throughout the semester, mini-preparation sessions reinforce the multiple meanings of inquiry. Opportunities occur for reflective discourse with peers and a teaching mentor and are supplemented with mentor and peer teaching observations. The observations are conducted using an observational protocol designed to record both the frequency and quality of selected teacher and student behaviors associated with inquiry instruction and learning. (Note: For a detailed list of these teacher preparation program components and to understand how the teaching preparation impacted specific teaching characteristics of the GLAs, please refer to Miller, 2010.)

## PURPOSE AND RATIONALE

### Novice Teachers' Conceptualizations of Teaching Science as Inquiry

There is essentially no educational research literature regarding GLAs' preparation to teach, and especially preparation to teach science as inquiry, or even inquiry in general. There is, however, important information to be gleaned on these same topics from the more ample research literature available on K–12 preservice and inservice teachers. Studies, for instance, indicate that professional development efforts geared towards building a better understanding of the nature of science (NOS) among elementary school teachers who lack a science background are frequently not successful. The lack of success of such teachers was measured by the long-term retention of proper notions of this topic (Akerson, Abd-El-Khalick, & Lederman, 2000). If the professional development activities consist of only interventions such as a short duration part of a teaching methods course or a workshop, then significantly little occurs. These researchers, however, demonstrated that providing ongoing teaching assistance while these teachers are actually teaching allows them to build, reflect upon, and teach appropriate conceptions of NOS using "inquiry science methods" (Akerson & Abd-El-Khalick, 2003). A different approach to the inculcation of inquiry teaching was reported by Keilborn and Gilmer (1999), who engaged preservice teachers in investigations involving the processes of science so as to successfully enable them to "internalize and transform new information for their own use and understanding" (p. 93) and, therefore, have a greater ability to teach science as inquiry to their students.

Even when preservice teachers have in-depth science background, however, conceptualizations of inquiry for instruction do not necessarily emerge with ease. In a study of 14 master's degree students in a secondary science teaching program where all participants held science degrees, Windschitl (2004) found that when these students had to plan inquiry investigations and later do the inquires as though they were students, they struggled with enacting an empirical inquiry study that had theoretical grounding. He concluded that despite their science backgrounds, these students struggled with asking a good question and designing a study to test that question. These struggles with enacting "authentic" inquiry in the classroom must be taken into account when constructing courses for preservice teacher:

> From a constructivist perspective, we know that pre-service experiences can be best designed only if we first understand the frameworks of knowledge that pre-service teachers bring with them to the program and the broader

> culturally reinforced models that maintain everyday ways of thinking about the disciplinary activities of scientists. (Windschitl, 2004, p. 508)

Windschitl also reported a number of additional factors that may influence how preservice teachers conceptualize inquiry including emulation of their previous instructors in K–12 education, how their undergraduate laboratory courses were taught and the roles student took within these labs, and their undergraduate coursework in a teacher education program (Blanchard, Southerland, & Granger, 2008; Windschitl, 2003).

## THEORETICAL FRAMEWORK

Our study was framed using the epistemology of constructionism. Constructionism, as defined by Crotty (1998), "is the view that all knowledge, and therefore all meaningful reality as such, is contingent upon human practices, being constructed in and out of interaction between human beings and their world, and developed and transmitted within an essentially social context' (p. 42).

Methods for this research study, additionally, utilized the lens of symbolic interactionism. This research design approach allowed us to assume a role of participant observer in order to collect data from GLA participants throughout their interactions with each other, with the researcher, and with their students. Symbolic interactionist perspectives often are used in research studies having a constructionist philosophy, as both are centered on the idea that knowledge is built through a social, interactive process between individuals and their surroundings (Blumer, 1969, as cited in Crotty, 1998). In this research study, the GLAs' experiences as novice teachers in introductory, undergraduate biology laboratories are characterized by symbolic interactionism as posited by Benzies and Allen (2001): The meanings constructed and assumptions developed by the GLAs were often revisited, reinterpreted, and then redefined.

## METHODS

### Data Collection Instruments and Analysis

The focus of this reported research is how GLAs came to conceptualize teaching biology as inquiry in their first teaching of introductory biology laboratories for non-science majors. All research was conducted in accordance with the University of Georgia's Human Subjects Institutional Review Board, Project # 2008-10060-2. The primary source of data was transcripts from three semi-structured GLA participant interviews. The first interview was conducted within two weeks of the teaching preparation workshop given prior to the start of any teaching. The second occurred approximately halfway through the 15-week semester, and the third took place during the last two weeks of the semester. Qualitative statistical software (ATLAS.ti.6) was used to assist with coding data from interview transcripts. The coding process followed the three-step data analysis model presented by Strauss and Corbin (1998). Member checking was used to establish trustworthiness when collecting and analyzing data related to interviews. Themes from the data analysis were triangulated with additional data sources including pre- and post-semester GLA demographic surveys, behavioral observations of GLAs, GLAs' reported relative comfort levels with the content they had to teach, GLAs' reactions to an inquiry teaching vignette, and informal discussions with GLAs.

### GLA Participants

Participants in this study included a subset of four GLAs from a population of 11 who were all assigned to teach BIOL 1103L, a one-credit introductory biology laboratory course for non-science majors with a content focus on cellular and molecular biology. While all 11 GLAs took part in the teaching preparation program and agreed to participate in the study, the study participants (Tara, Cameron, John, Evan [pseudonyms]) were purposely selected based on variance in the responses to a demographic survey that included self-reports of characteristics including gender, past teaching experience, past experience with inquiry, reported likelihood of teaching in the future, and initial "buy-in" to inquiry teaching methods (i.e., response to the teaching vignette previously mentioned). None of the participants previously had taught BIOL 1103L. We hoped that this range of responses would result in data whose analysis could produce an understanding of the impact of the teaching preparation program that could be generalized to a broader sample of novice graduate student teaching assistants.

## FINDINGS

The findings reported here are an attempt to examine how novice GLAs respond to teaching biology through inquiry when prepared via myriad methods presented throughout the duration of their teaching assignment. The first section of the chapter describes the findings related to GLAs' views on differences between science laboratory courses taught using a traditional format and those taught with an inquiry format. This is followed by findings arising from an analysis of GLAs' descriptions of what students need to "do" inquiry in section two. Next, in section three, the analysis of GLAs' recognition of instructional activities as conforming to their model of inquiry science is briefly presented. In the fourth section, we conclude with findings describing how GLAs define a "good" teacher of science as inquiry. Please refer to Table 14.1 for a summary of all findings.

**Table 14.1. A Summary of the Development of Conceptualization of Teaching Science as Inquiry**

| **Recognition of differences between traditional and inquiry laboratory activities** | |
|---|---|
| All GLAs articulated differences between traditional and inquiry laboratory activities throughout the semester of their teaching assignment. | |
| 1. All GLAs shared some commonalities in these described differences. | |
| 2. All GLAs developed individual ideas about what constitutes an inquiry laboratory exercise throughout the semester of their teaching assignment. | |
| 3. All GLAs' views of what constitutes a traditional laboratory exercise stayed the same throughout the semester of their teaching assignment. | |
| **What students need to "do" inquiry** | |
| Tara | Students need certain "tools" to do inquiry, but these are not necessarily the same tools that they need from her as instructor (e.g., need her to be prepared). |
| Cameron and John | Students need certain "tools" to do inquiry, and students need their instructors to provide those tools. |
| Evan | No information given on what students need to do inquiry, but they need him as their instructor to work on classroom management skills (e.g., be fair grader, not be "lax"). |
| **Recognition of inquiry laboratory activities** | |
| *BIOL 1103L laboratory series* | |
| 1. All GLAs agreed that particular laboratory series either had no or very little inquiry (e.g., *Case It!*) or a lot of inquiry (e.g., *enzymes, Mendelian genetics*). | |
| 2. All GLAs found that particular laboratory series were more "middle ground" inquiry activities (e.g., *antibiotic resistance, water quality*). These laboratory series elicited a range of individual variation in reference to 'levels' of inquiry experienced as well as when during the laboratory series the inquiry was experienced. | |
| **Defining a good teacher of science as inquiry** | |
| 1. All GLAs expressed multiple commonalities in references to characteristics of a good teacher of science as inquiry including approachability, strong content knowledge, taking active roles in student learning, showing excitement for student learning, and getting students to participate. | |
| 2. All GLAs individually described additional teacher attributes important to be a good teacher of science as inquiry. | |

## Traditional Science Laboratories Versus Inquiry Science Laboratories

The first activity of the academic year was the teaching preparation workshop. At the outset of the workshop, all BIOL 1103L GLAs were asked to examine the course's laboratory manual and verbally state any reactions they had to it as compared to laboratory manuals they had used when *taking* introductory biology laboratory courses. Among the most common statements was the observation that the BIOL 1103L manual had a limited number of defined experimental protocols. This observation initiated a discussion between us and the GLAs regarding the roles of teachers and students in "traditional" laboratory teaching, including the typical structure of laboratory manuals that accompany this instructional methodology. This conversation naturally blended into comparisons between "traditional" teaching of science in a laboratory setting and teaching "inquiry" science laboratories. Shortly after this workshop, the participants engaged in their first of three individual interviews. In those interviews, all participants described their experiences taking introductory science laboratory courses as traditional in how they were taught, how students were expected to demonstrate learning, and how laboratory activities were presented in manuals. Tara exemplified everyone's responses in her first interview:

> For basic bio and basic chem, it was "here's a protocol"… like a little 10-minute spiel about here's a little background and then it was, follow your instructions, do what you need to do, and ask me if you have any problems.… Then, you write up a lab report, turn it in next week. (08/09)

When asked to suggest why students and teachers took the roles assumed in these traditional laboratory courses, the GLAs provided several reasons. One individual felt that traditional and older faculty didn't want to change (John). Another felt that professors didn't know anything about what actually was happening in the laboratory class unless it contributed to the lecture course grade. He also felt that TAs (teaching assistants) only knew enough to write on the board, but could not expand on anything. This GLA (Cameron) also made the assumption that some TAs didn't want to be there because they didn't seem to like it. Another GLA (Tara) thought that the professors wanted to give students practice following directions (a skill to use in graduate school). One exception to this traditional experience was described by Cameron, who excitedly pointed out that his introductory biology laboratory course was inquiry-based (this could not be said for his other introductory science laboratory courses). He explained that he never knew that style of teaching and learning laboratory activities "had a name" (first interview, 08/09).

The idea of what was, or was not, an inquiry lab persisted through much of the data collection process. A question posed to the study participants in all interviews and during instructional preparation sessions asked them to differentiate between science laboratory courses taught using a traditional format from those taught with an inquiry format. We sought to ascertain how the GLAs' perceptions of the benefits or detriments of teaching science within an inquiry-based structure changed over the course of the semester.

## Tara

Consider this example of Tara, who, in her first interview, described a traditional instructional laboratory format as teachers "telling you what to do." She further stated that, in the traditional laboratory, students were given a protocol to follow, and she contrasted this with an inquiry format where students must figure out what to do and design their own protocol. She continued by saying that in order to do this, students must "understand enough about it to decide how you test it" (first interview, 08/09). By halfway through the semester, Tara had broadened her ideas of what constituted an inquiry science laboratory experience. This expanded conception included group discussions and explicit student participation; beyond this, students "had to put the work in" in order to finish the laboratory activities (second interview, 10/09). For Tara, traditional instructional laboratory courses were more passive for teachers and students: "You are up at front, talking at them, answering questions along the way. But most students don't ask questions because they want to go [i.e., leave the laboratory to go somewhere else]" (second interview, 10/09).

Tara's end-of-the-semester interview reiterated her statement that inquiry laboratory courses required more student responsibility and more interaction between students and teachers. By this point in her teaching experience, however, Tara also had developed as a teacher to a deeper level as she was able to describe, in a way that had been missing in previous interviews, how the differences between traditional and inquiry laboratories are related to student learning. She explained at length how the interactive nature of inquiry laboratories allowed teachers to acquire a greater knowledge of student conceptual learning. Positing her idea in terms of the more passive, traditional teaching role, she said:

> You don't know what students are learning until they take a test.... With inquiry-based you sort of gauge their progress along the way, whether they are actually understanding, and you can help them understand because that is the way the lab is set up. (12/09)

## John

John's sentiments regarding differences between traditional and inquiry instructional laboratories often mirrored Tara's, but he was able to express and demonstrate a deeper understanding of what teaching science as inquiry actually requires from teachers and students. His initial impressions of differences between the laboratory formats included the belief that traditional laboratories have everything laid out for students, so they simply follow a protocol to get through an exercise. He contrasted this method with inquiry laboratories that required thinking on the part of the student, including personal reflection on previous experiences in school and in life and how those experiences related to the science that were encountering in their laboratory courses. John further believed that the information revealed through students' reflections helped fill in gaps in the students' knowledge that the laboratories were not providing.

In his second interview, the growth in John's teaching knowledge was evidenced by his ability to expand on his understandings of differences between traditional and inquiry instructional laboratories. He noted that having students create laboratory research methodology actually made the process of learning science easier because it is their own work on which they can reflect. He saw that act of doing science to learn was a productive avenue to understanding the processes of science. John spoke to this avenue in a mid-semester interview:

> The biggest difference is students basically make up their own methodology versus just being given everything.... In some respects that makes it easier on the students later on...when stuff doesn't work then you can say "you made up this experiment, how you can do it better?" (10/09)

By the end of the semester, John had shifted his concerns and was describing them in terms of his frustrations in trying to teach science as inquiry. He described how student and teacher expectations in traditional laboratories tend to persist throughout the class at a level of expecting a "right answer" (third interview, 12/09). In this traditional instructional setting, he felt that the chance of getting an unexpected result was slim unless steps are completely skipped. Beyond this concern, the assurance of seeking correct answers created a low-stress environment for both the GLA and the student. John's idea of a low-stress environment contrasted sharply with the inquiry biology laboratories in which John engaged with students. He struggled with how to dispel students' misconception that there is always a "perfect" answer in science. He was observed to tell students several times during the semester of data collection: "Sometimes it [science] just fails miserably."

## Cameron

Cameron presented a third view of how a GLA becomes a teacher of inquiry. In his first interview, Cameron indicated that defining an instructor in inquiry laboratories should start with removing the word "instructor." He felt that this term was "misleading" because it implies that students are told how to do everything (first interview, 08/09), which is actually a characteristic of teachers in traditional instructional laboratories. For Cameron, the learning experience in an inquiry laboratory should be more of a conversation among teachers and students, so that teachers are not seen as overbearing authorities. By the end of his teaching assignment, he explained in detail the monotony of traditional laboratories where, even though the content topic changes each week, the methodological procedures never do. In Cameron's view, inquiry laboratories require critical thinking about the entire process of science: "you actually have to think about it and each week your mind is only on one problem [but] you have to get all aspects of it" (third interview, 12/09).

## Evan

Finally, we consider Evan's views on the distinction between inquiry and traditional laboratories. In his first interview, Evan did not try to distinguish between traditional and inquiry instructional laboratories because he reported that he had not experienced any kind of inquiry in such laboratories. By the mid-point of the semester, he began to report differences between the two types of laboratories. For Evan, traditional laboratories followed a method of students receiving a summary/overview from their teacher and then being allowed to do the activities. Inquiry labs were interactive between teachers and students, and teachers needed to be proactive about this interactive nature. By the mid-point of the semester, Evan also had begun to describe his frustrations, pointing out that inquiry laboratories were a "headache" (second interview, 10/09). He felt that introductory biology students were not ready to engage in activities such as designing experiments. Among his reasons for this claim were students' lack of confidence, lack of knowledge about the biology content or experimental design, and inadequate preparation to predict results of their experiments. Evan felt that students could not deal with the mistakes in their experimental plans, although he expressed the view that this was a frequent and important part of doing science. For him, these introductory biology students, who were not science majors, should either engage in an inquiry activity and then write a structured laboratory report on the experience or "go back to traditional laboratories where it is cut and paste at the top end and freely write about it" (second interview, 10/09). By the end of the semester, Evan emphatically decided that inquiry was not an appropriate instructional strategy for introductory biology laboratories.

### Summary

Overall, these GLAs were able to put forward ideas to differentiate a traditional laboratory from an inquiry one, but the degree to which they felt the inquiry laboratory was a superior instructional setting varied widely. From this issue, we now turn to examining student needs related to inquiry labs.

## WHAT STUDENTS NEED TO "DO" INQUIRY

Across the semester, the GLAs began to conceptualize students' needs in terms of what was required from instructors in order for inquiry learning to proceed. In interview discussions of this topic with us, most of the GLAs also shared valuable information about what students needed to *have* in order to *do* inquiry, or engage in the process of science. The following findings emerged as important pieces to the puzzle of what it means to teach science as inquiry.

### Tara

Tara reported that content knowledge was critical for students to be able to design an experiment and then test it, but she did not report that that her students needed *her* to provide that content. Rather, when asked to clarify what she felt her students needed her to provide them to successfully engage in inquiry, her responses focused on skills of two types: science writing and critical thinking. For the former, although she was teaching non-science majors, Tara articulated that since science writing is concise, and concise writing is essential for the business world (first interview, 08/09), her students needed to learn science writing mechanics. Tara also spent extensive time in interviews describing that critical thinking skills were key to students being able to successfully engage in inquiry labs because they needed to be able to make choices for themselves and justify the decisions that they made throughout the experimental design process. She emphasized how important it was for her to guide students to learn to articulate their reasoning for decisions made: "They need to know *why* they are doing things. I need to ask them why and get them to explain it and if they don't get it quite right I can clarify it. So I think that's what I am there for…to guide them" (08/09). For Tara, this *"why"* aspect of the inquiry biology laboratories was essential to the success of the laboratory experience; she consistently revisited this theme of higher-level thinking in all of her interviews. In a parallel fashion, she also stated that since that the laboratory activities emphasized questioning to such a large degree, her students needed her to be prepared for those questions.

## John

While John also reported that content knowledge and critical thinking skills were key components of students' abilities to be able to engage in design and testing of experimental methodology, decision making, and argumentation of choices, he added that "minor methodology" background was necessary in order to even approach inquiry as presented in the BIOL 1103 laboratory course (second interview, 10/09). However, John diverged from the student characteristics that Tara reported as necessary in order to do inquiry and the directives that students need to acquire from the instructors of inquiry based science courses. He felt that the attributes that students needed in order to do inquiry were the same ones he identified relative to what students needed from him. John felt that he needed to give students content knowledge and structure in how to approach the inquiry aspects of the laboratories, but he also felt that students need reassurance and confirmation that they are doing the "right thing." It was frustrating to him that they could not seem to grasp, even by mid-semester, that he was not going to give them the answers to solve the problems at hand. John wanted students to understand that he would guide them in certain directions to allow them to think critically and make choices, but would not explicitly tell them what to do. These views persisted throughout the semester.

## Cameron

Cameron, like John, reported that the attributes students needed to do inquiry were acquirable from him as an instructor: a safe and comfortable teaching and learning environment. He that felt important aspects of this environment were to be open and ready for questions at any time, to provide an interactive instructor–student atmosphere so students would be willing to ask questions, to be open-minded about what the students were asking, and to provide encouragement and instill confidence in students that they could "do" science (summarized by Cameron, third interview, 12/09). He also expressed that he had to guide them more at the beginning of the semester, but then he could back off from that role as the semester progressed.

### Evan

While Evan expressed two qualities that his students needed to be able to engage in inquiry labs that were identical to Cameron's view, encouragement and confidence, these were the only two similarities he shared with any GLA participants when describing what students need to do inquiry. In fact, Evan reported entirely pragmatic elements of what students needed from him as their instructor. His students, for instance, needed him to be a fair grader, to not be "lax" in the classroom, to challenge them, and to stick to the rules (third interview, 12/09). He likened his students to high school students: immature, not ready to take responsibility for being in college, and unable/unwilling to follow directions. Evan mentioned, in his last interview, that instructor content knowledge was important in inquiry laboratories. He specifically pointed toward his own deficits when he described having difficulties getting through the Mendelian genetics laboratory activities because of his weak content knowledge in this area.

In summary, consensus could not be drawn across these four GLAs' understanding of what students needed in order to do inquiry. While, to a large extent, John and Cameron expressed that what students need most from instructors are the same tools that they need to do inquiry (e.g., content knowledge, interactive atmosphere), Tara's perceptions were less aligned with those of her peers, suggesting that she viewed her instructor responsibilities as somewhat separate from providing all the tools students need to do inquiry. Evan did not provide a sense of what students needed to do inquiry, but suggested that he needed to focus on classroom management issues with his students. The findings in this section of data analysis parallel previously described findings from all interviews of the GLAs' descriptions of what an inquiry laboratory experience entails for both students and teachers.

## RECOGNITION OF INQUIRY LABORATORY ACTIVITIES

At the end of the semester, in final interviews, one of the main focal points of our discussion with the study participants was their overall impression of the laboratories they taught. Despite receiving the same ongoing teaching preparation and teaching the same laboratory sequences, they only partially agreed on the depth and nature of inquiry experienced in the laboratory exercises.

Three of the four GLAs agreed that two of the five laboratory series taught (enzymes, Mendelian genetics) offered the most inquiry opportunities for students and teachers because these series demonstrated the greatest emphasis on students fully exploring the experimental design process and therefore engaging in critical thinking skills. In these series, students engaged in interpreting unexpected data, using previous knowledge to help solve problems, and being able to complete laboratory activities with only teacher guidance. Two other laboratory series (water quality, antibiotic resistance), and their inquiry characteristics, were more debatable among the GLAs. These laboratories held varying degrees of characteristics of inquiry as described by the GLAs but included such characteristics as interpreting unexpected data, articulation of experimental design choices, and justification of conclusions. The remaining laboratory series, titled *Case It*, was determined by all GLAs to have few to no inquiry characteristics. Overall, the GLAs appeared to agree about the end points of the range that demarked the presence or absence of inquiry in a laboratory activity. In other words, they agreed upon what constitutes little or no inquiry as well as what constitutes extensive inquiry, but laboratory activities that were determined to be more "middle ground" resulted in variations of interpretation of what inquiry was experienced by teachers and students.

## DEFINING A "GOOD" TEACHER OF SCIENCE AS INQUIRY

When teaching science as inquiry, teachers must engage students in activities that involve experiencing the process of science. Such involvement can be achieved through partial inquiries, those that engage students in some of the "five essential features of classroom inquiry" (NRC, 2000, p. 25) or full inquires that meet all five features (p. 28). For the purposes of this study, notions of qualities of a "good" teacher of science as inquiry were based on these five essential features of classroom inquiry. While we never held specific conversations or instructional sessions that discussed these qualities in reference to being an effective teacher of science as inquiry, the mini-preparation sessions held throughout the semester, post-observational feedback, and instructional teaching notes described qualities that would enable GLAs to achieve some or all of the essential features. Instructional teaching notes given to GLAs, for example, often included questioning strategies to use, so that students would have to communicate their process of thinking to the GLAs. These strategies were guidelines and thus suggested avoidance of telling students the correct answer to problems but instead pushing them to articulate their thought processes.

In some instances of reflective discourse, our discussions with the GLAs revealed frustration students expressed with inquiry-based laboratory activities. Our discussions, conversely, also revealed how satisfying it was for the GLAs and students when the students were able to reach the stated objectives of a laboratory exercise. To reach these points of success necessitated that the GLAs remind students of their capabilities of developing and using higher-order thinking skills in a field that is not their preference. To accomplish inquiry, the GLAs had to allow students to make some mistakes; one of Evan's comments synthesizes the comments of all four GLAs: "Let's face it, in science there are failures. We learn from them" (second interview, 10/09). Actually allowing failure to happen in the classroom, however, might prove to be a difficult teaching task. Reflective discourse allowed us to discuss these teacher qualities with the GLAs in a general setting, rather than only through a specific discussion of characteristics of a good teacher of science as inquiry.

As the GLAs experienced teaching science as inquiry in BIOL 1103L, we attempted to capture how they developed their conceptions of what they needed to do, or have, for successful implementation of the laboratory activities. This information was critical to understanding how the GLAs approached conceptualizing teaching science as inquiry because their responses could potentially provide insight into what they deemed important to offer as a teacher of science as inquiry. The following section of findings illustrates analysis of GLAs' conceptions of a "good" teacher of science as inquiry. The findings presented as a result of this analysis indicate high agreement among all GLAs regarding specific personality characteristics, class environment characteristics, and content knowledge needed to be a successful teacher of science as inquiry.

## Tara

One theme emerging across three GLAs was that a good teacher of science as inquiry needed to have the personality characteristic of *being approachable*. For Tara, this was important for two reasons. One reason was that students would be comfortable asking questions, and the second was so that they would see she was comfortable answering these questions: "I guess [instructors] have to be someone … students can feel comfortable talking to [and must] feel comfortable responding to [student] questions" (first interview, 08/09). In this discussion of teacher personality characteristics, Tara reiterated the importance of being comfortable with content in order to answer questions. At the end of the semester, one of her reflections on the entire semester demonstrated that she fulfilled this last characteristic of a good teacher of science as inquiry: "I was already pretty confident as a teacher. But I definitively feel more like a teacher now, but I think that's more the whole basis of the lab…I feel pretty prepared in concepts" (third interview, 12/09).

By mid-semester, Tara described a good teacher of science as inquiry as someone who could effectively strategize questioning techniques in order to ask students questions that didn't give away answers. These probing questions needed to be carefully worded so that students would have to actively think through the steps they had taken to that point; this often would allow for students to answer their own questions. She also reemphasized that getting students to participate was critical in inquiry laboratories, and suggested that this might be easier for some GLAs than others due to personality characteristics: "Cameron is good at that because he's so friendly and so nice. He's all over the place, but they love it. I'm not that way" (second interview, 10/09).

In her final interview, Tara reiterated that a good teacher of science as inquiry needed to be able to ask the "right" questions to students so that they could critically assess their work (third interview, 12/09). As Lederman, Abell, and Akerson (2008) explained,

> At the same time that students are developing abilities for doing inquiry, they are building understanding about scientific inquiry…. The understandings are often the result of metacognition in relation to inquiry-based activities…. That is, good teachers help students think about the thinking processes that they have used and focus students on knowing how to improve. (p. 15)

Also, student participation was a necessity for the laboratories to flow well. Her discussion of this aspect of teaching BIOL 1103L indicated that the class environment might affect student confidence and therefore willingness to participate: "I think it's easier to do it in small groups … because I think a lot of people are afraid in front of a class but, when you are face to face, they have to at least respond" (third interview, 12/09).

## John

At the beginning of the semester, John listed a variety of characteristics exhibited by a good teacher of science as inquiry, and some of these overlapped with Tara's responses. First, he indicated that a good teacher had to be "passionate about the [students'] work and take interest in it" (first interview, 08/09). This personality characteristic conjures up images of a teacher showing excitement and overt enthusiasm for what is taking place in class, but we noticed throughout the semester that John's personality tended to be reserved and, at times, stoic. Through his questions to students and in discussions with us, he clearly showed interest in his students' work. He asked thoughtful questions when teaching, for example, which allowed students to consider multiple avenues for their experimental plans. His reserved outward personality, however, may not have given students confidence that he was interested in their work.

For John, a good teacher of science as inquiry could not be passive. Noting his experience with hands-off instructors who left him feeling alone in the learning process, he had few memories of support received through mentoring. He could only rarely recall being able to approach his professors with questions or problems. He talked about one of his past instructors who took an active role in discussing, planning, and evaluating John's work. John wanted his students to see him as approachable, so that they would come and ask questions. He felt it was important to stay up to date with students' progress to keep them on track. John initially reported that many students came to him with questions, but as noted earlier, his students often wanted him to just tell them the answers to problems. Many eventually stopped coming to ask him for help, but the ones who persisted were those whom John described as "really wanting to know how to get to the right answer" (second interview, 10/09).

A difficult role to take, but a necessary one in John's opinion, was to let students make mistakes along the way of engaging in the process of science "because that is where students learn" (first interview, 08/09). John felt confident that if students were to experience authentic inquiry, they needed to go through the reality of making errors in science and then finding ways to fix those errors. For an inexperienced teacher, John took a bold step by consciously creating a learning environment where students might resent his not telling them the "right way" to do things. By mid-semester, John described how good teachers of science as inquiry needed patience because inquiry can be "trying on students, and they take that frustration out on the TA to some extent" (second interview, 10/09). This student frustration aspect of teaching science as inquiry is something he reported having talked about with all the GLAs and with which they were all in agreement. He also noted that content knowledge and familiarity of the premise of laboratory activities, experienced through doing the same laboratory in a different course/setting or experiencing it in a research laboratory, was critical so that a teacher can troubleshoot problems and be able to answer questions. John emphasized, "even though you understand the context, you may not be able to answer some of the problems that arise without having actually done something like it" (second interview, 10/09). Another important issue for John was the significance of being able to rephrase questions. He came to see that not everyone understands an idea in the same way. He perceived peer learning as a means to achieve understanding if a teacher was unable to convey information to a particular student.

## Cameron

Cameron, like Tara and John, perceived good teachers of science as inquiry as having both content knowledge and contextual understanding of the laboratory activities 'because the minute you go up there and you really don't know what they are talking about, they'll see right through it" (first interview, 08/09). He admitted, like Tara, that this level of instruction takes extra effort on the teacher's part. We observed Cameron putting both of these teaching characteristics into play throughout the semester. In his introductions for each laboratory section, Cameron used different ways to present an overview of the laboratory activities, and these examples almost always pertained to the students' lives. He brought in terminology from different non-science disciplines and made parallels between those terms and the processes they involved and the terminology and processes the students were working within laboratory activities. For example, in a discussion of the cumulative writing assignment associated with the enzymes laboratory series, a news article that would be read by college students, Cameron drew parallels between the sections of a traditional laboratory report that students typically write in a science laboratory course and the sections typical of a news article. Calling on the journalism majors in the class, Cameron had them describe the term "lead" to the class and then ask what part of a lab report would be akin to the lead in a news article.

Similar to John's assertion that good teachers of science as inquiry should show passion for students' work, Cameron also discussed the importance of expressed enthusiasm for learning and questions. In his view, students should see that as the acts of a teacher: He enjoys questions and is willing to answer them. Cameron echoed Tara's stated sentiments that he needs a deep content base to answer students' questions.

By mid-semester, Cameron felt that to be a good teacher of science as inquiry, one had to listen to students carefully to make sure that their questions are being heard. He said it was easy to only half listen and to answer students with preconceived answers. Cameron also reported that, in order to support the interactive format that is inherent to inquiry laboratories, a teacher cannot talk down to students but instead should create conversations with students, keeping everyone on equal levels. The end of the semester found Cameron reflecting that patience is a virtue, and without it, the laboratory would fall apart: "Students are stressed and if you get flustered it all falls apart" (third interview, 12/09). Content knowledge was still deemed critical, as was being open-minded to seeing that, to teach science as inquiry, an instructor should take part in collaborative discussion with students: "It's not just me talking to the class. It's students who are talking back" (third interview, 12/09).

## Evan

Evan initially focused his thoughts on a good teacher of science as inquiry in reference to teaching characteristics noted by other GLA participants: not being introverted, showing excitement about what students are doing, and getting students to participate by taking on the role of a "game-show" host in order to motivate them (first interview, 08/09). He thought it was important to make sure students in these laboratories understand that they have to participate. By mid-semester, Evan said it was important to not give answers to students because they don't end up learning anything: "Don't give answers.... If I give it to you it's worthless.... I have been there ... you are telling me and I can do this on a test but I don't know what it means" (second interview, 10/09). He also noted, like Cameron, how important it was to listen to students' questions and not "write them off'" (second interview, 10/09). Evan likened questioning strategies to the profession of his sister, a defense attorney; in order to pull out what students know so that they can answer questions, you have to act as a lawyer and cross-examine them. This "brings students along" (second interview, 10/09) to the point where they can begin to start answering questions. Interestingly, these two interviews revealed two of several cases where Evan could verbally attribute aspects of the teaching preparation that the GLAs received to what needs to happen in his teaching but, when in the classroom, we frequently observed Evan telling students answers to questions and taking a passive role in student interactions. Here is one example of this dichotomy in his intended instruction versus his practice as recorded in notes from observations:

> Evan tells students all possible plate set-up choices and why each set-up would test certain parameters of antibiotic resistance. He needs to let student figure that out, or at least try. That is the point of the lab. (first set of observations, week 7 of semester; first week of antibiotic resistance laboratory series)

At the end of the semester, Evan reemphasized the need for a good teacher of science as inquiry to actively listen to student questions, citing himself as an example of what not to do: answering students' questions with preconceived answers when it could be conceptual issue they don't understand. The following excerpt from Evan's final interview described this experience when teaching the Mendelian genetics lab series:

> One of the students asked me a question and I answered with the most obvious.... I mean it wasn't a stupid answer. I was like "well let me look at your Punnett squares." There was no reason for me to look at their Punnett squares. I'm like, why did I even say that? I didn't listen. I had already in my head what he asked, even though it's not what he really asked. So there was no help of inquiry there. Where could I have led him to answer his own question? (12/09)

Evan also noted that content knowledge was critical to being a good teacher of science as inquiry, noting his problems with answering questions during the three-week laboratory series on Mendelian genetics, which he attributed to his own lack of strength in that content area. He further explained that his lack of knowledge in this area was because he had memorized everything in his undergraduate class in order to pass tests, but never felt he really understood what he was doing. Finally, Evan discussed that making the laboratory activities relevant to students' lives was important for effective teaching science as inquiry because that was the job of a teacher in general, to make things relevant to students. Both he and Cameron identified the antibiotic resistance and the water quality laboratory series as excellent examples of labs that provided students with experiences of the everyday relevance of inquiry science.

To summarize, approachability, strong content knowledge, taking active roles in student learning, showing excitement for student learning, and getting students to participate were teacher characteristics described by all GLAs as necessary to be a good teacher of science as inquiry. In this regard, it appears that there were common understandings among the four GLAs that developed from their teaching science as inquiry in the introductory biology labs. Beyond these characteristics, the GLAs described additional teacher attributes important to be a good teacher of science as inquiry that may reflect individual struggles that the GLAs faced based on their personalities or those of students.

## IMPLICATIONS

While science graduate students who are required to teach might be offered teaching education, it is often not discipline or pedagogically specific (Prieto, Yamokoski-Maynhart, & Meyers, 2006). This is a monumental issue considering that graduate students teach as many as two-thirds of undergraduate courses at some higher education institutions (Nelson, 1995). How can we assume that these novice teachers are accurately conceptualizing teaching science as inquiry, and, therefore, potentially translating this into their teaching, if they are not offered opportunities to develop these conceptions? A recent study by Park Rogers and Abell (2008) echoed this notion: "Not understanding inquiry teaching can make it difficult to translate one's beliefs about the nature of scientific inquiry into the practice of inquiry teaching" (p. 594).

We have presented an analysis of how four science graduate students teaching introductory biology laboratory courses to non-science majors came to conceptualize teaching science as inquiry over a one-semester experience. Many of the factors that the GLAs described as shaping their beliefs were elements of the teaching preparation program that they received throughout their teaching assignment (e.g., discussion of questioning strategies, discussion of consideration, and validation of student ideas). This research study suggests two hypotheses that should be studied with a larger sample size over a longer period of time: (1) factors (for instance, knowledge of content inherent to lessons being taught, views of how science should be learned, past experience with inquiry, emulation of former teachers, etc.) outside of a pedagogically specific teaching preparation brought to a teaching assignment can affect how conceptualizations of teaching science as inquiry are developed; and (2) the conceptualizations of teaching that develop during the teaching preparation and the enactment of teaching may affect the actual teaching in which the GLAs engage.

In light of both existing preservice and inservice teacher education literature on the numerous factors that may affect how teachers come to conceptualize inquiry (Akerson & Hanuscin, 2007; Varma, Volkmann, & Hanuscin, 2009; Windschitl, 2003, 2004) as well as limited research on this topic as it is relevant to graduate student instructors (Roehrig, Luft, Kurdziel, & Turner, 2003; Savage & Sharpe, 1998; Shannon, Twale, & Moore, 1998), we did not attempt to pinpoint the GLAs' understandings of teaching science as inquiry to one or even two particular characteristics. Rather, doors were left open for the GLAs to provide as much data as possible about factors influencing their conceptualizations.

The semester in which this study took place began with the four study participants reporting no or limited conceptions of "inquiry," but by its end, three of the four had come to conceptualize what teaching science as inquiry in introductory biology laboratories meant to them as individual instructors. While examination of multiple data sources revealed that all participants described numerous similar characteristics of what it meant to teach science as inquiry, analysis also revealed that three of the four GLAs' conceptions continued to develop throughout the semester while one stayed more steadfast. This latter GLA (Evan) was never able to clearly articulate ideas of teaching science as inquiry. Instead, he continued to relate inquiry to what he did as a scientist. One possibility for this is that he may have decided that when a particular teaching strategy worked, there was no need to change it. Considering the myriad responsibilities that each GLA carried in addition to the teaching assignment, it is possible to propose that a GLA would hold on to each and every success and try to rock the boat as little as possible.

The results of the analysis of the data indicated that commonalities existed between the four GLAs' descriptions of laboratory activities as offering either minimal or extensive opportunity for students to engage in inquiry. And yet, for these laboratory activities that did not have these clearly distinguishable "extremes" of involvement in inquiry, the GLAs identified different aspects of what constitutes an inquiry activity and differently interpreted how those parts of the laboratory exercise are actually experienced by students and teachers. This finding suggests that despite undergoing the same teaching preparation emphasizing understanding of pedagogy, the teaching that GLAs enact is a product of three variables that are largely out of the control of university faculty seeking to shift the nature of teaching in introduction science laboratory classes toward an emphasis on inquiry. These three factors are: (1) the knowledge and experiences that GLAs bring to their teaching assignment, (2) the meaningful learning about teaching that is taken from the teaching preparation, and (3) their experiences in the classroom as they teach the laboratory.

The findings addressed in this study suggest direction for further research into graduate student teacher education, not only in the considerations of what should factor into the design and implementation of a preparation program, but also what and how to assess what the graduate students are learning from that teaching education.

## REFERENCES


Abrams, E., Southerland, S. A., & Silva, P. (2008). Inquiry in the classroom: Identifying necessary components of a useful definition. In E. Abrams, S. A. Southerland, & P. Silva (Eds.), *Inquiry in the classroom. Realities and opportunities* (pp. xi–xiii). Charlotte, NC: Information Age Publishing.

Akerson, A. L., & Hanuscin, D. L. (2007). Teaching the nature of science through inquiry: The results of a three-year professional development program. *Journal of Research in Science Teaching, 44*(5), 653–769.

Akerson, V. L., & Abd-El-Khalick, F. (2003). Teaching elements of nature of science: A yearlong case study of a fourth-grade teacher. *Journal of Research in Science Teaching, 40*(10), 1025–1049.

Akerson, V. L., Abd-El-Khalick, F., & Lederman, N. G. (2000). Influence of a reflective explicit activity-based approach on elementary teachers' conceptions of nature of science. *Journal of Research in Science Teaching, 37*(4), 295–317.

Anderson, R. A. (2002). Reforming science teaching: What research says about inquiry. *Journal of Science Teacher Education, 13*(1), 1–12.

Benzies, L. K., & Allen, M. N. (2001). Symbolic interactionism as a theoretical perspective for multiple method research. *Journal of Advanced Nursing, 33*(4), 541–547.

Blanchard, M. R., Southerland, S. A., & Granger, E. M. (2008). No silver bullet for inquiry: Making sense of teacher change following an inquiry-based research experience for teachers. *Science Education, 93*(2), 322–360.

Blumer, H. (1969). *Symbolic interactionism: Perspective and method*. Englewood Cliffs, NJ: Prentice Hall.

Chiappetta, E. (1997). Inquiry-based science. Strategies and techniques for encouraging inquiry in the classroom. *The Science Teacher, 64*(7), 23–26.

Crotty, M. (1998). *The foundations of social research: meaning and perspective in the research process*. Thousand Oaks, CA: Sage.

Flick, F. B. (2003, March). *Inquiry as cognitive process*. Paper presented at the annual meeting of the National Association for Research in Science Teaching, Philadelphia, PA.

Gray, P. L., & Buerkel-Rothfuss, N. (1991). Teaching assistant training: A view from the trenches. In J. D. Nyquist, R. D. Abbott, D. H. Wulff, & J. Sprague (Eds.), *Preparing the professorate of tomorrow to teach* (pp. 40–51). Dubuque, IA: Kendall/Hunt.

Harris, D., & McEwen, L.A. (2009). A graduate teaching assistant workshop in a faculty of science. *Canadian Journal of Higher Education, 39*(2), 101–120.

Keilborn, T. L., & Gilmer, P. J. (Eds.). (1999). *Meaningful science. Teachers doing inquiry + teaching science*, Tallahassee, FL: SERVE.

Lederman, N. G., Abell, S. K., & Akerson, V. (2008). Students' knowledge and skill with inquiry. In E. Abrams, S. A. Southerland, & P. Silva (Eds.), *Inquiry in the classroom. realities and opportunities* (pp. 3–33). Charlotte, NC: Information Age Publishing.

McKeachie, W. J. (1990). Research on college teaching: The historical background. *Journal of Educational Psychology, 82*(2), 189–200.

Miller, K. (2010). *Advancing inquiry teaching among graduate teaching assistants in biology* (Unpublished dissertation). The University of Georgia, Athens, GA.

Minner, D. D., Levy, A. J., & Century, J. (2010). Inquiry-based science instruction—what is it and does it matter? Results from a research synthesis years 1984 to 2002. *Journal of Research in Science Teaching, 47*(4), 474–496.

National Research Council [NRC]. (1996). *National science education standards*. Washington, DC: National Academy Press.

National Research Council [NRC]. (2000). *Inquiry and the national science education standards. A guide for teaching and learning*. Washington, DC: National Academy Press.

Committee on Conceptual Framework for the New K–12 Science Education Standards; National Research Council [NRC]. (2012). *A framework for K–12 science education: Practices, crosscutting concepts, and core ideas*. Washington, DC: National Academy Press.

Nelson, C. (1995). Lessons from the job wars: What is to be done? *Academe, 81*(6), 18–25.

Park Rogers, M. A., & Abell, S. K. (2008). The design, enactment, and experience of inquiry based instruction in undergraduate science education: A case study. *Science Education, 92*(4), 591–607.

Prieto, L. R., Yamokoski, C. A., & Meyers, S. A. (2007). Teaching assistant training and supervision: An examination of optimal delivery modes and skill emphases. *Journal of Faculty Development, 21*(1), 33–43.

Roehrig, G. H., Luft, J. A., Kurdziel, J. P., & Turner, J. A. (2003). Graduate teaching assistants and inquiry-based instruction: Implications for graduate teaching assistant training. *Journal of Chemical Education, 80*(10), 1206–1210.

Savage, M. P., & Sharpe, T. (1998). Demonstrating the need for formal graduate student training in effective teaching practices. *Physical Educator, 55*(3), 130–138.

Shannon, D. M., Twale, D. J., & Moore, M. S. (1998). TA teaching effectiveness: The impact of training and teaching experience. *The Journal of Higher Education, 69*(4), 440–466.

Strauss, A., & Corbin, J. (1998). *Basics of qualitative research: Techniques and procedures for developing grounded theory.* Thousand Oaks, CA: Sage.

Sunal, D. W., Hodges, J., Sunal, C. S., Whitaker, K. W., Freeman, L. M., Edwards, L., Johnston, R. A., & Odell, M. (2001). Teaching science in higher education: faculty professional development and barriers to change. *School Science and Mathematics, 101*(5), 246–257.

Tanner, K., & Allen, D. (2006). Approaches to biology teaching and learning: On integrating pedagogical training into the graduate experiences of future science faculty. *CBE Life Sciences Education, 5*(1), 1–6.

Varma, T., Volkmann, M., & Hanuscin, D. (2009). Preservice elementary teachers' perceptions of their understanding of inquiry and inquiry-based pedagogy: Influence of an elementary science education methods course and a science field experience. *Journal of Elementary Science Education, 21*(4), 1–22.

Windschitl, M. (2003). Inquiry projects in science teacher education: What can investigative experiences reveal about teacher thinking and eventual classroom practice? *Science Education, 87*(1), 112–143.

Windschitl, M. (2004). Folk theories of "inquiry": How preservice teachers reproduce the discourse and practices of an atheoretical scientific method. *Journal of Research in Science Teaching, 41*(5), 481–512.

CHAPTER 15

# LESSON STUDY WITH GRADUATE TEACHING ASSISTANTS

## Three Comparative Cases in the Sciences

**Sharon Dotger and Deborah Barry**

### ABSTRACT

This chapter overviews three cases of graduate teaching assistants engaging in lesson study in the earth and life sciences. The chapter reviews the lesson study process, discusses how the process is applicable to the post-secondary context, and presents data from analysis of videotaped planning meetings and instructional events to generate claims about how GTAs learn to teach through lesson study. Evidence from these cases indicates that GTAs are challenged to justify their instructional decision making during lesson study. They notice that the design of groups impacts the interactions undergraduates have during class. They question the reliability of a students' right answer as an indicator of conceptual understanding and they begin to consider students' prior knowledge when planning instruction. Implications for GTA learning opportunities for teaching are discussed.

*Research Based Undergraduate Science Teaching*, pp. 463–488

Undergraduate science courses are taught by experts in the field; faculty lecture one to two times per week and graduate teaching assistants (GTAs) lead the associated laboratories. Currently, GTAs are responsible for delivering over 90% of college laboratory instruction (Sundberg & Armstrong, 1993). This general structure is widespread in the United States of America, and efforts to reform teaching practices in these situations have met several barriers. One notable barrier is GTAs' knowledge of teaching and sufficient opportunities for them to develop that knowledge (Wright & Sunal, 2004). Explicit training for teaching is important because, while "experts know their disciplines thoroughly, this does not guarantee that they are able to teach others" (National Research Council, 1996, p. 31). Providing explicit training, however, is challenged by the degree to which teaching is valued in the academy.

How teaching is valued in the academic hiring process in bachelor's, master's, and doctoral degree-granting institutions was investigated by Meizlisch and Kaplan (2008). The results of their survey indicated that bachelor's and master's degree-granting institutions rated teaching as more important than research, while there were no significant differences in the ratings between teaching and research potential in doctoral-granting institutions. The data indicate that teaching is at least as important as research in obtaining an academic position. Teaching and research, however, do not receive equal attention in the preparation of doctoral students for academic careers in the sciences. Teaching assistantships during graduate school often are viewed as less prestigious in comparison to research assistantships (Austin, Campa, Pfund, Gillian-Daniel, Mathieu, & Stoddart, 2009). The tension between research and teaching also was documented by Golde and Dore's (2001) survey of 4,000 doctoral students at 27 universities. The authors stated that equal attention should be paid to preparing doctoral students as teachers and researchers.

Graduate education in the sciences historically has followed an apprenticeship model where students learn to be proficient researchers. The research experience often is heavily supervised at first. As students progress in their studies, they become capable of conducting independent research to produce new scientific knowledge (Feldman, Divoll, & Rogan-Klyve, 2009). Although the apprenticeship model exists for teaching graduate students to conduct independent research, rarely does a similar, mentored experience enable them to become educators. According to Feldman et al., professors do not explicitly plan for the way they teach their graduate students to conduct research. They rely instead on observation, proximity, direct training on equipment, and the graduate students' ability to learn from the departmental culture. It seems that similar approaches are used to teach GTAs how to teach, and teaching has often been portrayed as a lesser pursuit in science departments (Addy & Blanchard, 2010). If improving the quality of teaching in undergraduate science is an important issue, then faculty will need to become more explicit in the training they provide to GTAs. Although GTA training programs exist and have become more popular in recent years, Luft, Kurdziel, Roehrig, and Turner (2004) suggest they should be modified to be discipline specific, draw upon current literature of learning to teach science, encourage reflective practice, and increase pedagogical content knowledge.

Pedagogical content knowledge (PCK) is knowledge of the "the most powerful analogies, illustrations, examples, explanations, and demonstrations—in a word, the ways of representing and formulating the subject that make it comprehensible to others" (Shulman, 1986). It is recognized as one of three essential forms of knowledge for teaching by educational researchers who study teaching across disciplines in a variety of contexts, including post-secondary ones. The two other significant forms of knowledge are knowledge of subject and knowledge of curriculum (Luft et al., 2004; Shulman, 1986, p. 8). Graduate teaching assistants' expertise in their content area qualifies them as having a strong depth and breadth of knowledge of the subject. Their strong content knowledge does not automatically translate to knowledge of curriculum or the development of PCK (Dotger, 2011). The development of PCK, and of knowledge of curriculum, must be explicitly taught and thus requires a change in the typical way graduate students learn to teach. Improving teachers' PCK is one of the pathways that may lead to changes in teachers' practices (Loughran, Mulhall, & Berry, 2004).

Several models have been developed to provide discipline specific training for GTAs, including semester-long courses providing pedagogical training, mentoring programs in teaching, and professional learning communities where graduate students learn to address student learning problems through action research (Austin et al., 2009; Baumgartner, 2007; Gaia, Corts, Tatum, & Allen, 2003). Several challenges for GTAs are recognized by Dotger (2010) in addition to a lack of formal training in teaching. An important challenge is that graduate assistants often lead undergraduate students through laboratory activities, discussions, and assessments that they did not help design or construct. These challenges make it difficult for GTAs to develop PCK in the context of their normal teaching duties.

In the research reported here, lesson study was used as a model of professional development with GTAs in order to develop their PCK (Cerbin & Kopp, 2006) and provide an opportunity to participate in the development of lessons taught in the undergraduate science laboratory setting. Lesson study may be particularly advantageous for graduate students and novice faculty members to develop their teaching skills (Alvine, Judson, Scheine, & Yoshida, 2007). The purpose of this chapter is to describe three different examples of lesson study implementation, focusing particularly on the structure of the lesson study cycles. We begin by reviewing the key components of professional learning for teaching and connecting them to lesson study. Previous research regarding lesson study use and effectiveness in both K–12 and undergraduate settings also will be reviewed prior to the presentation of the examples.

## JAPANESE LESSON STUDY

A longstanding assumption regarding the improvement of teaching practice has been that in order to improve teachers' practices in classrooms, their knowledge must become deeper, more nuanced, and contextualized. Research about teacher learning in the K–12 realm suggests a set of conditions that maximize the likelihood that teachers will learn from their professional development experiences and thereby improve their knowledge for teaching. First, for teachers to commit to improvement, they need sustained, collaborative professional interactions that focus on content and pedagogy (Briscoe & Peters, 1997). An extensive review of the literature for developing science teachers builds on this idea. Effective teacher professional development has four characteristics, according to Roth (2007, p. 1206), as it should "(a) engage teachers actively in collaborative, long-term problem-based inquiries, (b) treat content as central and intertwined with pedagogical issues, (c) enable teachers to see these issues as embedded in real classroom contexts, and (d) focus on the content and curriculum teachers will be teaching." While Roth developed these conditions by studying effective K–12 teacher learning, we believe that they are applicable to developing "knowledge-for-practice" (Cochran-Smith & Lytle, 1999) with teachers in the undergraduate setting. We argue here that lesson study meets each of these criteria effectively.

Japanese lesson study is a form of collaborative teacher research providing teachers with an opportunity to deepen their content knowledge and understanding of pedagogy through examining, discussing, and learning from their daily work and colleagues (Fernandez, 2002; Lewis, Perry, & Hurd, 2006, 2009; Lewis, Perry, & Murata, 2006; Stigler & Hiebert, 1999). Via lesson study, teachers increase their content and pedagogical knowledge, improve their ability to observe and assess students' learning, create stronger connections between their daily practice and long-term goals, and forge stronger collegial networks (Lewis, Perry, & Hurd, 2004). By engaging in lesson study, teachers can change their interpretive systems for thinking about teaching (Zawojewski, Chamberlin, Hjalmarson, & Lewis, 2008). This would follow from their improved abilities to make their thinking explicit, test the rationales underlying their thinking, analyze the outcomes of their teaching, and revise their future practices. Such teacher skills would, in turn, foster student learning.

To begin lesson study, a small group of teachers articulate goals for students' learning and study curricula in order to determine potential tasks that would embody those goals for instruction. Teachers study representations of content in curriculum documents and instructional materials (see Fernandez & Yoshida, 2004). They then design, modify, or use already existing materials that they believe will help students meet previously specified learning goals for the lesson. Next, the team of teachers creates detailed lesson plans, which identify teacher actions and articulate points to look for (Lewis, 2002; Lewis & Hurd, 2011) during the lesson. These points assist teachers in building their filter for what they will attend to, or notice, as the live lesson is taught. Theoretically, these goals and instructional examples frame what teachers' notice during the implementation of the lesson. One of the teachers on the team will facilitate the lesson with a group of students, while the other team members observe and collect data. In order to collect data to evaluate the research lesson, they must design instruction such that students' thinking during the lesson is evident. Thus, instruction needs to provide opportunities for students to talk, problem-solve, or write using processes that the observers can see and hear.

During the lesson, often referred to as a "research lesson" (Lewis & Tsuchida, 1998), teachers have an opportunity to intensively study student learning. Following the lesson, when teachers collaboratively study students' work, they develop a shared understanding of what occurred during the lesson and can draw conclusions for that particular lesson and for teaching more generally, thus reevaluating their goals and the curricula they studied during the early stages of the process. As lessons are redesigned, retaught, and reevaluated, teachers generate data they use to articulate claims about teaching, which are captured in a report of the process. Through repeated lesson study cycles over time, teachers can refine the tasks they develop for their students and their skills for noticing student learning during live lessons. This process is beneficial for teachers because it allows them to focus on the development of their knowledge in the context in which they work every day, providing them a forum for articulating and testing the rationales that guide their practice. The teachers' learning is evident in their post-lesson discussion, as well as in the report of learning written in the following days (see Lewis, 2002).

## LESSON STUDY IN THE UNDERGRADUATE CONTEXT

Several articles describe the implementation of lesson study in North American undergraduate settings. Lesson study and its implementation among 150 instructors within 25 disciplines on 10 campuses of the University of Wisconsin system were described by Cerbin and Kopp (2006). At the time the article was published, nearly one-quarter of the full-time instructors at one institution participated in at least one lesson study cycle. Most of the faculty who participated in lesson study had minimal previous training in teaching. In their lesson study work, several features of the traditional model are maintained: Learning goals were developed, a research lesson was designed, data were gathered with a focus on *how* students were learning (rather than only *what*), the evidence was analyzed, and the entire process was documented. The faculty on the team determined the timing of the lesson study work. They typically met for 15 to 20 hours per lesson study cycle over the course of the academic year. As a result of working through this process, the authors claimed that lesson study brings about the same benefits for university faculty as it does for K–12 teachers (Cerbin & Kopp, 2006).

Colorado State University and Southwestern University are two additional universities with active lesson study projects (Kamen et al., 2011). At Southwestern, four faculty teams representing five academic departments were established. Additionally, a faculty member at Southwestern partnered with teacher educators at five additional universities to use lesson study in their preservice teacher education courses. Given the geographic distribution of the participating faculty, they taught the research lesson at one university twice on one day to two different groups of students, with debriefing and revision of the lesson occurring between these two events. A second professor taught the third iteration of the lesson at a different university the following day. Six total iterations of the lesson were taught, with the final iteration occurring publicly at the 2005 School Science and Mathematics Association national conference. Participants explained that the experience helped them rethink their own approaches to mathematics teacher education in the remaining lessons within their course (Kamen et al., 2011).

In the two previous examples, lesson study participants were faculty. Two other examples exist in the literature describing lesson study with graduate students. Lesson study has been implemented as a graduate student teacher training program at Harvard University for first-year calculus (Alvine et al., 2007) and at a second research-intensive university in the northeast with graduate students teaching earth science (Dotger, 2011). In both cases, lesson study was a mechanism for providing graduate teaching assistants with opportunities to learn about teaching in the context of their content and with their students, which directly addresses the concerns other GTAs have expressed about general, decontextualized training for teaching (Baumgartner, 2007; Luft et al., 2004). In the calculus case, experienced calculus faculty worked with the GTAs who received course credit for their participation. In the earth science case, a science education faculty member worked with the Earth Science GTAs. Earth science faculty members were generally supportive of the project but did not attend any of the planning sessions, the research lesson, or the debriefing meetings. The GTAs participated in that lesson study cycle voluntarily and were not given course credit or stipends for their efforts.

The literature currently contains a few examples of how lesson study has been conducted in undergraduate settings. More examples need to be described, as well as the features of these structures that must be included in order to maximize the likelihood that participants will learn from the cycle. The lesson study examples presented in this chapter adhere to the Japanese lesson study model described in the higher education literature to varying degrees (Alvine et al., 2007; Cerbin & Kopp, 2006; Dotger, 2010). These examples provide us with the opportunity to better understand why this protocol is important and what aspects of the protocol are particularly beneficial for graduate teaching assistants in the sciences.

## METHODOLOGY

Each of these lesson study cycles was conducted at the same northeastern research-intensive institution with the same facilitator. The cycles, however, occurred in different semesters and with different samples from our GTA population. A total of thirteen GTAs participated in this study. These GTAs were recruited from a population of 33 GTAs in earth science, biology, and science teaching. The GTAs, recruitment process, and the context of their participation are further described in each example. In each case, the facilitator first introduced lesson study to the GTAs, provided background reading materials, and generally oriented the participants to each aspect of the Japanese lesson study model as it was described in the literature. In the findings below, we detail aspects of this model that were altered by the participants. The facilitator asked clarifying questions during meetings to develop the research lessons, recorded notes concerning the research lesson and the behavior of graduate teaching assistants, and provided brief explanations of lesson study procedures when necessary. In the first two cases, the graduate teaching assistant participants formed a voluntary lesson study group for introductory courses in their field (spring 2009 and fall 2008, respectively). In the third case, lesson study was completed in the context of a graduate level course in science education (spring 2010).

Data for the present study included observational notes written by the facilitator during lesson study planning meetings and the research lessons. In the second example described below, planning meetings were videotaped, and the debriefing session following the research lesson was audio recorded. These data were coded using Studiocode qualitative coding software, and illustrative quotes were transcribed. In the third case, additional data included written reflections by the graduate teaching assistants and interviews that were conducted six months after the lesson study cycle was completed.

These data were coded independently by each of the researchers using a constant comparative method. A coding scheme was developed, and the entirety of the data was recoded once the complete coding scheme was developed. The codes developed were compared among each of the examples of lesson study. These comparisons are discussed below, with evidence drawn directly from the data. The following sections describe three different examples of lesson study with GTAs, focusing on the structures of study events and how these structures contributed to GTAs' learning. The examples are identified by the subject matter that was the focus of the lesson study cycle.

## EXAMPLE ONE: PLATE TECTONICS

Four graduate teaching assistants and one science education faculty member formed a voluntary lesson study group to design a research lesson for an introductory earth science course. These GTAs were recruited via email, which was distributed to all 14 GTAs in the earth science department. Each of the graduate students participated in one previous lesson study cycle, described elsewhere (Dotger, 2011) and thus entered into the second cycle with some familiarity with the process. Prior to the initiation of this cycle, the earth science course only addressed the topic of plate tectonics during lecture. Inserting a laboratory about plate tectonics was a new addition to the course and was done with the permission of the course instructor.

The GTAs and the science educator met to plan the research lesson four times. Each meeting was two hours long, with one meeting per week. During the initial meeting, conversations focused on the goals for the course and how a laboratory focused on plate tectonics would contribute to those goals. The science educator was familiar with a previously published plate tectonics lesson (see Sawyer, Henning, Shipp, & Dunbar, 2005) and brought that lesson to the group during the second planning meeting. In the meeting, the group read the article that described the lesson design, studied the four maps providing data for students to analyze, and began discussing the merits of the existing lesson.

During the third meeting, the GTAs discussed how they would implement the lesson in their own context. They had a few constraints that were different than the ones the original lesson designers experienced. First, the duration of the lesson needed to fit within 120 minutes, rather than the

180 minutes planned for in the original design. Secondly, they understood that there would be no opportunity for following up with students' ideas in another laboratory section or during the lecture associated with the course. These constraints were compounded by their uncertainty that their students would be able to identify patterns within and among the maps as described in the original lesson. Thus, they altered the lesson to be more teacher-directed than the original lesson designers intended.

In the fourth and final planning meeting, the GTAs practiced the lesson with two other GTAs for the course who had not participated in the lesson study. During this meeting, the lesson study participants specified the introduction, the presentation of the problem, and the questions they would ask students to facilitate their thinking. Participating GTAs answered questions from their peers about the design of the lesson and its sequence. The faculty member teaching the course was invited to the meeting but did not attend. Differences of ideas among the GTAs who designed the laboratory via the lesson study process and those of the GTAs who would be adopting the designed lesson were settled among themselves.

The research lesson was taught by one of the participating GTAs to a group of 24 undergraduates. The other three GTAs attended the lesson to gather observational data. The facilitator and three science education graduate students attended the lesson to assist with data gathering. Two video cameras, set up in two corners of the classroom, also were used to capture the events within the lesson. Immediately following the research lesson, the planning team and the additional observers met for two hours to discuss the data and students' work. Discussion during the lesson debriefing focused on how the lesson might be further modified for other groups of students. As the discussion unfolded, implications for other laboratory experiences in the course and for undergraduate science teaching in general, were discussed.

As compared to the research literature describing lesson study, this cycle was missing two crucial elements. First, there was no written plan to guide learning. When the GTAs taught all other lessons with the course, they had no lesson plans to which they could refer. Their own lesson planning consisted of writing a bulleted list of vocabulary or ideas they would introduce to students at the opening of the laboratory. Their plan for the lesson was represented by the copies of the quiz they used to open the laboratory session, the bulleted list of words they would address during their introduction, and the handouts they would give the students to complete the lab. Writing a more detailed plan as is described in the lesson study literature, and common to teaching in the K–12 setting, was very foreign to the GTAs. While the lesson study facilitator suggested that the GTAs develop the plan, she had no leverage to require it. She was also anxious that pushing GTAs to write down their ideas would reduce their interest in participating in lesson study at all, even though the processes that are typically included in a lesson study cycle were made clear to the participants. The purpose of the plan to guide learning is partially to document the lesson plan. It also is useful in guiding the observers as they attend to students' actions during the lesson. Without the plan, the observers were left to interpret the lesson based on their own viewpoints, gathering data on student actions, rather than focusing their observations on aspects of students' thinking that were considered as the lesson was planned.

Secondly, and related to the first constraint, although note taking and summary report writing were discussed during the lesson study process, the group did not write about what they learned from observing students. While they discussed their observations of the students during the lesson, they did not record the substance of these discussions. Only the facilitator and the science education graduate students recorded notes during the research lesson, suggesting that their prior experiences in observing instruction assisted them in understanding the importance of this task. One possible reason for the lack of note taking and report writing was that the participants did not have experience in writing about teaching, nor did they believe they had an audience that would value what they learned. On one hand, they verbally stated that they understood more about the importance of teaching then they had before completing the lesson study process, yet they knew that relative to the knowledge of experienced science teachers, their knowledge was novice at best. They were uncertain that their understandings about teaching would have the potential to impact the practices of other GTAs with whom they worked, or be helpful in convincing faculty members in charge of the course of the changes they thought were necessary in the structure of the course or the other laboratories.

## EXAMPLE TWO: HARDY–WEINBERG EQUILIBRIUM

The lesson study group for example two, the Hardy–Weinberg Equilibrium, was comprised of six GTAs who met five times to plan and debrief their research lesson. Of these six GTAs, five were PhD students in biology, and one was a PhD student in college science teaching. All of the participants were assigned as graduate teaching assistants to an introductory biology course and developed this lesson to teach in that context. These participants were recruited from the population of 10 GTAs who teach the introductory laboratory in biology. They represent approximately half of the GTAs in the biological sciences at the study institution. Participation in the lesson study was voluntary, and not for course credit. These planning meetings, lasting approximately 1½ hours each, were videotaped and led by a science educator to guide the group through the lesson study process.

The videotaping provided a more complete record of what occurred during the cycle than was recorded during example one, where we relied on note taking. We are able to include quotes from the participants in this example to illustrate the kinds of conversations the GTAs had. For a more complete discussion of the content of their discussions and the GTAs' understandings of Hardy–Weinberg Equilibrium (please see Dotger et al., 2012).

Unlike the situation described in the first example, the course instructor attended the meetings. He commonly held weekly meetings to train the GTAs for the upcoming week's laboratory and extended these weekly meetings for lesson study. During the planning sessions, the science educator's primary role was to take notes, ask clarifying questions, and provide brief explanations concerning the lesson study process. The course instructor facilitated the discussion about the purposes of the laboratory and its content. Since the course instructor partly facilitated the group discussion, the science educator was able to take more complete notes about the lesson study process than had been possible in the first example's context. Although all GTAs who participated in the lesson study process taught the research lesson to their assigned laboratory sections, only one section was designated as the observed research lesson. The lesson study group was constrained by time, with many of the opportunities to observe the research lesson in direct conflict with the science educator's own teaching schedule and other GTAs' course schedules. The participating GTAs were very nervous about teaching in front of their peers or were unwilling to commit additional time to attending another section of the laboratory. Common observation of a research lesson therefore, did not occur; instead, each GTA taught the designed lab to his or her students. The designated research lesson was video recorded. The course instructor and science educator attended the research lesson and recorded observational notes, although other members of the lesson study team were not present.

The research lesson developed focused on the concept of the Hardy–Weinberg Equilibrium and was modified from two existing laboratories the GTAs located on the internet. The GTAs combined different sections of these existing laboratory exercises in order to reach their instructional goals of providing concrete examples of each of the factors that can disrupt the Hardy-Weinberg equilibrium, such as non-random mating and migration.

In this lesson study example, GTAs constructed an original lesson handout for the students. The handout's design was similar to other handouts provided to students in other laboratory sessions. Like the previous example, this lesson study group did not generate a plan to guide learning apart from this handout for students. The development of the handout, however, provided an opportunity for the GTAs to explore their knowledge of this content area in relation to how they planned to teach the major concept. During these planning meetings, the GTAs regularly and repeatedly asked other GTAs and the course instructor for content clarification of the material included in the lab. For example, this exchange occurred between one GTA and the course instructor during a lesson study-planning meeting (Dotger et al., 2011):

GTA: So, you said something that was confusing to me conceptually. 2pq is the heterozygote, so that's big p and little p and q. Which is the same as..., so big p and little q, so it's showing up twice?

Course Instructor: So, let's think about any old trait. We'll just say that rolling your tongue is dominant. Big R. Little r is recessive, and if you have two of those you can't roll your tongue. We can do the same thing with P and Q. It's just different location. So if we were doing a punnet square and I had a heterozygote that was Rr, we can do a punnet square really easy. There are two ways of getting Rr, you can use this big R or this one. We can do the same thing with the p and q. Different notation, same traits. My punnet square would come out to look like this. You've all taken algebra, so we can just write it this way. So, if we are talking about the p's and q's, we want [the students] to be aware that they are the exact same things that we used last week with the letters.

GTA: But, why p's and q's?

Course Instructor: That is just what they used. For this population equilibrium model, it's not important that one of them is dominant and the other is recessive. But, when we are teaching about it, that is how we simplify the concept of genetics. But for so many of these traits that we look at, there are more than just two alleles.

This exchange between one of the GTAs and the course instructor demonstrates a lack of conceptual clarity on the part of the GTA. This type of conversation, related to clarifying the content of the laboratory, happened frequently throughout the lesson study planning meetings. It should be noted that the format of the lesson study experience allowed for several planning meetings to be spent on one topic. This extended time period, and careful discussion of the wording and activities to be completed by students during the laboratory experience, allowed GTAs to explore their own understanding of the Hardy–Weinberg Equilibrium and prompted them to ask for clarification in appropriate ways both to present the material and in the content.

This clarification of content also occurred between GTAs when trying to construct the laboratory handout for students. The following conversation between three GTAs presents an example of how GTAs worked together to refine the laboratory:

> GTA 1: Why are the last two sentences relevant?
>
> GTA 2: Because you still want to figure out the frequency of alleles in the offspring.
>
> GTA 1: So, like what do you mean?
>
> GTA 2: So, this is what we have so far. The offspring (draws a table on the chalk board), we've got three possibilities, and they will tally them up like we did before.
>
> GTA 3: So, could we call that offspring genotype? That would make more sense to me. If you had the genotype and then the total number next to it.
>
> GTA 2: The genotype frequency is how often this Bb shows up in the population, which is this column divided by the total, but we still want them to be able to calculate the frequency of each allele so, we would still need to include that question, or the last two sentences.

In contrast with the first example, the second example contains a crucial element: the course instructor. Similar to the first example, however, is minimal writing and observation of the live lessons. As a result, there is little institutional knowledge developed about the students' thinking processes during the lesson. As researchers, we were able to discern the GTA's intentions and rationales for the lesson they planned, but these rationales are not available to future GTAs since the creation of the laboratory handout only documents the work the students should complete during the lesson.

In this case, due to the minimal writing and observation of the live lesson, GTAs utilized written student work to debrief and reflect on the lesson. Due to the timing of the research lesson near the end of the fall semester, the debriefing session was held following winter recess. The amount of time that lapsed between the research lesson and the debriefing session proved to be detrimental to GTAs reflecting on the lesson. The GTAs each brought copies of student work completed during their assigned laboratories and identified areas of the laboratory handout that appeared to be problematic based on student responses. This information was then used to reword questions for future iterations of the course.

## EXAMPLE THREE: THE CRYOSPHERE

This lesson study group was comprised of three graduate students, one doctoral student in earth science, a master's degree student in earth science, and a doctoral student in science education. These students completed the lesson study process in the context of a three-credit graduate level course in science education, facilitated by the first author. A total of nine graduate students enrolled in the course, including master's students seeking certification in high school teaching, a doctoral student in college science teaching, a doctoral student in science education, and the previously mentioned graduate students. Three lesson study groups were formed during this course, and we describe one example here. This course met for three hours weekly and included readings regarding lesson study and undergraduate science teaching, online discussion groups, and written assignments. The written assignments and videotapes of the research lesson implementations provided the data we refer to in this example. Interviews with the two earth science graduate students were conducted approximately six months following the lesson study. The two graduate students in earth science were GTAs for an introductory earth science course whose laboratories were the focus on this group's work. This lesson study group began with an analysis of the general goals for the laboratory provided by the earth science lecture course instructor. These general goals were: to build upon and reinforce material covered in lecture, provide opportunities to perform "hands-on" earth science work, and ask questions regarding lecture material in a smaller class setting, encouraging student collaboration.

The lesson study process progresses from the designation of lesson goals to the development of a carefully crafted research lesson. As this was the only lesson about the cryosphere to be taught, determining the learning goals and objectives was extremely challenging for the GTA participants. The challenge was partly because there never had been a laboratory about the cryosphere in the course before, and none of the GTAs had ideas about what their students would think about the cryosphere. To gather some information about students' thinking regarding the topic, the GTAs conducted a clinical interview with two undergraduate students. These students were asked several questions about the cryosphere and glaciers.

From the findings of the interviews, the lesson study group brainstormed a list of important concepts for students to understand about the cryosphere and then narrowed the lesson objectives using that list. Three broad content goals were developed: (1) students will understand how glaciers move while in the solid state, (2) students will be able to conceptualize the rate of glacial flow in the real world, and (3) students will understand how glaciers can alter a landscape. This research lesson focused on a model of glacial movement students could manipulate. It served as a representation for how glaciers may interact with and impact local terrain. Few studies have focused on how models of scientific phenomena relate to the real-world process they represent (Mathews, 2007). As such, this became an important target for the research lesson. The study of models as representations of scientific phenomena is not unique to the earth sciences, and it has been shown to help to develop scientific thinking in students (Clement & Rae-Ramirez, 2008; Gilbert & Boulter, 2000).

The lesson study group did several materials trials in an attempt to build a working model of glacial movement. Finally, the group determined that the model of glacial movement was most realistic using Oobleck, a non-Newtonian fluid made with cornstarch and water, and a ramp made with tin foil. Students were given a small cup of Oobleck to poor down the foil ramp. Drops of gel food coloring were placed along the ramp to represent sediment. The team recognized several limitations of the Oobleck model of glacial advance and thought that it was important for students to focus on these limitations. This lesson was designed around comparing the Oobleck model to time-lapse videos of glacial advance. This allowed students an opportunity to confront their misconceptions of glacial movement and develop ideas about models as representations of scientific phenomena.

Students generated hypotheses in small groups for the interactions between the Oobleck, a tin foil "landscape," and gel food coloring distributed on top of the foil. Students then poured a small amount of Oobleck onto the sloped tin foil, calculated the rate of flow, and recorded their observations. Three short time-lapse videos of glacial movement were watched as a class (Extreme Ice Survey videos are available at http://www.extremeicesurvey.org). Students then were asked to make comparisons between their Oobleck model and the real-world glaciers in the videos. Students recorded similarities and differences on a poster that was shared with the class.

The research lesson was implemented and observed five times during a one-week period. The GTAs met after each observation of the research lesson to discuss student learning and any needed changes to the lesson format. Although the format of the recitation provided some challenges to lesson design, it also allowed for the completion and testing of lesson revisions within a single week because the recitation was taught a total of 14 times. The repeated teaching of the lesson, along with careful observation and reflection, allowed the lesson study group to make several iterative changes to the lesson format over the one-week time period.

The lesson study team drew several conclusions about what the undergraduate students learned about the cryosphere from the lessons. The process of observing the undergraduate students helped the GTAs build specific knowledge related to teaching about glaciers, or pedagogical content knowledge. Both of the earth science GTAs, for example, noted that they incorporated hypothesis development and testing into their other lessons. Given the difficulties they witnessed students having with recording observations, they reevaluated the impact of handout structure for students' answers. Having the students report their findings to the whole class using a large poster, GTAs commented that they learned new techniques for facilitating class discussion.

There were some very superficial comparisons drawn by students, such as that both the glaciers and Oobleck are white and they both move. One group of students also noted that there was a force below a glacier that made it move. This misconception may have stemmed from confusion related to the video, which showed water carrying large chunks of ice away from the terminus of the glacier, or from prior knowledge the students had. Most of the talk occurring between GTAs concerning their students centered on the wild answers some provided on quizzes or laboratory handouts prior to the lesson study experience, similar to the findings of Barry and Dotger (2011). These answers provided an easy chuckle and a way for GTAs to validate that their students are performing in way similar to other recitation sections. The framework of the lesson study experiment, however, also forced the GTAs to think about why their students made these claims about glaciers, such as the "underlying force" noted above and what they can and should do about these student misconceptions. The GTAs did not ignore or laugh about these student ideas; instead, they attempted to help clarify these ideas by discussing them with the students, providing evidence of a shift in their conceptions of their students as learners.

The GTAs developed general knowledge about teaching from participating in the lesson study group. They noted that this was the first time they had observed one of their peers teaching a recitation. Taking on the observer role in the classroom allowed them to note processes contributing to student learning of which they were not previously aware. The GTAs stated that they had never considered group interactions as having an impact on student learning. They noted a lot of variation in how students formed groups and assigned tasks to group members. Student talk provided a window into the learning process and helped identify any misconceptions students may have had. One of the graduate students noted, "I think there are so many good existing lessons that we could just change the order of steps or change the way the students report their answers that would get them out of their routine and make student learning more observable."

One of the most valuable and possibly one of the most frustrating insights the GTAs gained from participating in the lesson study group was the realization that learning is a complicated process. In the past, they had viewed student learning as relatively straightforward: They presented the content, and therefore, the students would learn it. Although this was not a very sophisticated view of teaching and learning, it was the one they held, with very little understanding as to why some students were unable to be successful. One GTA, for example, stated:

> I have learned that student learning is not as simple as I thought it was at the beginning or middle of the semester, or even right before the research lesson. It wasn't until I was looking through observations and student work that I realized I have no idea how student learning works.

Additional evidence of GTAs' realization that learning is a complicated process is that although students may have the "correct" answer on their worksheets, it is not always indicative of learning. One of the GTAs stated,

> I used to get so frustrated when I would explain pretty much each question at the beginning of recitation, the students would ask questions and then students would still deliver far-out answers. Looking back at these experiences, I realize that many of these cases were with students that may have been trying to learn the material and not just answer the questions. Then there were other students that had the "right answer" on their paper, but couldn't explain it.

## CONCLUSION AND IMPLICATIONS

Each of these three examples demonstrates various structures for implementing lesson study. In the first example, plate tectonics, the team focused on planning, observation, and post-lesson discussion. The lack of a written plan to guide learning limited the data collected by observers and did not facilitate the construction of a report about what they learned. In the second example, plans to guide learning were not constructed, either. The course instructor was involved but did not require observation of the live research lesson. The lack of shared observation prevented data collection regarding students' learning processes and outcomes: GTAs were left to rely on students' written work as an indicator of their learning. In the third example, GTAs learned the limitation of relying solely on students' written work as an indicator of their understanding. The cryosphere study was most complete and contained all of the aspects of the Japanese lesson study process as described in the literature. Thus, we have the most information about students' and GTA learning from this example.

The timing of the implementation, evaluation, redesign, and reimplementation of the research lessons is a significant feature of lesson study and its potential for faculty or graduate student participation. Although teaching multiple versions of a lesson within a week is intense, this bounds the lesson study experience to a relatively short implementation and study period. We believe that this makes the lesson study cycles feasible for graduate students and faculty to participate.

As lesson study is focused on the learning resulting from a teaching experiment; it helps modify the role of the teacher to include a researcher perspective (Fernandez, Cannon, & Chokshi, 2003). The researcher perspective often is missing from classroom teaching; however, its inclusion helps to bridge the gap between educational theory and practice. Explicit opportunities to study undergraduate learning often are not included in the doctoral education process for GTAs. Lesson study provided an opportunity for participating GTAs to engage in classroom research, centered on learning issues in their particular field. This allowed GTAs to develop ideas about best practice, resulting from their direct observations of students in action. It may be valuable to provide similar experiences to graduate teaching assistants in the sciences to leverage their understandings of research in order to improve teaching, as well as encourage a scholarly approach (Allen, Donham, & Tanner, 2004). The structure of graduate school requires GTAs to develop their identities as researchers ahead of their identities as teachers (Austin, 2010). The collaborative nature of this process can reduce isolation commonly felt by educators (Lewis, Perry, & Murata, 2006), providing a formal space for GTAs to create or modify exercises specifically for the students they teach.

While it is often assumed that GTAs have the depth and breadth of content knowledge to be adequately prepared to teach introductory level courses, these examples revealed that providing a forum such as lesson study for GTAs to explore their ideas in a specific content way allowed the GTA participants to clarify their understanding. Lesson study can lead to improved knowledge of content (Lewis, Perry, & Murata, 2006). In previous cases, teachers challenged one another to explain their thinking about mathematical patterns (Lewis, Perry, & Hurd, 2009). These challenges to explain thinking were also evident in these lesson study cases and are evidence of content knowledge.

The context of the lesson study planning meetings also allowed GTAs to develop their PCK related to teaching about each topic. As Dotger (2011) points out, the use of lesson study with GTAs supports the exploration of the *why* and *how* behind instructional decisions in addition to providing an avenue for GTAs to test their ideas in the classroom. As GTAs explored why they believed it was important to include or reword certain aspects of the laboratory, they also demonstrated some indications of the ability to discuss their pedagogical decisions. This type of discourse may be absent from the other formats of graduate student teaching preparation, which instead focus on providing a model of how to deliver information (Barry & Dotger, 2011). This development of teaching knowledge has the potential to transfer to other teaching situations, generally increasing GTAs pedagogical knowledge (Luft et al., 2004). The increase in pedagogical, content, and pedagogical content knowledge developed within the lesson study context can also work to increase GTAs self-efficacy, which may have positive influences on graduate students' scholarly development (Feldon et al., 2011).

The development of the research lesson in the cryosphere case proved to be a valuable form of professional development for participating graduate students, as indicated during their follow-up interviews. The analysis of the lesson plan indicated that these GTAs developed appropriate pedagogical content knowledge relevant to teaching aspects of the cryosphere. Participating GTAs became more aware of the importance of reflection during their regular teaching duties, which Loughran (2002) notes is an essential part of teacher development. For example, one of the GTAs stated:

> The biggest impact lesson study has had on me has been the reflective portion of my time. The majority of it has been through thought and discussion with the group and some notes from the week of the research lesson. I have already been planning on how to install a more regular time for reflection on my classes. As graduate students our schedules get so busy and typically after class I am immediately thinking of my next priority. Lesson study forced me to stop and think for a moment before the next step.

In addition, these GTAs were able to focus on student thinking and interaction as central to lesson design. As one GTA noted, "evaluating student learning is like evaluating a toddler with LEGOs or blocks. We give them all the pieces and can tell them to make a building, but we have no idea how they decide to put the pieces together." Participating in this process has allowed these GTAs the opportunity to start thinking about how the information they present to undergraduate students, the LEGOs or blocks, start to form an understanding of a science topic, the building.

There are many examples of teaching innovations in college science in a certain context (Krockover et al., 2002), but lesson study could be applied to science instruction in any location at the undergraduate level. We assert that this valuable experience has allowed these GTAs to develop their teaching practice. As one GTA noted:

> Empowering the professor with skills to design a lesson that yields immediate feedback regarding student learning and understanding is vital to the success and development of a teacher. Perhaps this idea is much more entrained in the thought process of education students, but this was completely new to me and appears to not have trickled into our department or the sciences in general. I think the way we've been trained is that there is a correct and incorrect answer in sciences, and there is no room for discussion.

An awareness of learners and learning is an important aspect in the development of undergraduate science educators (Donovan & Bransford, 2005). Lesson study provides one mechanism that facilitates this awareness and is an accessible method of professional development for graduate students that can be incorporated into GTA training programs. Although undergraduate student learning has always been the goal of graduate student-led recitations, making the explicit connection for GTAs encourages the development of an instructor who has the skills to focus on learning and teaching in addition to their research goals (Cerbin & Kopp, 2006; Dotger, 2011).

Lesson study may have potential for developing interdepartmental collaboration between graduate students throughout the various sciences taught at the undergraduate level and science education researchers. Lewis, Perry, and Murata (2006) discuss lesson study as a way to decrease the amount of isolation experienced by educators, instead fostering a connection to colleagues to approach issues in teaching and learning. We note that science education personnel do not necessarily need to be involved in the lesson study process. However, Krockover et al. (2002) documented that science content specialists often lack educational research tools. The content specialty graduate students participating in the current study had difficulty collecting and analyzing qualitative data because they had no prior experience with this type of data analysis. After consultation with the education researchers, they were able to complete this process. This suggests that some training in qualitative data collection and analysis may be beneficial for content specialty graduate students to successfully complete a lesson study cycle. As evident during the third case, graduate students who received additional training in data collection and analysis were able to complete the lesson study process more fully. In future studies, researchers could investigate GTAs' ideas about the kinds of data that are most relevant for evaluating teaching effectiveness, how to collect and analyze that data, and how these conceptions may change after participating in lesson study.

The lesson study collaboration has allowed these GTAs to focus on student thinking, which is not a task that they had formally done in the past, as noted by GTAs during their lesson planning meetings. Because lesson study is focused on observing student thinking, they were forced to design an experience for students with points of observation. This required the creation of an exercise that students could not complete without actually talking to each other. One of the graduate students noted in her personal reflection,

> Our intent was to have content that we could design an observable activity around. Without specific discussion about how the format of our lesson dif-

> fered from the usual lesson format, we designed something for the students that did get them out of their weekly habits.

The developed recitation exercises were taught on a large scale and can be used again in future iterations of the courses. One commonly recognized benefit of lesson study in the United States is the carefully developed lesson plans that result (Lewis, Perry, & Hurd, 2006). However, we believe that a greater benefit to undergraduate science instruction comes from the development of a new way to think about teaching in the GTAs that occurred as a result of their participation in lesson study. This new approach to teaching is more focused on the student as a learner and less on "covering material" and having students write "the right answer" in the appropriate area in their laboratory notebook. It is our hope as researchers that these ideas about teaching and learning will persist within these GTAs and become evident in their future teaching endeavors. Future research concerning the role of lesson study in the development of graduate students' pedagogies should include the collection of longitudinal data to understand the impacts of participation in lesson study over time.

The GTAs were allowed a lot of freedom in the lesson development process, which they had not experienced before as recitation and laboratory instructors. The GTAs noted that the large time commitment to the development of the research lesson encouraged them to become invested in its success, unlike their prior experience as GTAs leading discussions, laboratories, and recitations that they had no part in creating. One student recounted an experience with previous GTA duties, stating that:

> At the time I knew what was expected of me from the department and professor was generally show up, help the students work through the cookbook-style exercises each week, grade their exercises, have an office hour or two, and in some cases attend the lecture.

This type of experience for GTAs in the sciences is relatively commonplace (Dotger, 2010); however, the evidence provided here suggests that allowing GTAs to participate in lesson development may result in development of both PCK and PK.

A large part of faculty responsibility includes teaching, regardless of institution type and research commitment. We assert, therefore, that the professional development of GTAs in regards to both their teaching and research abilities is important. This study provides evidence that lesson study may have the potential to support the development of content specific teaching skills is GTAs. The same GTA quoted in the preceding paragraph wrote:

> We all teach at some level, whether it is in the form of presenting our research at a poster or oral session at a meeting, or in a classroom, and presumably we will continue to teach through either or both venues throughout our careers. The techniques and tools learned in [lesson study] provide an excellent opportunity for graduate student development and promote thinking about ways to convey concepts that I never would have encountered otherwise.

This process may help this GTA to become more thoughtful in the development of lessons, as well as to incorporate reflection into future teaching endeavors. We provide several indicators throughout these three cases of how these GTAs have become more thoughtful educators. Although we do not claim that every GTA participant has gained each of these skills, our data indicate that the following gains in some of these GTAs have resulted from their participation in lesson study. Graduate teaching assistants learned how to plan for effective group-work, how to justify why questions are asked on laboratory handouts, how to question whether the correct answer on a laboratory handout is an indication of student learning, and new ways to get students to talk about their ideas. It may be possible that lesson study could provide similar benefits to other graduate students in a variety of disciplines.

One of the lasting impacts of participation in a lesson study group for these GTAs has been to increase their confidence in lesson development (Allen et al., 2004). In the subsequent semester, both of the GTAs who participated in the cryosphere lesson study were assigned to teach different courses and worked in conjunction with the lecture instructor to design laboratory experiences that involved student exploration with fundamental concepts to earth science. These GTAs noted during their follow-up interviews that they have been more aware of their questioning practices while teaching, as well as developing lessons with specific learning objectives in mind. The lesson study experience was the first for both of these GTAs that required them to determine learning objectives for a laboratory experience and to think about whether or not the students actually met those learning objectives. These GTAs were careful to distinguish between "the right answer" on a laboratory hand-out and student learning even six months after the lesson study experience.

Alvine et al. (2007) assert that lesson study could be incorporated into GTA training programs that occur both at the beginning and during the course of the semester to allow GTAs to have more responsibility for the planning of laboratory experiments and feedback on their teaching practices. Participation in a lesson study group has provided these GTAs with an important opportunity to collaborate in the planning and execution of a recitation or laboratory session.

By comparing the structures of these three examples of lesson study, there are several indications that features of lesson study previously described in the literature provide important opportunities for graduate teaching assistants to benefit from the lesson study process. We suggest that the written lesson plan and shared observation of the research lesson are key features that allowed the graduate teaching assistant participants in the cryosphere case to benefit more fully from their lesson study cycle. Although we understand that lecture course instructors have limited time for engagement in the planning of recitations and laboratories due to their other academic responsibilities, our data support the notion that the participation of the course instructor throughout the process is beneficial, particularly to assist in content clarification. The inclusion of the course instructor allows for a more thorough understanding by GTAs of the placement of the material in the larger context of the course, and may provide an important model for GTAs of how faculty value teaching and student learning.

## REFERENCES

Addy, T. M., & Blanchard, M. R. (2010). The problem with reform from the bottom up: Instructional practices and teacher beliefs of graduate teaching assistants following a reform-minded university teacher certificate programme. *International Journal of Science Education, 32*(8), 1045–1071

Allen, D., Donham, R., & Tanner, K. (2004). Approaches to teaching and learning: Lesson study—building communities of learning among educators. *Cell Biology Education, 3*(001-007), 1–7.

Alvine, A., Judson, T. W., Scheine, M., & Yoshida, Y. (2007). What graduate students (and the rest of us) can learn from lesson study. *College Teaching, 55*(3), 109-113.

Austin, A. E. (2010). Reform efforts in STEM doctoral education: strengthening preparation for scholarly careers. In J. M. Smart (Ed.), *Higher education: Handbook of theory and research* (Vol. 25, pp. 91–128). New York, NY: Springer Scientific.

Austin, A. E., Campa, H., Pfund, G., Gillian-Daniel, D. L., Mathieu, R., & Stoddart, J. (2009). Preparing STEM doctoral students for future faculty careers. *New Directions for Teaching and Learning, 117*, 83–95.

Barry, D., & Dotger, S. (2011, June). *Enhancing content knowledge in graduate teaching assistants through lesson study*. Paper presented at the NSEUS national conference on Research Based Undergraduate Science Teaching: Investigating Reform in Classrooms. University of Alabama, Tuscaloosa AL.

Baumgartner, E. (2007). A professional development teaching course for science graduate students. *Journal of College Science Teaching, 36*, 16–21.

Briscoe, C., & Peters, J. (1997). Teacher collaboration across and within schools: Supporting change in elementary science teaching. *Science Education, 81*(1), 51–65.

Cerbin, W., & Kopp, B. (2006). Lesson study as a model for building pedagogical knowledge and improving teaching. *International Journal of Teaching and Learning in Higher Education, 18*, 250–257.

Clement, J. & Rae-Ramirez, M. (Eds.). (2008). *Model based learning and instruction in science* (Vol. 2). Dordrecht, The Netherlands: Springer.

Cochran-Smith, M., & Lytle, S. (1999). Relationships of knowledge and practice: Teacher learning in communities. *Review of Research in Education, 24*, 249–305.

Dotger, S. (2010). Offering more than "here is the textbook": Teaching assistants perspectives on introductory science courses. *Journal of College Science Teaching, 39*(3), 71–76.

Dotger, S. (2011). Exploring and developing graduate teaching assistants' pedagogies via lesson study. *Teaching in Higher Education, 16*(2), 157–169.

Dotger, S., Barry, D., Wiles, J., Benevento, E., Brzozowski, F., Hurtado-Gonzales, J., ... Wisner, E. (2012). Developing graduate students' knowledge of Hardy-Weinberg Equilibrium through lesson study. *Journal of College Science Teaching, 42*(1), 20–24.

Donovan, S., & Bransford, J. D. (2005). *How students learn: History, mathematics, and science in the classroom*. Washington, DC: National Academies Press.

Feldman, A., Divoll, K., & Rogan-Klyve, A. (2009). Research education of new scientists: Implications for science teacher education. *Journal of Research in Science Teaching, 46*, 442–459.

Feldon, D. F., Peugh, J., Timmerman, B. E., Maher, M. A., Hurst, M., Strickland, D., Gilmore, J. A., & Stiegelmeyer, C. (2011). Graduate students' teaching experiences improve their methodological research skills. *Science, 333*(6045), 1037–1039.

Fernandez, C. (2002). Learning from Japanese approaches to professional development. *Journal of Teacher Education, 53*(5), 393–405.

Fernandez, C., Cannon, J., & Chokshi, S. (2003). A US–Japan lesson study collaboration reveals

critical lenses for examining practice. *Teaching and Teacher Education, 19*(2), 171–185.

Fernandez, C. & Yoshida, M. (2004). *Lesson study: A Japanese approach to improving mathematics teaching and learning*. Mahwah, NJ: Lawrence Erlbaum Associates.

Gaia, A. C., Corts, D. P., Tatum, H. E., & Allen, J. (2003). The GTA mentoring program: An interdisciplinary approach to developing future faculty as teacher-scholars. *College Teaching, 51*, 61–65

Gilbert, J., & Boulter, C. (Eds.). (2000). *Developing models in science education*. Dordrecht, The Netherlands: Kluwer Academic.

Golde, C. M., & Dore, T. M. (2001). *At cross-purposes: What the experiences of today's doctoral students reveal about doctoral education*. Philadelphia, PA: Pew Charitable Trust.

Kamen, M., Junk, D. L., Marble, S., Cooper, S., Eddy, C. M., Wikerson, T. L., & Sawyer, C. (2011). Walking the talk: Lessons learned by university mathematics methods instructors implementing lesson study for their own professional development. In L.C. Hart, A. S. Ashton, & A. Murata (Eds.), *Lesson study research and practice in mathematics education* (pp. 164–174). New York, NY: Springer Science.

Krockover, G. H., Sheparson, D. P., Adams, P. E., Eichinger, D., & Nakleh, M. (2002). Reforming and assessing undergraduate science instruction using collaborative action-based research teams. *School Science and Mathematics, 102*(6), 266–284.

Lewis, C. (2002). *Lesson study: A Handbook of teacher-led instructional change*. Philadelphia, PA: Research for Better Schools.

Lewis, C., & Hurd, J. (2011). *Lesson study step by step: How teacher learning communities improve instruction*, Portsmouth, NH: Heinemann.

Lewis, C., Perry, R., & Hurd, J. (2004). A deeper look at lesson study. *Educational Leadership, 61*(5), 18–23.

Lewis, C., Perry, R., & Hurd, J. (2006). Lesson study comes of age in North America. *Phi Delta Kappan, 88*(4), 273–281.

Lewis, C., Perry, R., & Murata, A. (2006). How should research contribute to instructional improvement? The case of lessons study, *Educational Research, 35*(3), 3–14.

Lewis, C., Perry, R., & Hurd, J. (2009). Improving mathematics instruction through lesson study: A theoretical model and North American case. *Journal of Mathematics Teacher Education 12*(4), 285–304.

Lewis, C., & Tsuchida, I. (1998). A lesson is like a swiftly flowing river: How research lessons improve Japanese education. *American Educator, 22*(4), 12–17, 50–51.

Loughran, J. J. (2002). Effective reflective practice. *Journal of Teacher Education, 53*(1), 33–43.

Loughran, J., Mulhall, P., & Berry, A. (2004). In search of pedagogical content knowledge in science: Developing ways of articulating and documenting professional practice. *Journal of Research in Science Teaching, 41*, 370–391. doi: 10.1002/tea.20007.

Luft, J., Kurdziel, J. F., Roehrig, G., & Turner, J. (2004). Growing a garden without water: Graduate teaching assistants in introductory science courses at a doctoral/research institution. *Journal of Research in Science Teaching, 41*, 211–233.

Mathews, M. R. (2007). Models in science and in science education: An introduction. *Science and Education, 16*, 647–652

Meizlisch, D., & Kaplan, M. (2008). Valuing and evaluating teaching in academic hiring: A multidisciplinary, cross-institutional study. *Journal of Higher Education, 79*(5), 489–512.

National Research Council. (1996). *National science education standards*. Washington, DC: National Academies Press.

Roth, K. (2007). Science teacher knowledge. In S. Abell & N. Lederman (Eds.), *Handbook of research on science education* (pp. 1205–1250). Mahwah, NJ: Lawrence Erlbaum Associates.

Sawyer, D. S., Henning, A. T., Shipp, S., & Dunbar, R.W. (2005). A data rich exercise for discovering plate boundary processes. *Journal of Geoscience Education, 53*(1), 65–74.

Shulman, L. S. (1986). Those who understand: Knowledge growth in teaching. *Educational Researcher, 15*(2), 4–14.

Stigler, J., & Hiebert, J. (1999). *The teaching gap: Best ideas from the world's teachers for improving education in the classroom*. New York, NY: Simon & Schuster.

Sundberg, M. D., & J. E. Armstrong. (1993). The status of laboratory instruction for introductory biology in U.S. universities. *The American Biology Teacher, 55*, 3144–146.

Wright, E. L., & Sunal, D. W. (2004). Reform in undergraduate science classrooms. In D. W. Sunal, E. L. Wright, & J. B. Day (Eds.), *Reform in science teaching for the 21st century* (pp. 33–51). Greenwich, CT: Information Age Publishing.

Zawojweski, J., Chamberlin, M., Hjalmarson, M., & Lewis, C. (2008). Developing design studies in mathematics education professional development: Studying teachers' interpretive systems. In A. Kelly, R. Lesh, & J. Baek (Eds.), *Handbook of innovative design research in science, technology, engineering, and mathematics (STEM) education* (pp. 216–245). Mahwah, NJ: Lawrence Erlbaum Associates.

CHAPTER 16

# STOICHIOMETRY'S PCK OF UNIVERSITY CHEMISTRY PROFESSORS

**Kira Padilla and Andoni Garritz**

## ABSTRACT

The purpose of this chapter is to document the pedagogical content knowledge (PCK) of a set of four university chemistry professors teaching stoichiometry, the study of the mass and amount of substance ratios between two or more substances undergoing a chemical change or, in brief, the science of chemical calculations. This topic can be taught with a simple algorithmic purpose going for immediate procedures without much understanding about what to do and/or why one is doing it, or it can be used to reinforce crucial concepts on the chemical reaction or even the particulate constitution of matter. A discussion is presented on the approach given by these four professors in their general chemistry classes, classified as conceptual, representational, contextual, and procedural. Results are conclusive on the various pedagogical foci of three of the approaches (representational, contextual, and procedural), and the equivalence of the four professors' conceptual approach. Results also reveal a link between conceptual and procedural knowledge.

*Research Based Undergraduate Science Teaching*, pp. 489–512

The term *pedagogical content knowledge* (PCK) was introduced by Shulman (1986, 1987) in order to draw attention to the value of the special amalgam of content and pedagogical knowledge that a teacher needs to be an outstanding one. Stoichiometry is a specific topic of the college general chemistry course which PCK deserves to be documented and commented, as has been pointed out by De Jong, Veal, and Van Driel (2002). A survey on the literature on stoichiometry is given and the contrast between algorithmic problem solving and conceptual understanding is included.

With the intention of documenting stoichiometry's PCK, Loughran, Mulhall and Berry's (2004) Content Representation (CoRe) method has been employed. We have used this methodology in previous research and found it an interesting and appropriate method of documenting, portraying, and analyzing PCK (Garritz, Porro, Rembado, & Trinidad, 2007; Padilla, Ponce de León, Rembado, & Garritz, 2008). We have chosen the proposal of Magnusson, Krajcik, and Borko (1999) as a PCK model. This one has five elements that every professor should know: (1) orientation to teaching science, (2) knowledge of science curricula, (3) knowledge of students' understanding of science, (4) knowledge of instructional strategies, and (5) knowledge of assessment of scientific literacy. So the questions in the CoRe frame were adapted to this model.

Once the four general chemistry teachers' CoRe was completed in order to characterize them, we used four categories of the sentences given therein: conceptual (if understanding concepts is the central goal), contextual (if he/she uses context as a motivational intention), procedural (if he/she simply utilizes problem solving as an algorithmic objective), or representational (if her/his aim is to make use of historical, analogical, metaphorical, demonstrational, experimental, digital, visual and other kinds of representations). In the next section this classification is explained in detail.

## STOICHIOMETRY TEACHING CATEGORIES

Stoichiometry has played a key role in the evolution of chemistry as a science, marking the difference between qualitative and quantitative chemistry. As it was pointed out by Kolb (1978), the term "stoichiometry" comes from the Greek *stoicheion* (element) and *metron* (measure). It was devised by the German chemist Jeremias Benjamin Richter (1762–1807), as a concept designed to quantify the mass proportions of several combined substances. Richter found that the proportions of reagent masses were constant, so the equivalent quantities of an acid and a base in a neutralization reaction were always constant. Richter was a mathematician interested in chemistry, and he believed that chemistry should be considered a branch of mathematics. As Partington (1961, p. 675) wrote: "he busied himself in finding regularities among the combining proportions." Richter was graduated as a Philosophiae Doctor in 1789, writing his thesis on the use of mathematics in chemistry. At that point in history, chemists were interested in making chemistry more mathematical, in the way that physicists had done starting with Galileo and Kepler.

As it was pointed out by Padilla and Furió-Mas (2008), Ernst Fischer (1754–1831) in 1802 called attention to Richter's results saying that they could be presented in a table to show the equivalent weights of an acid and a base when they are compared with one thousand parts of sulphuric acid as the standard substance. Just after the equivalentist paradigm was settled, it came the atomic hypothesis by Dalton, who established an interpretation for the equivalent masses in terms of atoms and its amounts in compounds. The equivalentist paradigm belonged to a tradition of matter theory (continuity) that did not believe in the fundamental existence of the smallest particles (atoms). The atomistic paradigm belonged to a tradition of matter theory (discontinuity) that asserted the existence of discrete atoms and molecules.

The first special booklet, designed to specifically teach stoichiometry to beginning students of chemistry, was written in 1865 by Frickhinger and Cooke (Jensen, 2003), who used the equivalent weights instead of the atomic weights, despite Cannizzaro's work presented in Karlsruhe in 1860. Today, the literature related to stoichiometry can be classified in two categories focusing on:

1. problem solving, where we can find contextual problems (Pinto, 2005a, 2005b); analogies (Arce de Sanabia, 1993; Fortman, 1993; Haim, Cortón, Kocmur, & Galagovsky, 2003; Merlo & Turner, 1993); conceptual approaches (Chandrasegaran, Treagust, Waldrip, & Chandrasegaran, 2009; Krieger, 1997; Taasoobshirazi & Glynn, 2009); and visual representations (Arasasingham, Taagepera, Potter, & Lonjers, 2004; Arasasingham, Taagepera, Potter, Martorell, & Lonjers, 2005; Ault, 2001; Evans, Yaron, & Leinhardt, 2008; Sanger, 2005)
2. students' conceptual understanding of fundamental ideas about the structure of matter before doing calculations (Lekhavat & Jones, 2009; Nakhleh, 1993; Nurrenbern & Pickering, 1987)

The research related to teachers' conceptions about stoichiometry is usually investigating preservice or secondary school teachers, and there is almost nothing related to university professors. This lack of research is why our main subject is college chemistry professors' stoichiometry PCK. Based on the literature, we have chosen four different ways to classify stoichiometry teaching: conceptual, contextual, procedural, and representational.

## Conceptual Stoichiometry Teaching

We will refer in conceptual stoichiometry teaching to the construction of a holistic view of the content by inductive and deductive critical thinking. This term has been coined by Arons (1997), who mentions as the objective of Chapter 13 of his book to "unpack the term critical thinking to list a few simpler underlying processes of abstract logical reasoning that are common to many disciplines and that can be cultivated and exercised separately in limited context accessible to the students" (p. 375). Conceptual knowledge is assembled by using different kinds of representation forms of the concept (especially verbal, graphic, and symbolic), while procedural knowledge has rather a mathematical expression and meaning in applied actions. The aim is to get a full authentic comprehension of the underlying concepts and theories, to reorganize that knowledge using evidence, and to maintain a critical and more objective view of the subject. There are two main complementary trends guiding the purpose of teaching in this new century. One guide is a focus on is the critical thinking ability to reason, which involves the dominion of specific contents, conceptual understanding of frameworks and of the processes of science. The other guide is a focus on developing the problem-solving/decision-making capacity to become an effective citizen. We want to point out that our definition of conceptual understanding emphasizes breadth and depth of knowledge (Alao & Guthrie, 1999). Breadth is related to "the extent of knowledge that is distributed and represents the major sectors of a specific domain" and depth to "the knowledge of scientific principles that describes the relationship among concepts" (Alao & Guthrie, 1999, p. 244).

Different authors have pointed out the importance of helping students to get a better comprehension of what stoichiometry means, further than just memorizing the steps required to make a calculation (Agung & Schwartz, 2007; Akkus, Gunel, & Hand, 2007; BouJaoude & Barakat, 2003; Chandrasegaran et al., 2009; Niaz & Lawson, 1985; Taasoobshirazi & Glynn, 2009; Yarroch, 1985). This conceptual process, however, is not far away from problem solving; the point is what kind of problems should be proposed to students to let them have a better comprehension of stoichiometry, like chemical equations, balancing, limiting reagent, chemical formulas, and so on. In this sense, BouJaoude and Barakat mention that the problem solving process goes from "algorithmic, when conceptual knowledge is missing, to conceptual, when conceptual knowledge is available and when algorithms are stored meaningfully in memory" (2003, p. 2). An investigation was conducted by Yarroch (1985) to identify how students understand balancing of chemical equations. Yarroch reported that many students could balance an equation but, when asked to represent it in molecular terms, many of them cannot do it. In this sense, Yarroch says, students could balance chemical equations in an algorithmic way without showing any evidence of understanding. Niaz and Lawson (1985, p. 49) did a similar investigation, noting that "it is not recommended that students be given algorithmic solution strategies because this would allow them to correctly balance equations without the need for formal reasoning, thus depriving them of an opportunity for its development." According to Ramsdem (1983) (as cited by BouJaoude & Barakat, 2003; Entwistle & Ramsden, 1983) Understanding student learning. New York: NY: Nichols Publishing Company.), meaningful learners have a deep approach to learning when they "build a holistic description of content, reorganize new content by relating it to prior knowledge and/or to personal experiences, are inclined to use evidence, and maintain a critical and a more objective view" (p. 3). In the same way, Akkus et al. (2007) say that "conceptual scientific knowledge is an understanding of the ideas and theories that form the backbone of the scientific community's knowledge and includes the application of knowledge in novel problem-situations" (p. 1748).

## Contextual Stoichiometry Teaching

Many other proposals to teach stoichiometry suggest the importance of contextualizing exercises and lab work to make it interesting and motivational to students.

Contextualizing may include several strategies (Crawford, 2001), such as learning:

1. In the context of one's life experiences or preexisting knowledge (relating)
2. By doing: through exploration, discovery, and invention (experiencing)
3. By putting the concepts to use (applying)
4. In the context of sharing, responding, and communicating with other learners (cooperating)
5. By using knowledge in a new context or novel situation: one that has not been covered in class (transferring)

We go beyond Alao and Guthrie's description, as mastery of concepts in a specific area of science is not of main importance, but include their relationships and interactions discussed within everyday life phenomena (e.g., burning of a candle, tarnish of silver cutlery) and topics in the area of science, technology, society, and environment (STSE) (e.g., greenhouse effect, waste management, and recycling).

In the context of this study, contextual learning is interpreted as students' ability to apply the learned scientific concepts to scientific phenomena in everyday life situations. This application includes, for example, the ability to recognize new information as something different from one's current understanding and beliefs, to identify inconsistencies, and to construct explanations to reconcile knowledge conflicts, or to seek connections among diverse pieces of information.

Contextualization can help students not just with stoichiometry problems but, besides, to think critically and to realize the relevance of chemistry in their daily lives (Pinto, 2005a, 2005b). In this sense, some topics are: boron in fertilizers, mineral waters, calcium and physiology (Pinto, 2005a, 2005b), gas chamber as the production of HCN from polyacrylonitrile (Hunter, Wilkins, & Pearson, 1992), green chemistry (Cacciatore & Sevian, 2006), amino acid complementary (Vitz, 2005), kinetics in chemical reactions (Toby, 2000; Toby & Tobias, 2003), and so on. We think the contextualization of concepts that are very abstract is an important tool for teaching. We should, however, think about the purpose of context. We agree with Pinto in what implies contextualization, but we also think that, in many cases, despite the use of contextual exercises, many proposals are focusing also on the algorithmic process without considering a meaningful concepts' understanding.

## Representational Stoichiometry Teaching

One of the most interesting strategies reported in literature is relating different kinds of representations to improve learning of stoichiometry (Arce de Sanabia, 1993; Chebolu & Storandt, 2003; DeMeo, 2002; Fortman, 1993; Haim et al., 2003; Kashmar, 1997; Krieger, 1997; Rohring, 2000; Roser & McCluskey, 1999; Witzel, 2002). We have found the following kind of representations: historical, analogical, visual, analogue maps, lab experiments or demonstrations, molecular models, and material models. In almost all of the above papers, the authors pay more attention to the relations among substance and its molecular representations and how these relationships can help students understand some stoichiometry ideas, like limiting reagent, mass conservation, amount of substance, and so on. One example of analogy uses hamburger sandwiches (Haim et al., 2003), where students reflect about formulas, chemical equations, mass conservation, and limiting reagent. The general idea is to let the students identify those simple mathematical procedures needed to solve stoichiometric problems, and lead them to acknowledge the need for new vocabulary. Another analogy is the connection of particles with seeds or clips (Arce de Sanabia, 1993), where the key concept is the relative mass (Fortman, 1993) of the seeds to arrive to samples with the same number of them. Some authors also include in this category the use of historical cases as a framework for students' understanding (Giunta, 1998; Holton, 2003; Masson & Vázquez-Abad, 2006; Niaz & Rodríguez, 2001).

## Procedural Stoichiometry Teaching

Hereafter, we will call "procedural knowledge" as the knowledge that requires the use of a memorized set of procedures for the solution of a problem, which denotes dynamic and successful utilization of particular rules or algorithms within relevant representation forms. Most of the literature reports many different strategies to teach stoichiometry, however, almost all of them are focused on the procedure or algorithmic process (Arasasingham et al., 2005; Ault, 2001; DeMeo, 2005; DeToma, 1994; Figueira, Coch, & Zepica, 1988; Kolb, 1978; Murov & Stedjee, 2001) without considering if students achieve a meaningful learning. In all of these reports, authors emphasize in the steps students should follow solving stoichiometry exercises in a correct way. Some of them focus on the use of graph strategies, dimensional analysis, formulas, or maps that let students memorize some constant values like Avogadro's number or molar volume. Ault (2001) presents several units used to measure an amount (mass, amount of substance, volume, and number of elementary entities) and how to convert one into another; after that, he gives the way to create a visual representation for the solution of several typical stoichiometric problems (amount of substance to amount of substance, mass to mass, mass to volume, etc.) and the different transformation factors that can be employed in each case.

The law of conservation of matter is a cornerstone in the development and advancement of modern chemistry, as expressed by Paixão and Cachapuz (2000). These researchers propose a teaching strategy based in history and philosophy that departs from the combustion reactions and their contemporary economic, environmental, social, and political contexts, exploring STSE perspectives in the teaching of science. Exploration is centered upon the context of oxygen theory discovery. On the other hand, Özmen and Ayas (2003) analyze some misconceptions of the conservation of matter of 150 high school students concerning this topic during a chemical reaction in open and closed systems. Agung and Schwartz (2007) examined Indonesian high school students' understanding of conservation of matter, balancing of equations and stoichiometry, in 22 schools with 19 teachers that validated the 25-question survey (which was a concept questionnaire constructed at the Harvard Smithsonian Center for Astrophysics to identify misconceptions) used with 877 students. In general, students' understanding of the fundamental principles in chemistry was low.

## CONCEPTUAL LEARNING VERSUS ALGORITHMIC PROBLEMS

In this section, we center on the supposed dichotomy between conceptual vs. procedural knowledge (in mathematics learning it has been summarized by Haapasalo & Kadijevich, 2000). There have been a large number of terms referring to those two kinds of knowledge, as described by Haapasalo and Kadijevich in the following set of pairs of knowledge:

- Conceptual vs. practical
- Knowing that vs. knowing how
- Declarative vs. procedural
- Facts vs. skills
- Understanding vs. algorithmic
- Theological vs. schematic
- Deductive vs. empirical
- Meaningful vs. mechanical
- Logical/relational vs. instrumental
- Structural vs. operational

One has to recognize that this listing represents certain polarities of the two knowledge types and can, therefore, lead to oversimplifications. In the conclusions, we will describe our views of this alleged dichotomy.

In a study by Yarroch (1985), only half of the 14 high school students he interviewed were able to represent the correct linkages of atoms in molecules. That linkage represents the difficulties of changing from one chemical level of representation—macro, submicro, or symbolic—to the others (Gilbert & Treagust, 2009). Davidowitz and Chittleborough (2009) consider that stoichiometric problems can be used to tackle misunderstandings in relation to the constitution of molecules and their formulas. But building such relationships implies going further than the algorithmic nature implicit in them. They say that students "find more difficult [solving stoichiometric problems] using sub-micro representations despite the opportunities to practice using the sub-micro representations in tutorials" (Davidowitz & Chittleborough, 2009, pp. 182–184).

It has been pointed out that students' views of the particulate nature of matter are cause for concern (Gabel, Samuel, & Hunn, 1987). Instructors of introductory courses know that many students do not understand how to solve problems and frequently resort to algorithmic solutions. In order to solve a problem correctly the concepts involved in the problem must be understood and must be recalled without prompting. After a preliminary description of the problem is given, it needs to be re-described according to the problem solver's frame of reference. In chemistry, depicting the physical phenomena in terms of the particulate nature of matter is helpful. We conclude that the ability to represent matter at the particulate level is very important in explaining phenomena such as chemical reactions, changes in state, the gas laws, stoichiometric relationships, and solution chemistry. It is fundamental to the nature of chemistry itself.

A series of papers started by Nurrenbern and Pickering (1987) have been appearing in the *Journal of Chemical Education* related to the handicap that good stoichiometric problem solvers face with conceptual problems in basic chemistry. Nurrenbern and Pickering applied some problems of algorithmic nature and some that require conceptual understanding to be solved. They found students answering problems about gases without knowing anything much about the nature of a gas, or solving limiting-reagent problems without understanding the nature of chemical change. This result is consistent with the work of Yarroch (1985) and Gabel et al. (1987). Pickering (1990) goes further, asking what happens to the students when they go to other courses in chemistry—for example, organic chemistry. Are there two kinds of students, some who possess an ability to do conceptual problems and some who can solve mathematical-algorithmic problems without molecular understanding? Is the distinction between the groups a difference of ability or just a gap in knowledge? He stresses that, presumably, the instructor's and the textbook's emphases have caused students to direct their efforts toward problem solving. The ability to solve a problem, while desirable in itself, does not seem to imply much real understanding of microscopic reality, and it is this understanding that is at the heart of chemical science. Sawrey (1990) repeated the Nurrenbern and Pickering experiment with a larger and more uniform sample of university students. She found that students viewed the traditional type of questions as mere exercises, but the pictorial concept questions as true problems.

The literature contains evidence that novice problem solvers in chemistry usually have greater success with solving problems in an algorithmic mode than problems with a more conceptual base (Bunce, 1993; Nakhleh, 1993). Niaz and Robinson (1992) concluded that student training in algorithmic-mode problems did not guarantee successful understanding of conceptual problems: "Algorithmic and conceptual problems may require different cognitive abilities" (p. 54). Mason, Shell, and Crawley (1997, p. 906) worked on the following research question:

> How do the general problem-solving procedures used by high-ability algorithmic/high-ability conceptual, low ability algorithmic/high-ability conceptual, high-ability algorithmic/low-ability conceptual, and low-ability algorithmic/low-ability conceptual students compare to each other and to the general problem-solving procedures used by the faculty expert in solving paired algorithmic and conceptual problems?

They conclude that regardless of the students' problem-solving ability, algorithmic-mode problems always required more time and a greater number of transitions for completion than did the paired conceptual-mode problems. Regardless of the topic, however, all students correctly solved the algorithmic-mode problems more frequently than the corresponding paired conceptual-mode problems.

The influence of prior knowledge, use of learning strategies, interest, and learning goals on conceptual understanding and the contribution of each one of the factors was analysed by Alao and Guthrie (1999). These authors developed and used an 18-item knowledge test to measure conceptual understanding and the *Learning Goals, Interest and Strategy Use Questionnaire* to assess students' intentions to try to understand ecological science concepts; this last was also developed by them from previous instruments. They conclude that all factors are important to knowledge acquisition, but prior knowledge accounted for a significant portion of the variance in conceptual understanding after the contribution of interest, learning goals, and strategy use were controlled.

The prevailing practice at the university level when teaching chemistry consists of lectures by the professor, follow-the-recipe laboratory activities, exercise-solving recitation sessions, and examinations oriented toward algorithmic or lower-order cognitive skills. The lecture format for instruction is incompatible with most higher-order cognitive skills and conceptual learning; so success in solving algorithmic problems does not indicate mastery of the relevant chemical concepts (Zoller, Lubezky, Nakhleh, Tessier, & Dori, 1995). Science education researchers indicate that many novice learners in chemistry (Nakhleh, 1993; Nakhleh & Mitchel, 1993) are able to apply algorithms without significant conceptual understanding. We want to elucidate whether this is due to those who teach introductory chemistry placing more value on algorithmic learning than on conceptual understanding, giving the learners the impression that science is "math in disguise" (Pushkin, 1998).

The results of a project to reform the way undergraduate chemistry is taught is presented by Nakhleh, Lowrey, and Mitchel (1996). This project set out to narrow the gap between conceptual and algorithmic understanding in freshman chemistry, using the Generative Learning Model of Wittrock (1986). The nature of the assessment in the course moved from a heavy emphasis on mathematical problem solving to a mix of conceptual questions and more traditional problem-solving questions involving the use of algorithms. The results indicate that special sessions and conceptual exam questions can improve students' abilities to work successfully with both concepts and algorithms. The special sessions provided diagnostic assessment of strengths and weaknesses for both students and professor. In another study, Lin, Kirsch, and Turner (1996) applied Nakhleh's (1993) paired type questions (one with conceptual emphasis and the other with an algorithmic objective) related to several topics of the general chemistry course: gas laws, equations, limiting reagents, empirical formulas, and density. The researchers' focus was on the selection of conceptual versus algorithmic strategies by students belonging to minorities, arriving at the conclusion that these kinds of students are more interested in concepts than in algorithmic aspects of chemistry problem solving.

It has been stressed by Nieswandt (2007) that conceptual understanding of science is a complex phenomenon. It incorporates an understanding of single concepts such as "mass" or of more complex concepts such as "stoichiometry"—declarative or factual knowledge—which, following certain rules and models, combines multiple individual concepts (e.g., particle model, mass conservation, amount of substance, equivalent, etc.), results in a new concept. Thus, conceptual understanding comprises declarative knowledge; procedural knowledge, including concepts, rules, and algorithms; and conditional knowledge, including the understanding of when to employ procedural knowledge and why it is important to do so (Paris, Cross, & Lipson, 1984).

Recently, Salta and Tzougraki (2011) investigated more than one thousand students' (grades 9 and 11) performance with problems of conservation of matter during chemical reactions. The researchers classified the problems in three types: "algorithmic-type," "particulate-type," and "conceptual-type." All students had a far better performance in particulate-type problems than in the other two. Although students' ability in solving algorithmic-type problem increases as their school experience in chemistry progresses, their ability in solving conceptual-type problems decreases. Four different ways of teaching stoichiometry have been discussed here, including a general survey of papers reported in the literature. These are more than just simple strategies, but teaching options that could be used in different moments in the classroom. In our work we tried to identify these four ways of teaching among in college chemistry professors.

## METHOD

The participants in this study were two female and two male professors. All were working full time in either a Mexican or an Argentinean university. We arbitrarily selected as their names Alex, Ana, Anthony, and Alice. The first one has 15 years of teaching experience and a PhD in inorganic chemistry with a postdoctoral work at a renowned European university. The second and third earned BS.c. degrees in chemical engineering and each had more than 30 years of teaching experience. Finally, the fourth has a PhD degree in biochemistry and almost 30 years of teaching experience. All are considered excellent professors by their peers and their pupils.

The documenting of pedagogical stoichiometry knowledge of four university professors was developed using Loughran, Mulhall, and Berry's (2004) proposal of Content Representation (CoRe). The CoRe tries to find out teaching objectives; knowledge of alternative student conceptions; problems that commonly appear when learning; effective sequencing of topic elements; important approaches to the framing of the ideas; use of appropriate analogies, demonstrations, and examples; and insightful ways of testing for understanding, among others.

The questions of the CoRe frame that we have selected and adapted are presented in Table 16.1. We, the professors and researchers, first discussed which could be the central concepts or ideas related to teaching stoichiometry, a crucial component of the CoRe). We understand the central ideas as those at the core of understanding and teaching the theme; there are topics that belong to the disciplinary knowledge and that are used by professors as end point of one theme and, at the same time, as initial point to the following theme. The clue is that those ideas sharply reflect the most important aspects of the topic and may include some of its precedents.

**Table 16.1. Core Questions Used to Document Chemistry Professors' Stoichiometry Pedagogical Content Knowledge**

1. Why is it important for students to learn this idea and what do you intend teaching it?
2. From STS and historical context, why is it important for students to learn this?
3. Difficulties/limitations connected with learning this idea
4. Difficulties/limitations connected with teaching this idea
5. Knowledge about students' thinking which influences your teaching of this idea
6. What representations do you use to engage students with this idea (analogies, metaphors, examples, demonstrations, reformulations, etc.?)
7. Specific ways of ascertaining students' understanding or confusion around this idea

After a long set of conversations, professors and the researchers arrived at consensual agreement that the six central ideas involved in teaching stoichiometry are:

1. Ratios and proportions
2. Purity of substances
3. Composition
4. Empirical and molecular formulas
5. Balancing chemical equations
6. Expressions of concentration

Then, the professors received the frame of Table 16.1 and were asked to answer the questions for each one of these central ideas, and to do it at home, without any pressure.

Based on research reported in the literature (Mortimer, 1995; Padilla et al., 2008), we decided to use the classification of four conceptual profile zones that were the same as those mentioned in the section "Stoichiometry Teaching Categories" of this chapter to start our analysis of what professors mentioned in their CoRes:

- **Conceptual**: Phrases related to the importance given by teachers to try to help students understand the fundamental concepts before students start doing problems, to employing inductive and deductive reasoning, and to the recognition that some ideas generate confusion among students because they are difficult to understand.
- **Contextual**: Sentences using everyday problems or references that help students contextualize the subject and make it closer to them. It also includes learning by doing or by applying and cooperating.
- **Procedural.** This zone is characterized by remarks on the use of algorithms and mathematical formulae as analytical tools applied without a complete understanding of the conceptual relationships involved.
- **Representational.** Comments on the use of ways to represent the topic, such as historical narratives, analogies, demonstrations and laboratory work, metaphors, stories, web-based teaching, controversies, and so on.

Each one of us classified phrases in the CoRe answers to the questions in Table 16.1 for the main ideas fundamental to teaching stoichiometry, by marking them in four different colors, each corresponding to a conceptual profile zone, and discussing and solving the differences existent between their viewpoints. Then, we counted the number of times that each one of these profile zones appeared for each one of the professors, characterized them, and expressed it as percentages.

## RESULTS

The results of counting each of the responses belonging to each of the conceptual profile zones are presented in Figure 16.1 for our four professors.

All of the professors show a similar percentage of use of conceptual strategies, despite not having a similar complete profile (except perhaps Alex and Alice). It is interesting, because we have said how important it is that students learn in a meaningful way, which means students should understand those ideas in a qualitative way. The general profile of the four professors is quite different if we analyse each profile zone; for example, it seems that Anthony points out the importance of procedural knowledge in teaching stoichiometry ideas, in spite of his use of a conceptual way of teaching. At the same time, Anthony is representational and contextual. Alice and Alex have a very similar profile because both of them are cognitive and representational. They make use of procedural knowledge almost in the same proportion (Alex a little more than Alice, but as we will discuss below, in a different way). Ana uses the same proportion of cognitive and procedural knowledge, and at the same time, she uses contextual and representational ways of teaching, giving more importance to the first. What it is important to notice is that, despite some seeming to have almost the same profile, the main differences are in the kind of phrases they show in their CoRe, and that will be discussed below. An analysis of each one of the four professors' answers is now developed.

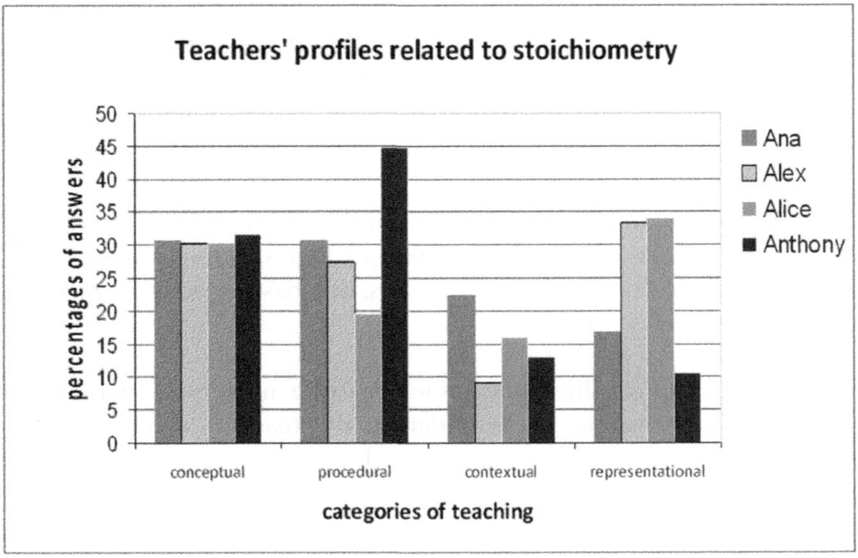


**Figure 16.1.** Professors' profiles related to stoichiometry teaching.

## Ana

To start our discussion of the analysis, we have selected four sentences from Ana, each belonging to one of the profile zones, just to give examples of how they were selected. We highlight these with quotation marks and italicize some portions of the professor's CoRes that led to our decision of categorizing the whole phrase in a given profile zone.

Ana's procedural sentence is: "*It is fundamental* that students *know how to calculate* substances elemental composition from the chemical formula and *vice versa*. What I want is that *students learn how to do the process*, understanding each mathematical step involved." She also mentions the following conceptual phrase, which alludes to the students' "difficulties to *understand the meaning of formula subscripts*, because they change them while making the chemical balancing, without being conscious that those changes affect the *nature of the substances involved.*" We selected the following sentence by Ana as included in the representational profile zone: "The *difficulties are based on the superposition of representational levels*: macroscopic, microscopic and symbolic." One of her sentences in the contextual profile zone was: "In the STS (Science, Technology, and Society) context, *those concepts can be applied to food, medicines and cleaning products*."

Ana recognizes the importance of mathematical calculations, but she emphasizes that it is quite important that *students understand each mathematical step* taken, which does not mean that students already have meaningful learning. In many cases students just learn algorithmic procedures for some style of problem and, if they have to solve a slightly different one, they do not know how to proceed. In her conceptual sentence Ana points out the importance of students understanding the chemical formula and the meaning of the subscripts, which implies that they must comprehend the concept of amount of substance. In the representational category, Ana was the only one who emphasized in the three representational levels proposed by Johnstone (1993). It has been demonstrated that the relationships among them are the most difficult ideas to be understood by students in all educational levels (Gilbert & Treagust, 2009). Finally, she pointed out to her students that stoichiometry is a subject used in many other matters related to chemistry, and mostly in the chemical industry.

## Alex

Alex was quite consistent in his teaching strategies. He makes use of almost the same percentage of conceptual and procedural strategies. It seems, however, that he emphasizes representational strategies but pays little attention to contextual ones. Examples of Alex's answers for each category are the following:

> (Procedural) "In general, the process of *calculation and unit conversions in concentration problems could be mechanical*. Students could be efficient to do calculations in some way; however, *the logic behind the process is still dark to them.*"
>
> (Representational) "At this point, to illustrate the idea of percentage mass/mass and mole fractions I always *use the traditional analogy of cakes* (with different masses) cut in slices sometimes of the same size and other times different."
>
> (Contextual) "The concentration idea is something quite intuitive for students, because they have made lemonade at least once; that is why *I tried to represent those many ways to quantify the amount* of lemon juice, water and sugar using different ways to express chemical concentrations."
>
> (Conceptual) For balancing chemical reactions, Alex said, "This is important not just from a conceptual view (like those factors that could affect the chemical reaction yield), but also when students have to study complicate subjects like chemical equilibrium."

In these phrases, Alex is recognizing that students could be very efficient in stoichiometry calculations, but at the same time, sometimes this problem-solving process could be unfathomable to them because they do not understand the logic behind it. When we analyzed his CoRe it seems that Alex does not emphasize students' reflections related to the qualitative comprehension of these ideas. In the representational sentence, Alex makes use of analogies or material models to teach stoichiometry concepts. This does not mean that Alex's strategies were unimportant, but we think the levels of representation presented by Johnstone (1991, 1993) should be taught to students in a comprehensive way at the same time rather than by the use of other models. One strategy used by Alex to teach relative masses has been implemented as a practical experience in the general chemistry lab work at our School of Chemistry (during the first semester). This representational strategy makes use of nails, nuts, screws, and the like to try to help students develop a better comprehension of relative masses and why they are used in chemistry. With respect to the conceptual category, Alex considers the importance of students constructing a meaningful understanding of balancing chemical reactions. Such an understanding conceptualizes what amount of substance is and why it is used in chemistry. Such an understanding is one of the most complicated and important in chemistry, because teachers have different conceptions of the amount of substance, as Padilla et al. (2008) have shown.

## Alice

Alice is the professor with more phrases in the representational profile zone, because she uses a lot of historical comments on her CoRe:

> I know the transformations that these concepts have had, from *two visions: equivalentist and atomist.* I understand that mole concept first appeared in the equivalentist conceptual framework, with Ostwald, a denier of the atomic hypothesis.
>
> A great problem to understand these concepts is the frequent changes they have had, so a *deep knowledge of history is necessary to understand them* until what we know now. It is a strange case this in which *the unit (mole) is first defined and explained and afterwards appears the magnitude (amount of substance).*
>
> I know that amount of substance is accepted as a fundamental unit of the International System of Units, first by IUPAF and later on, in 1965, by IUPAQ. This moment was a breakthrough that started in Richter's times at the end of the 18$^{th}$ century who thought *in stoichiometry as a way to "mathematize" chemistry to quantify chemical reactions.*

To us, the historical evolution of chemical ideas is quite important for teaching, and in some cases, it is fundamental that students recognize them because it will lead them to understand qualitative ideas and comprehend them much better. In our CoRe, the second question is about STS and historical ideas; however, only Alice uses the historical ideas to represent how this subject has evolved from its origins using the equivalentist paradigm to now, where atomism is the predominant paradigm. It is interesting to analyze the last sentence given by Alice in this category where we could reflect about how stoichiometry was conceived as a way to mathematize chemistry, which is taken so literally by some teachers.

Alice also makes use of analogies, where everyday objects are always present:

> Usually we go to the market to buy grapes by their weight not by their number. Of course that is the same with rice or beans, which are not bought by the number of grains. Only the great fruits can be bought by their number.
>
> I use an analogy between the mass magnitude, its unit the kilogram, and the magnitude amount of substance and its unit mole

She also uses demonstrations: "Classroom demonstration that allow students to understand the difference between to measure amounts or masses of diverse objects or substances, for example to have a dozen of flowers or 10g of copper."

In these analogies and demonstrations Alice is trying to help her students understand the differences between measuring big objects and tiny objects. In this way, she wants to exemplify differences among mass and amount of substance, helping students comprehend these differences. One problem occurs in her last phrase, "to have a dozen of flowers or 10g of copper" because the chemistry dozen is a return to considering the amount of substance's unit mole, as a number, which is mistakenly used by teachers as well as in textbooks.

## Anthony

Anthony has a dominant procedural profile zone. Here, we have some examples of sentences classified in as procedural in his CoRe (we emphasize the procedural portion of the phrases with italics):

> I first let the students use the *procedure they feel experts on* and then I make them use *conversion factors* to solve the same examples.
>
> *It is the mathematical model*, besides the conservation of mass law and the mole concept *what makes possible balancing equations to coincide with what happens in a real chemical process*.
>
> I propose them to solve *a lot of exercises* of all kinds. This is enough to achieve good results.
>
> The main difficulty in teaching stoichiometry is to make students understand the relation between concentration and density,—in physics or chemistry units [here Anthony is doing a distinction between mass and volume, as physical units, and amount of substance, which he consider a chemical unit, as is the case with some other teachers that make a distinction between physics and chemistry magnitudes and units of measure]. The second is to *convince* [students] that these *concentration expressions* are intensive magnitudes, calculated from quotients of extensive ones. Once these two obstacles are surpassed understanding goes better. In reactions where there is not change in oxidation state of the substances involved *it is enough for balancing the trial or algebraic methods*.

All these phrases emphasize how Anthony teaches stoichiometry. His students do a lot of exercises. If they get a correct result, they are considered to have learned stoichiometry. He focuses on *convincing students* instead of *helping them* to meaningfully understand these ideas. There are many teachers like Anthony. These consider leaving students to do exercises implies that they are doing "problem solving," when what they are really doing is solving algorithmic problems. In this sense, it could be interesting to reflect on what problem solving means. Solving problems goes much further than just following a sequence of steps. It really implies that students can take decisions, use the information in a correct way, as well as have the capacity to interpreting the results. We perceive this process as quite related to a conceptual way of teaching. While teaching stoichiometry, teachers often pay more attention to the procedural process without considering the importance of students conceptualizing basic ideas like amount of substance, concentration, limiting reagent, chemical balancing, and chemical formulas.

Anthony has lower percentages (see Figure 16.1) in the representational profile zone; nevertheless he, like Alice, makes use of historical representations; one of his phrases of this type is the following: "The processes to purify substances come from alchemists' time, which in their eagerness of transforming metals into gold developed almost all purification processes that are used until now". In his profile, Anthony almost doesn't show sentences related to the contextual profile zone, however, in the next sentence we could distinguish contextual and conceptual ideas: "I asked questions to know if they could distinguish among substances and mixtures, I used daily life products like food, drinks, medicines, etc. [contextual]. To bring misconceptions from everyday world is almost always the reason of their confusion [conceptual]." In this last phrase, Anthony said that some ideas, brought by students from their everyday context, make them confused. This could be explained in terms of chemistry as a subject, which is present in all everyday activities; however, it is not so easy to explain chemical facts, so students may build some explanations using the knowledge learned in previous courses.

## CONCLUSIONS AND POSSIBLE IMPACT ON TEACHING

After a deep review of the literature, we could categorize the kinds of stoichiometry teaching. We take advantage of this to make our analysis so that we could build conceptual teaching profiles. We think that the profiles constructed in this investigation are very particular, because all teachers have almost the same level of conceptual profile zone, yet they have different percentages (see Figure 16.1) in the other categories. Alex and Alice use the same percentage (33%) of representational phrase; however, the kinds of "representations" used by them are quite different. Alex is more analogical, and Alice is more historical. What we notice through the literature and in this study that stoichiometry teaching tends to be more procedural because of the ontological meaning and origin of this subject. As Alice said, this subject came from a "mathematization" of chemistry, and this idea has permeated chemistry education.

We considered that it is centrally important to understand how procedural knowledge and conceptual knowledge relate to each other. It seems appropriate to underline that these two types of knowledge must be somehow related when the learning process is our focus. The variables in the assessment of this process, however, promote or obstruct possible qualitative and quantitative links between the two knowledge types. One must take into account the complementary presence of both kinds of knowledge while learning—that is, the necessity of having both, procedural and conceptual components, in teaching science—a perspective similar to the "complementary" considered in the middle-American and Eastern worldviews.

The pedagogical approaches that derive from the enhancement of procedural vs. conceptual knowledge (or vice versa) cannot construct a modern view of teaching and learning, because both extremes mean a conventional teacher-based, behaviorist instruction of concepts and/or procedures. Which factors in our education—or perhaps in the whole of society— are important for the development of our thinking abilities and multimodality in human brains? This issue basically calls upon and considers the representations taught to address the questions: do I know that (conceptual), do I know why (contextual and representational), do I know how (procedural), and do I know how I know (metacognitive).

## REFERENCES

Agung, S., & Schwartz, M. S. (2007). Students' understanding of conservation of matter, stoichiometry and balancing equations in Indonesia, *International Journal of Science Education*, *29*(13), 1679–1702.

Akkus, R., Gunel, M., & Hand, B. (2007). Comparing an inquiry-based approach known as the science writing heuristic to traditional science teaching practices: Are there differences? *International Journal of Science Education*, *29*(14), 1745–1765.

Alao, S., & Guthrie, J. T. (1999). Predicting conceptual understanding with cognitive and motivational variables. *The Journal of Educational Research*, *92*, 243–254.

Arasasingham, R. D., Taagepera, M., Potter, F., & Lonjers, S. (2004). Using knowledge space theory to assess student understanding of stoichiometry. *Journal of Chemical Education*, *81*(10), 1517–1523.

Arasasingham, R. D., Taagepera, M., Potter, F., Martorell, I., & Lonjers, S. (2005). Assessing the effect of web-based learning tools on student understanding of stoichiometry using knowledge space theory. *Journal of Chemical Education*, *81*(10), 1517–1523.

Arons, A. B. (1997). *Teaching introductory physics*. Hoboken, NJ: Wiley.

Arce de Sanabia, J. (1993). Relative atomic mass and the mole: A concrete analogy to help students understand these abstract concepts. *Journal of Chemical Education, 70*(3), 233–234.

Ault, A. (2001). How to say how much: amounts and stoichiometry. *Journal of Chemical Education, 78*(9), 1347–1349.

BouJaoude, S., & Barakat, H. (2003). Students' problem solving strategies in stoichiometry and their relationships to conceptual understanding and learning approaches. *Electronic Journal of Science Education, 7*(3).

Bunce, D. M. (1993). Introduction: Symposium: Lecture and learning: Are they compatible? *Journal of Chemical Education, 70*(3), 179–180.

Cacciatore, K. L., & Sevian, H. (2006). Teaching lab report writing through inquiry: a green chemistry stoichiometry experiment for general chemistry. *Journal of Chemical Education, 83*(7), 1039–1041.

Chandrasegaran, A. L., Treagust, D. F., Waldrip, B. G., & Chandrasegaran, A. (2009). Students' dilemmas in reaction Stoichiometry problem solving: deducing the limiting reagent in chemical reactions. *Chemistry Education Research and Practice, 10*, 14–23.

Chebolu, V., & Storandt, V. C. (2003). Stoichiometry of the reaction of magnesium with hydrochloric acid. *Journal of Chemical Education, 80*(3), 305–306.

Crawford, M. L. (2001) *Teaching Contextually. Research, Rationale, and Techniques for Improving Student Motivation and Achievement in Mathematics and Science*, Waco, TX: CCI Publishing.

Davidowitz, B., & Chittleborough, G. (2009). Linking the macroscopic and sub-microscopic levels: Diagrams. In J. K. Gilbert & D. Treagust (Eds.), *Multiple representations in chemical education* (pp. 169–191). Secaucus, NJ: Springer.

De Jong, O., Veal, W. R., & Van Driel, J. H. (2002). Exploring chemistry teachers' knowledge base. In J. K. Gilbert, O. De Jong, R. Justi, D. F. Treagust, & J. H. Van Driel (Eds.), *Chemical education: Towards research-based practice* (pp. 369–390). Dordrecht, The Netherlands: Kluwer Academic Publishers.

DeMeo, S. (2002). Using limiting-excess stoichiometry to introduce equilibrium calculations: A discrepant event laboratory activity involving precipitation reactions, *Journal of Chemical Education, 79*(4), 474–475.

DeMeo, S. (2005). Mass relationships in a chemical reaction: Incorporating additional graphing exercises into the introductory chemistry laboratory. *Journal of Chemical Education, 82*(8), 1219–1922.

DeToma, R. P. (1994). Symbolic algebra and stoichiometry. *Journal of Chemical Education, 71*(7). 568–570.

Entwistle, N., & Ramsden, P. (1983). *Understanding student learning.* New York: NY: Nichols.

Evans, K. L., Yaron, D., & Leinhardt, G. (2008). Learning stoichiometry: A comparison of text and multimedia formats. *Chemistry Education Research and Practice, 9*, 208–218.

Figueira, A. R., Coch, J., & Zepica, M. (1988). Teaching stoichiometry. *Journal of Chemical Education, 65*(12), 1060–1061.

Fortman, J. J. (1993). Pictorial analogies IV: Relative atomic weights. *Journal of Chemical Education, 70*(3), 235–236.

Gabel, D. L., Samuel. K. V., & Hunn, D. (1987). Understanding the particulate nature of matter. *Journal of Chemical Education, 64*(8), 695–697.

Garritz, A., Porro, S., Rembado, F. M., & Trinidad, R. (2007). Latin-American teachers' pedagogical content knowledge of the particulate nature of matter. *Journal of Science Education, 8*(2), 79–84.

Gilbert, J. K., & Treagust, D. (Eds.). (2009). *Multiple representations in chemical education.* Seacaucus, NJ: Springer.

Giunta, C. J. (1998). Using history to teach scientific method: the case of argon. *Journal of Chemical Education, 75*(10), 1322–1325.

Haapasalo, L., & Kadijevich, D. (2000). Two types of mathematical knowledge and their relation. *Journal für Mathematik-Didaktik, 21*(2), 139–157.

Haim, L., Cortón, E., Kocmur, S., & Galagovsky, L. (2003). Learning stoichiometry with hamburger sandwiches. *Journal of Chemical Education, 80*(9), 1021–1022.

Holton, G. (2003). What historians of science and science educators can do for one another. *Science and Education, 12*, 603–616.

Hunter, N. W., Wilkins, C. C., & Pearson, E. F. (1992). Gas chamber stoichiometry. *Journal of Chemical Education, 69*(5), 389–390.

Jensen, W. B. (2003). The origin of stoichiometry problems. *Journal of Chemical Education, 80*(11), 1248.

Johnstone, A. H. (1991). Why is science difficult to learn? Things are seldom what they seem. *Journal of Computer Assisted Learning, 7*, 75–83.

Johnstone, A. H. (1993). The development of chemistry teaching. *Journal of Chemical Education, 70*(9), 701–705.

Kashmar, R. J. (1997). The use of cut-out molecular models on the overhead projector to illustrate stoichiometry and limiting reactants. *Journal of Chemical Education, 74*(7), 791–792.

Kolb, D. (1978). The mole. *Journal of Chemical Education, 55*(11), 728–732.

Krieger, C. R. (1997). Stoogiometry: A cognitive approach to teaching stoichiometry. *Journal of Chemical Education, 74*(3), 306–309.

Lekhavat, P., & Jones, L. L. (2009). The effect of adjunct questions emphasizing the particulate nature of matter on students' understanding of chemical concepts in multimedia lessons. *Educación Química, 20*(3), 351–359.

Lin, Q., Kirsch, P., & Turner, R. (1996). Numeric and conceptual understanding of general chemistry at a minority institution, *Journal of Chemical Education, 73*(10), 1003–1005.

Loughran, J. J., Mulhall, P., & Berry, A. (2004). In search of pedagogical content knowledge in science: Developing ways of articulating and documenting professional practice. *Journal of Research in Science Teaching, 41*(4), 370–391.

Magnusson, S., Krajcik, J., & Borko, H. (1999). Nature, sources, and development of the pedagogical content knowledge for science teaching. In J. Gess-Newsome & N. G. Lederman (Eds.), *Examining pedagogical content knowledge* (pp. 95–132). Dordrecht, The Netherlands: Kluwer Academic Publishers.

Mason, D. S., Shell, D. F., & Crawley, F. E. (1997). Differences in problem solving by nonscience majors in introductory chemistry on paired algorithmic–conceptual problems. *Journal of Research in Science Teaching, 34*(9), 905–923.

Masson, S., & Vázquez-Abad, J. (2006). Integrating history of science in science education through historical microworlds to promote conceptual change, *Journal of Science Education and Technology, 15*(3), 257–268.

Merlo, C., & Turner, K. E. (1993). A mole of M&M's. *Journal of Chemical Education, 70*(6), 453.

Mortimer, E. F. (1995). Conceptual change or conceptual profile change? *Science & Education, 4*, 267–285.

Murov, S., & Stedjee, B. (2001). Analysis of zinc tablets: An extension to a stoichiometry experiment. *Journal of Chemical Education, 78*(10), 1389.

Nakhleh, M. B. (1993). Are our students conceptual thinkers or algorithmic problem solvers? Identifying conceptual students in general chemistry. *Journal of Chemical Education, 70*(1), 52–55.

Nakhleh, M. B., Lowrey, K. A. and Mitchel, R. C. (1996). Narrowing the gap between concepts and algorithms in freshman chemistry. *Journal of Chemical Education, 73*(8), 758–762.

Nakhleh, M. B., & Mitchel, R. C. (1993). Concept learning versus problem solving. *Journal of Chemical Education, 70*(3), 190–192.

Niaz, M., & Lawson, A. E. (1985). Balancing chemical equations: The role of developmental level and mental capacity. *Journal of Research in Science Teaching, 22*(1), 41–52.

Niaz, M., & Robinson, W. R. (1992). From 'algorithmic mode' to 'conceptual gestalt' in understanding the behavior of gases: an epistemological perspective. *Research in Science & Technological Education, 10*, 53–64.

Niaz, M., & Rodríguez, M. A. (2001). Do we have to introduce history and philosophy of science or is it already 'inside' chemistry? *Chemistry Education: Research and Practice in Europe, 2*, 159–164.

Nieswandt, M. (2007). Student affect and conceptual understanding in learning chemistry. *Journal of Research in Science Teaching, 44*(7), 908–937.

Nurrenberm, S. C., & Pickering, M. (1987). Concept learning versus problem solving: is there a difference? *Journal of Chemical Education, 64*(6), 508–510.

Özmen, H., & Ayas, A. (2003). Students' difficulties in understanding of the conservation of matter in open and closed-system chemical reactions, *Chemistry Education Research and Practice, 4*(3), 279–290.

Padilla, K., & Furió-Mas, C. (2008). The importance of history and philosophy of science in correcting distorted views of 'amount of substance' and 'mole' concepts in chemistry teaching. *Science and Education, 17*, 403–424.

Padilla, K., Ponce de León, A. M., Rembado, F. M., & Garritz, A. (2008). Undergraduate professors' pedagogical content knowledge: The case of "amount of substance." *International Journal of Science Education, 30*(10), 1389–1404.

Paixão, M. F., & Cachapuz, A. (2000). Mass conservation in chemical reactions: The development of an innovative teaching strategy based on the history and philosophy of science. *Chemistry Education Research and Practice, 1*(2), 201–215.

Paris, S. G., Cross, D. R., & Lipson, M. Y. (1984). Informed strategies for learning: A program to improve children's reading awareness and comprehension. *Journal of Educational Psychology, 76*, 1239–1252.

Partington, J. R. (1961). *A history of chemistry* (Vol. III). New York, NY: Martino Publishing.

Pickering, M. (1990). Further studies on concept learning versus problem solving. *Journal of Chemical Education, 67*(3), 254–255.

Pinto, G. (2005a). Stoichiometry of calcium medicines. *Journal of Chemical Education, 82*(10), 1509–1512.

Pinto, G. (2005b). Stoichiometry problems in context, *Education in Chemistry, 42*, 108–109.

Pushkin, D. B. (1998). Introductory students, conceptual understanding, and algorithmic success. *Journal of Chemical Education, 75*(7), 809–810.

Rohring, B. (2000). Fizzy drinks: Stoichiometry you can taste. *Journal of Chemical Education, 77*(12), 1608A–B.

Roser, C. E., & McCluskey, C. E. (1999). Pressure and stoichiometry. *Journal of Chemical Education, 76*(5), 638–639.

Salta, K., & Tzougraki, C. (2011). Conceptual versus algorithmic problem-solving: Focusing on problems dealing with conservation of matter in chemistry. *Research in Science Education, 41*(4), 587–609.

Sanger, M. J. (2005). Evaluating students' conceptual understanding of balanced equations and stoichiometric ratios using a particulate drawing. *Journal of Chemical Education, 82*(1), 131–134.

Sawrey, B. A. (1990). Concept learning versus problem solving: Revisited. *Journal of Chemical Education, 67*(3), 253–254.

Shulman, L. S. (1986). Those who understand: Knowledge growth in teaching. *Educational Researcher, 15*, 4–14.

Shulman, L. S. (1987). Knowledge and teaching: Foundations of the new reform. *Harvard Educational Review, 57*(1), 1–22.

Taasoobshirazi, G., & Glynn, S. M. (2009). College students solving chemistry problems: A theoretical model of expertise. *Journal of Research in Science Teaching, 46*(10), 1070–1089.

Toby, S. (2000). The relationship between stoichiometry and kinetics. *Journal of Chemical Education, 77*(2), 188–190.

Toby, S., & Tobias, I. (2003). What is the overall stoichiometry of a complex reaction? *Journal of Chemical Education, 80*(5), 520–523.

Vitz, E. (2005). Amino acid complementarity: a biochemical exemplar of stoichiometry for general and health sciences chemistry. *Journal of Chemical Education, 82*(7), 1013–1016.

Wittrock, M. C. (1986). Students' thought processes. In M. C. Wittrock (Ed.), *Handbook of Research on Teaching* (3rd. ed., pp. 754–772). New York, NY: Macmillan.

Witzel, J. E. (2002). Lego stoichiometry. *Journal of Chemical Education, 79*(3), 352A–B.

Yarroch. W. L. (1985). Student understanding of chemical equation balancing. *Journal of Research in Science Teaching, 22*(5), 449–459.

Zoller, U., Lubezky, A., Nakhleh, M. B., Tessier, B., & Dori, Y. J. (1995). Success on algorithmic and lower-order cognitive skills vs. conceptual chemistry exam questions. *Journal of Chemical Education, 72*(1), 987–989.

# ABOUT THE AUTHORS

**Iris Alkaher** received her PhD in science education from Technion, the Israeli Institute of Technology. She is a lecturer in the faculty of science at the Kibbutzim College of Education in Israel. Prior to her current appointment she was awarded a postdoctoral research fellowship at the Virginia Tech, USA. As a postdoctoral researcher, she conducted research related to inquiry teaching and learning in higher education. Dr. Alkaher's research interests are environmental and biology education, research-based science in higher education, and sociocultural aspects of science education. She currently teaches undergraduate and graduate students in the areas of biology and environmental educatand serves as the coordinator of the graduate Master of Education program in environmental education. She has recently published articles in high-level journals such as the *Cultural Studies of Education, Environmental Education Research* and *International Journal of Science Education* and presented papers at the annual meetings of the American Educational Research Association and the National Association for Research in Science Teaching.

**Kayla Anzalone** received her BS in cell biology from Western Washington University in 2011. She is currently enrolled in the Washington State University College of Pharmacy.

**Varda Bar** of the Hebrew University of Jerusalem received a PhD in physics. She has participated in astronomical research at the Solar Observatory in Capri, Italy, as well as the Sacramento Peak and Kitt Peak observatories in the United States. She has also worked in science education at the Tel Aviv and Hebrew University. Her educational research focuses on cognitive development and curriculum design. She has developed and implemented science education curriculum for special education learners at the elementary and middle school levels. Dr. Bar has completed several research projects on the scientist's image in different societies, as well as the conceptual development of basic physics concepts such as electricity and mass. Her many publications include a book about nuclear physics for educated nonprofessional adolescents and adults. She is currently completing another book about the sun.

**Deborah Barry** is a Research Associate with The Center for Advanced Study of Teaching and Learning in Higher Education in the Curry School of Education at the University of Virgina. Her research interests include determinants of career aspiration in STEM doctoral recipients, and exploring avenues that lead to a scholarly approach to teaching in undergraduate science settings.

**Emily Borda** received her PhD in organic chemistry and an MEd in educational leadership and policy studies from the University of Washington in Seattle, WA. She is currently an associate professor of chemistry and science education at Western Washington University in Bellingham, WA. Dr. Borda's research focuses on incorporating constructivist-based curriculum and teaching methods in large introductory science courses, and studying how undergraduates learn scientific modeling skills. Dr. Borda has published refereed articles in *Science & Education, Journal of College Science Teaching,* and the *Electronic Journal of Science Education.* She has presented her research at various regional and national venues, including meetings of the National Association for Research in Science Teaching (NARST), Association for Science Teacher Education (ASTE), and the American Chemical Society (ACS).

**Gayle Buck** is a professor of science education at Indiana University. She has taught in higher education for over fourteen years and conducted many empirical studies on science teaching and learning. Her scholarship and teaching focus on relationships across learner diversity and science. To that end, she has developed a research agenda that is grounded in sociocultural theoretical perspectives and utilizes participatory research approaches. Specifically, her research focuses on student populations traditionally underserved by science education and neglected epistemological assumptions in science teaching and learning. Dr. Buck has over forty publications that have appeared in journals such as *Journal of Research on Science Teaching, Journal of Science Teacher Education, Science*

*Education, Educational Action Research, Middle School Journal*, and *Learning Environments Research International.* This work has been presented at the annual conferences for many organizations including the National Association for Research in Science Teaching, American Educational Research Association, and Association for Science Teacher Education.

**Mikel Ceberio** received a PhD in physics from the University of Basque Country, UPV / EHU. He is currently an associate professor of physics in the School of Industrial Engineering of Bilbao, at the Basque Country University (UPV/EHU). Since the beginning of his career, he has been interested in various issues related to teaching and learning methodologies, such as learning difficulties and epistemology of science within the field of physics, introductory physics concepts, and problem-solving in physics. Dr. Ceberio has authored one book, 9 chapters, and over 15 refereed publications in journals such as *Science Education, Research in Science Education, European Journal of Engineering Education, Science and Education, Enseñanza de las Ciencias*, and *Caderno Brasileiro de Encino de Fisica.* Among other professional activities, he is currently Vice Dean of Innovation at the School of Industrial Engineering of Bilbao.

**John Dantzler** is an assistant professor in the University of Alabama's Department of Educational Studies in Psychology, Research Methods and Counseling in the College of Education. Dr. Dantzler has a PhD in educational research with an emphasis in psychometrics and program evaluation. He currently teaches educational research methods courses for the university and has worked as a consultant for the *National Study of Education in Undergraduate Science* (NSEUS) project. Prior to his work in higher education, he was the owner of a program evaluation consulting firm. His fifteen years of work with this firm consisted of designing research and evaluation surrounding federally funded programs in engineering education, K–12 education, mental health, low-income housing, substance abuse prevention, criminal justice, HIV/AIDS treatment and prevention, and public health. Currently, Dr. Dantzler's research interests are in the areas of psychometric development, graduate student identity as researchers, and count-based data analysis.

**Lin Ding** is an assistant professor of science education in the Department of Teaching and Learning at The Ohio State University. His research work focuses on investigating and improving student conceptual understanding, problem-solving, and learning attitudes in the physics domain. Previously a postdoctoral researcher in the physics department at Ohio State, Dr. Ding has led a research group in developing an innovative clicker methodology to effectively promote an interactive learning environment in undergraduate physics classrooms. Dr. Ding also specializes in the development of assessment instruments and quantitative research methods. He has published over a dozen peer-reviewed journal articles, presented numerous conference papers, and organized a number of workshops at national and international symposia.

**Erin Dolan** earned a BA in biology at Wellesley College and a PhD in neuroscience from the University of California at San Francisco (UCSF). During her graduate work, she volunteered extensively with UCSF's Science and Health Education Partnership, which spawned her interest in science education. At this time this chapter was written, she was an associate professor in the department of biochemistry and molecular biology and senior scholar in biology education in the division of biological sciences at the University of Georgia. She is now Executive Director of the Texas Center for Science Discovery in the College of Natural Sciences at the University of Texas at Austin. Her research involves understanding science research as a learning context: specifically, how and in what ways research collaborations among students, instructors, and scientists serve as a context for students to generate and evaluate scientific arguments, engage in scientific discourse, and develop scientific identities. She is editor-in-chief of *CBE—Life Sciences Education*, which is published by the American Society for Cell Biology.

**Sharon Dotger** received a PhD in Science Education from North Carolina State University. She is currently an Associate Professor of Science Education and the Chair of the Department of Science Teaching at Syracuse Her research interests include studying the learning processes necessary for teachers to change their instructional practices. Dr. Dotger studies these because she believes that, with the improvement of the set of practices teachers engage in for instruction, student learning can improve over time. She utilizes lesson study as a mechanism for studying teachers' practices and for their impact on student learning.

**Deirdre Englehart** received an EdD in curriculum and instruction with a focus on early childhood, science, and literacy from the University of Central Florida. She is currently an instructor and program coordinator for the Early Childhood Development and Education Program at the Daytona campus for the University of Central Florida. Dr. Englehart's research interests include teacher dispositions, science teaching, online learning, and science nonfiction literature. She recently presented at meetings of the National Association for Research in Science Teaching and the Eastern Educational Research Association. Her recent publications are included in

*The International Journal of Educational and Psychological Assessment* and *The Association for Childhood Education International.*

**Andoni Garritz** is full-time professor and 40-year faculty member in the School of Chemistry, National University of Mexico (UNAM), where he has taught courses such as General Chemistry, Science and Society, Molecular Symmetry, Chemistry Didactics, and Matter Structure. His research interests include pedagogical chemistry knowledge based on an historical and STS perspective. Dr. Garritz studied chemical engineering at UNAM in 1971. In 1977, he received his PhD from UNAM following a sabbatical visit with the Quantum Chemistry Group at Uppsala University, Sweden, where he received the Hylleraas' Award in 1975. He served as a consultant for UNESCO, coordinating seven chemistry experimental education projects in Latin America. He has written seven chemistry textbooks in Spanish and Portuguese and is the director of the journal *Educación Química*. In addition, Dr. Garritz has published approximately 150 papers and book chapters and participated in 350 presentations, 120 of them at international meetings.

**Jenaro Guisasola** is professor of physics in the applied physics department and of physics education in the MA for secondary science teachers program at the University of the Basque Country. He received his BS in physics and an MS in theoretical physics from the University of Barcelona, as well as a PhD in applied physics from the University of the Basque Country. As a director of the research on physics, mathematics and technological education group at the University of the Basque Country, Dr. Guisasola has supervised ten PhD dissertations. He has published extensively in international journals, along with seven textbooks regarding science for secondary education, and three books on science education research. His research interests include teaching and learning conceptual physics knowledge at the precollege and college levels; the use of history and philosophy of science in the classroom; and the relationships among science museums, science classrooms, and scientific literacy.

**Mojgan Matloob Haghanikar** received her BS in solid state physics from the National University of Iran (Shahid Beheshti). After graduation, she worked for several commercial companies who were conducting development projects such as telecommunications, solar cells, laser and holography. Dr. Haghanikar also taught introductory physics labs in local universities and was involved in establishing science museums, informal methods of teaching, and curriculum development in Iran, along with conducting physics education research. In 2002, she received a CHEVENING scholarship for completing a master's degree research program in the United

Kingdom. Her master's degree research in astronomy education was conducted at the University of Glasgow, Scotland. Following that experience, she completed her PhD in physics education from the physics department at Kansas State University. She is now an assistant professor in the physics department at Towson University in Baltimore, MD.

**Sarah Tanya Heaston** is a science educator at California State University, Chico. Tanya has developed preK–12 science curricula and teaching materials. She currently teaches science courses to preservice teachers for the science education department. Ms. Heaston has a secondary education credential in science and a master's degree in science education.

**Gail Horowitz** is currently a lecturer of organic chemistry at Brooklyn College of the City University of New York. She has spent the past two decades teaching chemistry at the undergraduate level. She holds a master's degree in organic chemistry and a PhD in science education and utilizes action research as a way of combining her interests in research and teaching. Since joining Brooklyn College, Dr. Horowitz has focused her research efforts on determining strategies and methods to best help underrepresented students achieve success in organic chemistry and other introductory science courses.

**Sally G. Hoskins**, professor of biology at the City College of CUNY, is a science educator and developmental biologist. With NSF support, she and collaborator Dr. Leslie M. Stevens (University of Texas, Austin) developed C.R.E.A.T.E. (Consider, Read, Elucidate hypotheses, Analyze and interpret data, and Think of the next Experiment)—a way to use intensive analyses of sequentially published journal articles in undergraduate science courses to demystify research and humanize researchers. Piloted at CCNY, C.R.E.A.T.E. produced gains in students' critical thinking ability and understanding of content. Dr. Hoskins is currently teaming with Dr. Kristy Kenyon (Hobart and William Smith Colleges) to plan C.R.E.A.T.E. teaching workshops for faculty from a national cross-section of 2- and 4-year colleges and universities. A three-time CCAPP Teacher of the Year at City College, she is enthusiastic about the potential for building students' understanding of science as well as their interest in research careers.

**Pazit Koren** of the Hebrew University of Jerusalem earned her PhD in education and philosophy of science. Her research agenda focuses on understanding and addressing the reluctance of students to enter STEM professions. Specifically, she explores the students' and teachers' understandings of the social and moral context of the nature of science and how that affects students' motivation to be a scientist. Dr. Korn has taught for the past fourteen years in a wide variety of educational levels, from elementary school to university, the latter at the Hebrew University of Jerusalem. She has published her research in journals such as the *International Journal of Science Education*, and presented her work at various academic venues.

**Corinne Lardy** is currently a secondary science student teacher supervisor at San Jose State University. She holds a PhD in mathematics and science education from San Diego State University and the University of California–San Diego and an MS in biological sciences from San Jose State University. Dr. Lardy has taught a variety of undergraduate and postgraduate courses in biological sciences and science teaching and learning. In addition, she has taught biology, chemistry, and earth science at the high school and middle school levels. Dr. Lardy's current primary research interests center around teacher beliefs and the impacts of those beliefs on individuals' abilities to institute and sustain aspects of reformed inquiry-oriented science teaching at the elementary, secondary, and undergraduate levels.

**Mathew Lockett** graduated from Western Washington University in 2010 with a BS in biochemistry and a teaching certificate for secondary education. Mr. Lockett is currently teaching high school science in the state of Washington.

**Bev Marcum** received her PhD in biology from the University of Virginia. As a postdoctoral fellow in Switzerland and Japan, she studied developmental biology using hydra as a model system in pattern formation and stem cell determination. She joined the biology faculty at California State University–Chico to focus on her teaching, and she has taught a wide range of classes including tropical marine biology at the Gerace Research Centre in the Bahamas. Dr. Marcum has long had an interest in science education and is the director of the Inland Northern California Science Project, part of a statewide professional development network for K–12 teachers. In addition to NSF and NIH science research grants, she has received numerous grants for science education research. With NASA/NOVA funding she developed the highly successful Internship in Science Teaching course for future teachers in the Hands-on Lab at CSU–Chico promoting science education from Kindergarten students to university science faculty.

**Cheryl L. Mason** is professor emerita of biological sciences and science education at San Diego State University. She received her PhD in science education from Purdue University, and her bachelor's and master's degrees in biological sciences from Indiana University. Her teaching experiences include undergraduate and graduate courses in biological sciences and science education, along with 12 years at the precollege level. Supporting her research, Dr. Mason has received extramural funding that has resulted in publications in numerous journals and books and presentations to hundreds of audiences. She has also been an active leader of international science organizations and has received international and national awards for her extensive successes in teaching and directorship roles. Dr. Mason's research focus is on the relationship of cognitive and attitudinal factors regarding successful science teaching and learning. She is especially concerned with helping girls and persons of color succeed in scientific endeavors.

**Kristen Miller** received a PhD in science education from the University of Georgia (UGA). She currently is the director of introductory biology laboratory courses at UGA. Her research interests include professional development for science graduate student instructors and writing as a mode of learning in undergraduate science courses. Since 2008 she has authored five peer-reviewed publications and delivered fourteen workshops/presentations on this research at regional and national STEM (science, technology, mathematics, and engineering) and science education conferences. She has authored four biology laboratory manuals that support inquiry teaching and learning and developed an online resource that provides science educators with reflection-based pedagogical tools called *caselets*. Dr. Miller is a board member of the Association for Biology Laboratory Education (ABLE) and is active in the STEM teaching and research community at UGA. She also instructs professional development teaching courses for graduate students who are novice science instructors.

**Sytil Murphy** was a postdoctoral student in the physics department at Kansas State University, 2007–2011, and is now an assistant professor of physics at Shepherd University in Shepherdstown, WV. Her research is in the area of physics education. Current projects involve investigating students' perceptions of physics concepts as shown in videos developed by the students and other media, along with accompanying write-ups. Her most recent project involves developing a blended online/traditional course to serve non-science majors.

**Glenda L. Ogletree** received a PhD in curriculum and instruction/elementary science education from the University of Alabama, Tuscaloosa. She currently is an assistant professor at Armstrong Atlantic State University in Savannah, GA. Her recent research has been on using iPads in the elementary science classrooms. She has recently presented at the NSTA National Conference, the National Literacy Research Association Conference, the National Youth-At-Risk Conference, and the Georgia Association of Teacher Education (GATE) annual conference. Dr. Ogletree also serves on the review board for the journal entitled *Elementary Science Education* published by the Association of Science Teacher Education.

**Steve Oliver** is professor of science education and has been a faculty member of the department of mathematics and science education at the University of Georgia since 1990. In 2011, Dr. Oliver was recognized as one of five inaugural Distinguished Research Mentors in the UGA College of Education and also received the Yeany Research Award for his cumulative contributions to research in science education. Dr. Oliver has been an investigator on grants totaling approximately $5 million and has collaborated widely with scientists including Dr. Jim Moore, professor of large animal medicine at UGA. He and Dr. Moore received a SEPA grant from the NIH that established an effort to create state-of-the-art inquiry teaching materials featuring high quality 3-D animations of physiological processes. Dr. Oliver has published in a wide variety of professional venues and served as an associate editor of the *Journal of Research in Science Teaching*.

**Kira Padillla** is an associate professor at the faculty of chemistry at National Autonomous University of Mexico (UNAM). She earned a bachelor's degree in chemistry, a master's degree in physical chemistry, and then a doctoral degree in science education (2004) from the University of Valencia in Spain. Following that she conducted postdoctoral research at Leiden University in the Netherlands (2008). Her main research interest deals with teachers' conceptions and how these are transformed or are used to teach any subject. As a result, she has written several papers on the pedagogical content knowledge (PCK) of university instructors. Dr. Padilla has also collaborated with the National Education Council (SEP) in Mexico to modify the curriculum and teaching approach in elementary and secondary schools and has written chemistry textbooks for secondary curriculum.

**Neville W. Reay** earned a doctorate in experimental nuclear physics, which morphed into a high-energy physics career that began as a postdoctoral student at the University of Rochester and ended as a professor at the Ohio State University. Dr. Reay conducted experiments studying topics as diverse as neutrino oscillations and charm and beauty particle lifetimes, ultimately becoming a fellow of the American Physical Society. His next position was University Distinguished Professor at Kansas State University. To help future generations of scientists, Dr. Reay returned to the Ohio State University as a fulltime instructor and physics education researcher. This final decade of his physics career was spent developing and studying use of clicker question sequences that help students overcome the contextual dependence of their learning. Currently, Dr. Reay is a flight instructor in Nashville,TN, and he also works with Nashville schools to support the research efforts of Professor Lin Ding.

**Erika Steele** received her master's degree in biology in 2008. She is currently completing her PhD in biology science education at the University of Alabama–Tuscaloosa with December 2012 as a target date of graduation. Her dissertation examines the impact of undergraduate course reform on students' perceptions of the learning environment. Ms. Steele's research interests include university instructors' beliefs about pedagogy and how they impact the science classroom learning environment, as well as the impact of undergraduate students' perception of the learning environment has on their learning.

**Cynthia Szymanski Sunal** completed her PhD at the University of Maryland and is now professor and chair of curriculum and instruction at the University of Alabama and director of the office of Research on Teaching in the Disciplines. She primarily works with elementary and middle level teachers. Among her publications are numerous books, journal articles, and monographs. In addition, she serves as executive editor of two journals and publishes a research series. She has also been involved in several funded projects from the National Science Foundation, the Department of Energy, and other agencies. Recently, Dr Sunal received the University of Alabama's College of Education's Bryant Professor of Research Award.

**Dennis W. Sunal** received a PhD from the University of Michigan. He currently is professor of science education at the University of Alabama, Tuscaloosa. His university teaching experiences include undergraduate and graduate courses in physics, engineering, curriculum and instruction, and science education. He holds both secondary (6th–12th) and elementary (K–6$^{th}$) teacher certifications and has taught at both levels. His research interests include effective undergraduate science teaching, physics teacher education, and pedagogical content knowledge of science teachers and faculty. He has been project director and co-director on numerous grants (e.g., NSF, NASA, Department of Education, USIA, and U.S. Department of Energy). In addition, Dr. Sunal has published numerous articles and chapters in refereed journals and books. He is also co-editor of the Research in Science Education (RISE) series published by Information Age Publishing.

**Donna P. Turner** graduated from the University of Alabama, Tuscaloosa with a PhD in secondary science education. She received her master's degree in biology education from Alabama A&M University and her AA certification from the University of Alabama–Birmingham. She has

15 years teaching experience with the Birmingham City School System. Currently, Dr. Turner works as a postdoctoral research fellow for the National Study of Education in Undergraduate Science project at the University of Alabama, Tuscaloosa. Her research focus is science education reform and the pedagogical content knowledge of undergraduate faculty and inservice elementary teachers.

**Siri Wuotila** is a graduate of Western Washington University. She holds a BS degree in communication and a master's degree in teaching. She is also a Robert Noyce Scholar and has recently received her Washington State teaching credential in secondary science. Ms. Wuotila completed a research paper on the effectiveness of single-sex education to commence her master's degree and has future research interests in comparing standards-based instruction and meaningful learning environments. She is a co-author of four other submitted research papers and has the goal of conducting and submitting her own research studies in her role as a secondary teacher. Currently, Ms. Wuotila is seeking a middle school science teaching position in Washington State.

**Xinying Yin** is currently an assistant professor in science education at California State University San Bernardino, and previously a high school chemistry teacher in China. Her research interests are in formative assessment in science education in different sociocultural contexts. Ms. Xinying plans to study the use of formative assessment under the influence of test-based accountability systems, such as in the United States, and examination-oriented educational systems, such as in China. She will use action research for developing in-service science teachers' knowledge of formative assessment, and enhancing preservice science teachers' understanding of scientific inquiry and their ability to teach inquiry-based science. She has presented her work at the annual meetings of several professional organizations including the National Association for Research in Science Teaching, National Science Teacher Association, and Association for Science Teacher Education.

**Dean Zollman** is university distinguished professor of physics at Kansas State University. He also holds the title of distinguished university teaching scholar. From 2001 to 2011 he was William and Joan Porter Professor and Head of Physics. Dr. Zollman is part of the Kansas State University Physics Education Research group (KSUPER) that was founded in 1972. His current research concentrates on investigating the mental models and operations that students develop as they learn physics and the transfer of knowledge in the learning process. He also applies cutting-edge technology to teaching physics and providing instructional materials to physics teachers and students. He has received three major national awards—the National Science Foundation's Director's Award for Distinguished Teacher Scholars (2004), the Carnegie Foundation for the Advancement of Teaching Doctoral University Professor of the Year (1996), and American Association of Physics Teachers' Robert A. Millikan Medal (1995).

**Kristina Zuza** received a PhD in physics education from the University of the Basque Country (UPV/EHU). She is currently a lecturer in the department of applied physics at the Polytechnic School of San Sebastian in the University of the Basque Country. Her research focuses on the teaching and learning process of physics in university students. She also has an interest in the preparation of secondary school teachers in science and technology. In addition, Dr. Zuza has authored over 15 refereed articles in journals including *American Journal of Physics, European Journal of PhysicsI,* and *Physical Review Special Topics: Physics Education Research.* She is a member of the astronomy department of the Aranzadi Science Society.

Printed and bound by CPI Group (UK) Ltd, Croydon, CR0 4YY
15/09/2026
14962370-0007